Lecture Notes in Applied and Computational Mechanics

Volume 43

Series Editors

Prof. Dr.-Ing. Friedrich Pfeiffer
Prof. Dr.-Ing. Peter Wriggers

Lecture Notes in Applied and Computational Mechanics

Edited by F. Pfeiffer and P. Wriggers

Further volumes of this series found on our homepage: springer.com

Vol.43: Ibrahim, R.A.
Vibro-Impact Dynamics
312 p. 2009 [978-3-642-00274-8]

Vol. 42: Hashiguchi, K.
Elastoplasticity Theory
432 p. 2009 [978-3-642-00272-4]

Vol. 41: Browand, F., Ross, J., McCallen, R. (Eds.)
Aerodynamics of Heavy Vehicles II: Trucks, Buses, and Trains
486 p. 2009 [978-3-540-85069-4]

Vol. 40: Pfeiffer, F.
Mechanical System Dynamics
578 p. 2008 [978-3-540-79435-6]

Vol. 39: Lucchesi, M., Padovani, C., Pasquinelli, G., Zani, N.
Masonry Constructions: Mechanical Models and Numerical Applications
176 p. 2008 [978-3-540-79110-2

Vol. 38: Marynowski, K.
Dynamics of the Axially Moving Orthotropic Web
140 p. 2008 [978-3-540-78988-8]

Vol. 37: Chaudhary, H., Saha, S.K.
Dynamics and Balancing of Multibody Systems
200 p. 2008 [978-3-540-78178-3]

Vol. 36: Leine, R.I.; van de Wouw, N.
Stability and Convergence of Mechanical Systems with Unilateral Constraints
250 p. 2008 [978-3-540-76974-3]

Vol. 35: Acary, V.; Brogliato, B.
Numerical Methods for Nonsmooth Dynamical Systems: Applications in Mechanics and Electronics
545 p. 2008 [978-3-540-75391-9]

Vol. 34: Flores, P.; Ambrósio, J.; Pimenta Claro, J.C.; Lankarani Hamid M.
Kinematics and Dynamics of Multibody Systems with Imperfect Joints: Models and Case Studies
186 p. 2008 [978-3-540-74359-0]

Vol. 33: Nies ony, A.; Macha, E.
Spectral Method in Multiaxial Random Fatigue
146 p. 2007 [978-3-540-73822-0]

Vol. 32: Bardzokas, D.I.; Filshtinsky, M.L.; Filshtinsky, L.A. (Eds.)
Mathematical Methods in Electro-Magneto-Elasticity
530 p. 2007 [978-3-540-71030-1]

Vol. 31: Lehmann, L. (Ed.)
Wave Propagation in Infinite Domains
186 p. 2007 [978-3-540-71108-7]

Vol. 30: Stupkiewicz, S. (Ed.)
Micromechanics of Contact and Interphase Layers
206 p. 2006 [978-3-540-49716-5]

Vol. 29: Schanz, M.; Steinbach, O. (Eds.)
Boundary Element Analysis
571 p. 2006 [978-3-540-47465-4]

Vol. 28: Helmig, R.; Mielke, A.; Wohlmuth, B.I. (Eds.)
Multifield Problems in Solid and Fluid Mechanics
571 p. 2006 [978-3-540-34959-4]

Vol. 27: Wriggers P., Nackenhorst U. (Eds.)
Analysis and Simulation of Contact Problems
395 p. 2006 [978-3-540-31760-9]

Vol. 26: Nowacki, J.P.
Static and Dynamic Coupled Fields in Bodies with Piezoeffects or Polarization Gradient
209 p. 2006 [978-3-540-31668-8]

Vol. 25: Chen C.-N.
Discrete Element Analysis Methods of Generic Differential Quadratures
282 p. 2006 [978-3-540-28947-0]

Vol. 24: Schenk, C.A., Schuëller. G
Uncertainty Assessment of Large Finite Element Systems
165 p. 2006 [978-3-540-25343-3]

Vol. 23: Frémond M., Maceri F. (Eds.)
Mechanical Modelling and Computational Issues in Civil Engineering
400 p. 2005 [978-3-540-25567-3]

Vol. 22: Chang C.H.
Mechanics of Elastic Structures with Inclined Members: Analysis of Vibration, Buckling and Bending of X-Braced Frames and Conical Shells
190 p. 2004 [978-3-540-24384-7]

Vol. 21: Hinkelmann R.
Efficient Numerical Methods and Information-Processing Techniques for Modeling Hydro- and Environmental Systems
305 p. 2005 [978-3-540-24146-1]

Vibro-Impact Dynamics

Modeling, Mapping and Applications

Raouf A. Ibrahim

Raouf A. Ibrahim
Department of Mechanical Engineering
Wayne State University
5050 Anthony Wayne Dr.
Room 2119 Engineering Bldg
Detroit, MI 48202
USA
E-Mail: ibrahim@eng.wayne.edu

ISBN: 978-3-642-00274-8 e-ISBN: 978-3-642-00275-5

DOI 10.1007/978-3-642-00275-5

Lecture Notes in Applied and Computational Mechanics ISSN 1613-7736
e-ISSN 1860-0816

Library of Congress Control Number: 2009920692

Typeset & Cover Design: Scientific Publishing Services Pvt. Ltd., Chennai, India.

Printed on acid-free paper

9 8 7 6 5 4 3 2 1 0

springer.com

In Memory of my parents.

Preface

Vibro-impact dynamics has occupied a wide spectrum of studies by dynamicists, physicists, and mathematicians. These studies may be classified into three main categories: modeling, mapping and applications. The main techniques used in modeling of vibro-impact systems include phenomenological modelings, Hertzian models, and non-smooth coordinate transformations developed by Zhuravlev and Ivanov. One of the most critical situations impeded in vibro-impact systems is the grazing bifurcation. Grazing bifurcation is usually studied through discontinuity mapping techniques, which are very useful to uncover the rich dynamics in the process of impact interaction. Note the available mappings are valid only in the absence of non-impact nonlinearities. Complex dynamic phenomena of vibro-impact systems include subharmonic oscillations, chaotic motion, and coexistence of different attractors for the same excitation and system parameters but under different initial conditions.

Selected applications of vibro-impact dynamics. These include lumped and continuous systems. Lumped systems cover a bouncing ball on an oscillating barrier, mass-spring-dashpot systems, normal and inverted pendulums, the spherical pendulum, the ship roll motion against icebergs, joints with free-play, rotor-stator rubbing in rotating machinery, vocal folds, microactuators, strings, beams, pipes conveying fluids with end-restraints, nuclear reactors and heat exchangers, and plates. These applications are discussed within the framework of the deterministic theory. Under random excitation the treatment requires special tools. The techniques of equivalent linearization and stochastic averaging have been applied to limited number of problems. One of the most beneficial outcomes of vibro-impact dynamics is the development of impact dampers, which have witnessed significant activities over the last four decades and have been used in several applications. On the other hand, vibro-impact has detrimental effects on the operations of mechanical systems and damage of pipes and rods in nuclear reactors. Each Chapter includes extensive literature review for its major sections followed by analytical description and main results. The book is supported by an extensive bibliography which exceeds 1,100 references of technical journal papers, technical reports, research monographs, MS and Ph.D. theses, and conference proceedings.

Some of the presented results were the outcome of research grants from the National Science Foundation (NSF) and the USA Office of Naval Research (ONR). I like to express my gratitude to the Program Directors, particularly Dr. Devendra Garg (NSF) and Dr. Kelly Cooper (ONR). I like to thank Dr. Valery Pilipchuk, Dr. Mohamed El-Sayad, and Mr. Ihab Grace who collaborated with me in conducting these research projects. The original brief form of this book was in the form of a review article citing about 550 references. In view of its length, Professor Freidrich Pfeiffer, of the Technical University of Munich and the Editor of the Springer series of Lecture Notes in Applied and Computational Mechanics, suggested that I should modify it as a research monograph. For that reason, I am indebted to Professor Pfeiffer and would like to thank him for his encouragement. During the course of modifications, I encountered some problems in presenting the derivations of Zhuravlev non-smooth coordinate transformation and Nordmark discontinuity mapping in a format suitable for the beginners. I am indebted to Professor Victor Bedichevsky of Wayne State University, Professor Harry Dankowicz of the University of Illinois, Urbana-Champaign, and Professor Xiaopeng Zhao of University of Tennessee who provided me a lot of help.

It was my wife, Sohair, who has been always supporting and encouraging me to accept the challenge and to complete the task on my own pace. I thank her so deeply for her patience and long-suffering during writing this book. Indeed without her support I could not complete this book.

Detroit, Michigan Raouf A. Ibrahim

Contents

Introduction

Vibro-impact systems had been known to laymen and children before they were considered by the scientific community. A skipping stone on the water surface and woodpecker toy are common examples. According to Wikipedia: "*Stone skipping is a pastime which involves throwing a stone with a flattened surface across a lake or other body of water in such a way that it bounces off the surface of the water. The object of the game is to see how many times a stone can be made to bounce before sinking.*" It is amazing that this game has attracted many researchers to provide analytical and physical insight (see, e.g., [685], [358], [963], [126], [183], [706], [895], [914], [535]). The woodpecker toy operates by self-excited vibrations due to the combined effects of friction, impact and weight. The motion of this toy is greatly influenced by simultaneous impacts, which cause discontinuous bifurcations ([836], [364], [837], [838], [579], [581], [365], [578]). The dynamics of the woodpecker toy can be analyzed with a one-dimensional Poincaré map.

Generally, vibro-impact systems involve multiple impact interactions in the form of jumps in state space. In most cases, there is energy loss due to impacts and the coefficient of restitution usually measures the degree of energy dissipation associated with an impact event. The time scale involved during impact is much smaller than the time scale of the natural frequency of the oscillator. The motion of vibro-impact systems in the presence or absence of friction, is usually described by strongly nonlinear non-smooth differential equations. These systems together with the modeling of impact forces require special treatments and representations, which will be described in Chapter 1. Chapter 1 presents some modeling and analytical techniques of vibro-impact interaction. These include the phenomenological modeling, Zhuravlev and Ivanov non-smooth coordinate transformations, Hertzian contact, point-wise mapping, and saw-tooth-time-transformation. The early work of vibro-impact dynamics is believed to be made by Bespalova [109] and Bespalova et al [110]. The basic theory of vibro-impact dynamics is well documented in many research monographs (see, e.g., [369], [518], [519], [53], [60], [61], [132], [1135], [317], [135], [137], [456], [558], [964], [567], [697], [139]). Chapter

R.A. Ibrahim: Vibro-Impact Dynamics: Model., Map. & Appl., LNACM 43, pp. 1–5.
springerlink.com

1 will not address a subclass of non-smooth multibody systems with planar frictional-impact contacts ([837], and [140]). Wu et al [1085] and Brogliato and Acary [139] presented a unified formulation of the equations of motion for constrained multibody systems with Coulomb friction, stiction, and impact together with constraint addition-deletion.

One of the most powerful tools of describing the complex characteristics of vibro-impact dynamics is the mapping of grazing bifurcation. Chapter 2 describes discontinuities mappings such as grazing and C-bifurcation mappings. Dankowicz and Piiroinen [203] presented a rigorous mathematical technique for stabilizing periodic or other recurrent motions. The spectrum of Lyapunov exponents of attractors of mechanical systems with impacts was estimated by De Souza et al [224] and Twizell et al [1021]. They introduced transcendental maps that describe solutions of integrable differential equations between impacts. This was supplemented by transition conditions at the instants of impacts. De Souza and Caldas [221] applied a small and precise perturbation on a given control parameter to stabilize desired unstable periodic orbits, embedded in the chaotic invariant sets of mechanical systems with impacts. Later, De Souza et al [219] applied a feedback control technique to suppress chaotic behavior in dissipative mechanical systems by using a small-amplitude damping signal. The control signal was generated by varying the damping coefficient according to the velocity direction. Wang et al [1059] developed an impulsive control scheme to stabilize chaos in a class of vibro-impact systems. The control is implemented just when the impact occurs, i.e., in discrete time instants. The impact velocity measurement was required for designing the controller. In another work, Wang et al [1060] examined the dynamic behavior of a controlled two-degree-of-freedom vibro-impact system with a damping control law using numerical simulation. The control signals are expressed by two piecewise-linear absolute value functions. It was found that this control scheme can successfully suppress chaos to periodic orbit.

Controllability and stabilization issues associated with a class of non-smooth dynamical systems, namely complementarity dynamical systems were addressed by Brogliato ([136], [138]), Brogliato et al [141] and Leine and Nijmeijer [580]. Convex analysis and complementarity problems were claimed to be the main analytical tools for control related studies. Lee and Yan ([575], [576]) proposed the position control based on feedback control force of impact oscillator under asymmetric double-sided barriers. It was shown that the stable or unstable (chaotic) impact oscillators can be controlled and kept in a desired position using a synchronization scheme.

Vibro-impact systems are encountered in many engineering applications such as vibro-impact response of heat exchanger tubes to aerodynamic excitation, impact of floating ice with ships, slamming of ocean waves on off-shore structures, ships colliding against fenders, rubbing between the stator structure and rotor blades in turbomachinery, hand-held percussion machines, loosely fitting joints, gear-pair systems with backlash, a ball bouncing on a table, pile drivers, collision of human vocal folds, and automotive braking systems. Chapters 3 through 6 present different types of applications. For example, Chapters 3 and 4 are devoted for lumped single and multi-degree-of-freedom systems, respectively. In many situations, vibro-impact can be associated with the generation of undesirable noise and sound. Some studies have been carried out with the purpose of reducing impact noise. The prediction of impact noise was the major theme of a series of research papers ([887], [888], [889], [890]). Igarashi and Aimoto ([440], [441]) discussed the mechanism of impact sound generation when a ball collides against a plate. The ball was attached with a force transducer and collided with a freely suspended square steel plate. The frequency characteristics were not affected by the size of the ball, the ball material, or the impact velocity. However, they were dependent on the plate boundary conditions. Oppenheimer and Dubowsky [752] predicted the noise and vibration of machines and their support structures using a heuristic energy-based criterion. The criterion revealed that the mechanism-support coupling affects noise radiation. In some cases the coupling was found to significantly affect vibration and noise radiation of the support structure, while having a relatively minor effect on mechanism response. Chapter 5 deals with other applications such as mechanical joints, micro-actuators, vocal folds, vibration protection systems, and other applications. Impact interactions can excite complex nonlinear responses in structures subjected to simple periodic excitation ([501], [63]). Some specific features include concentration of high harmonics in corresponding waveforms and the dependence of their behavior on system parameters, initial conditions and synergistic spatial effects.

Chapter 6 addresses the vibro-impact dynamics of continuous systems such as strings, beams, constraint pipes conveying fluid, nuclear reactors and heat exchangers, and plates. Equally important to analytical and numerical techniques is experimental validations using small dynamical models. Experimental investigations are very valuable in revealing such complex phenomena that are not predicted analytically or numerically. It has been proven that there is a possible similarity of several important vibro-impact systems under certain specific limits or constraints [89].

A common feature of these systems is that there is a discontinuity in the stiffness at the onset of collision. The repetition of impact can be periodic or random, and thus each class requires special mathematical treatment. Such systems exhibit complex types of resonance, bifurcations and chaos. The problem becomes more complex when different types of nonlinear resonance conditions such as parametric and internal resonances occur simultaneously in vibro-impact systems. When friction is accounted for, a variety of highly complex characteristics including chaos are possible ([189], [1039], [1040], [102], [118]).

The stability of periodic motion of impact systems can be studied by developing recurrence equations describing the evolution of periodic motion disturbances from impact to impact [356]. Kalagnanam [487] applied the Ott-Grebogi-Yorke (OGY) algorithm [758] to impact oscillators and stabilized their chaotic attractor on period-1 and period-2 orbits using small time-dependent perturbations of the driving frequency. It was demonstrated that the ability to switch the chaotic system between period-1 and period-2 orbits was achieved by controlled time-dependent perturbations. However, before the system settles on the stabilized orbit, it exhibits a long chaotic transient.

Janin and Lamarque [464] studied some numerical methods that can handle the occurrence of discontinuities in single-degree-of-freedom vibro-impact systems. These methods include classical Newmark and Runge-Kutta schemes. Regularization techniques used with finite element usually avoid the exact computation of all discontinuities using some kind of smoothing scheme [297]. Some special non-standard finite-difference schemes were developed ([277], [278]) for solving the response of impact oscillators whose numerical treatment by traditional methods is not always successful. These schemes incorporated the intrinsic qualitative parameters of the system such as the coefficient of restitution and the structure inherent nonlinearity. These schemes were found unconditionally stable and replicate a number of important physical properties of the oscillator such as the conservation of energy between two consecutive impact times. Shaw et al [929] presented some results pertaining to chaotic motions in a periodically forced impacting system, which is analogous to the version of Duffing's equation with negative linear stiffness. They developed a general method for determining parameter conditions under which homoclinic tangles[1] exist, which is a necessary condition for cross-well chaos to occur. It was shown how one may manipulate higher harmonics of the excitation in order to affect the range of excitation amplitudes over which fractal basin boundaries between the two potential wells exist. Imamura and Suzuki [444] developed the steady–state response and periodic solution of an impact oscillator. A general form of the periodic solution was obtained from a steady–state response. Later, Imamura ([442], [443]) derived an exact global form of all periodic solutions appearing in one–degree–of–freedom forced impact oscillator based on pseudo-feedback approach. For zero stiffness, Imamura [443] obtained all periodic solutions of forced vibro-impact systems with damping. Specific initial conditions that lead to periodic solutions were defined using pseudo-feedback approach.

This monograph provides an assessment of the common analytical techniques, numerical algorithms, and experimental results under deterministic and random excitations. The treatment under random excitation is addressed in Chapter 7 and covers briefly the pertinent features of the stochastic complex characteristics associated with simple vibro-impact systems. Related to vibro-impact dynamics is the development of impact dampers described in Chapter 8. This monograph will not address the problem of differential inclusion associated with unilateral constraints. The concept of a standard

[1] Homoclinic tangles refer to the intersection of the stable and unstable manifolds at a hyperbolic saddle point under the Poincaré map. Each point of transversal intersection is called a transversal homoclinic orbit.

inelastic shock was introduced by Moreau [694]. This concept yielded a very consistent formulation for the dynamics of systems with frictionless unilateral constraints. Moreau introduced convex analysis and formulated the governing equations in terms of measure differential inclusions. Schatzman (1998) studied the existence and uniqueness of solutions of one-dimensional dynamics with impact. Paoli and Schatzman ([783], [784]) considered discrete dynamical systems with perfect unilateral constraints and employed Moreau's impact law to decompose the velocity on the normal and tangent cones to the set of admissible positions at the impact point. Vibro-impact systems with a finite number of degrees of freedom was described by Paoli ([780], [781]) by a second-order measure differential inclusion for the unknown position completed with a constitutive impact law. Another formulation of a frictionless sweeping process is possible such that the unknown velocity belongs to an appropriate functional space and satisfies a first-order measure differential inclusion. Later, Paoli [782] proposed a time-discretization of the measure differential inclusion to describe the system dynamics. The proposed scheme proved the convergence of the approximate solutions to a limit motion, which satisfies the constraints. As an application, Paoli and Schatzman [785] numerically estimated the motion of a slender bar dropped on a rigid foundation. The bar was discretized by a system of rigid bodies linked by spiral springs or by a pair of linear springs. It was assumed that the impact is frictionless modeled by Newton's law.

Every chapter includes an extensive assessment in the beginning of each major section. The book is closed by a list of over 1,100 references that are cited in this research monograph.

Chapter 1
Modeling and Analytical Approaches

1.1 Introduction

The analytical modeling of vibro-impact systems is very crucial in predicting their dynamical behavior. The system can be linear or weakly nonlinear in the absence of impact. However, in the presence of impact, or friction, or both, it becomes "strongly" nonlinear. This strong nonlinearity owes it origin to the fact that the velocity before and after impact experiences a sudden change in its direction, and thus resulting in what is known as "*non-smooth dynamics*." Three particular techniques have been developed over the years in order to transform the non-smooth models into smooth ones. These include the power-law phenomenological modeling, the Zhuravlev and Ivanov non-smooth coordinate transformations, and the Hertzian contact law. Vibro-impact dynamics of linear systems, known as piecewise linear systems, have been treated in the literature using point-wise mapping. This approach solves the linear differential equation in two stages. The initial conditions of each stage are taken as the values of the solution of the previous stage at the end of its period. Other techniques include the saw-tooth-time transformation and the Lie group transformation. The basic principles of these approaches are presented in this chapter. The analysis of different models of vibro-impact systems was considered by Babitsky and Krupenin ([59], [61]). Analytical models approximating purely elastic and inelastic impact using smooth functions have been reviewed by Manevich and Gendelman [636].

1.2 Power-Law Phenomenological Modeling

In order to understand this modeling we consider a particle moving between two walls positioned at $x = \pm x_i$ as shown in Fig. 1.1(a). For an assumed rigid impact one must have to introduce the constraint $x \leq |x_i|$, where x is the particle displacement. This modeling is similar to a great extent to the one used by Shaw and Shaw [924], who represented the impact by the momentum

R.A. Ibrahim: Vibro-Impact Dynamics: Model., Map. & Appl., LNACM 43, pp. 7–29.
springerlink.com

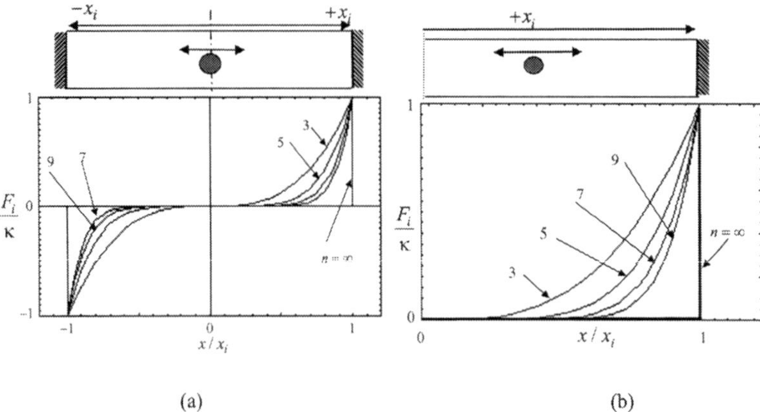

Fig. 1.1. Spatial dependence of impact for different exponent values for (a) double-sided barrier (b) single-barrier.

equation together with the definition of the coefficient of restitution. The collision was assumed as a discontinuous process. This constraint complicates the analysis since one must match solutions at points of interaction which are a *priori* unknown. In order to avoid operations with this type of constraints, one can introduce a phenomenological modeling that describes the interaction between the system and barriers with a special potential field of interaction, which is very weak in the region between the body surface and the barrier, but becomes fast growing in the neighborhood of the point, $x = \pm x_i$. The force of interaction can thus be phenomenologically represented by the power function

$$F_i = \kappa \left(\frac{x}{x_i} \right)^{2n-1}, \tag{1.1}$$

where $n >> 1$ is an integer and κ is a positive constant parameter usually estimated from experimental measurements. Such representation was proposed in vibro-impact problems by Hunt and Crossley [434] and in simulating liquid sloshing impact in moving containers by Pilipchuk and Ibrahim [852], El-Sayad, et al [289] and Ibrahim and El-Sayad [439]. Fig. 1.1 shows the dependence of the impact force on the spatial coordinate, x/x_i, for different values of n for the cases of double- and single-barrier. As $n \longrightarrow \infty$, we have the case of absolutely rigid body interaction, where the corresponding potential energy takes the square well form. For finite and large values of n, the interaction field is not absolutely localized at the points $x = \pm x_i$. This implies that both the system and barrier are not absolutely rigid, but admit a small deformation at the regions of impact, $x = \pm x_i$, for double-sided barrier. Nonlinear surface stiffness together with nonlinear surface damping was adopted in the literature [405] in studying the reliability and noise emissions from real mechanisms.

Alternatively, a smoothened tri-linear spring model may be adopted ([770], [772])

$$F(x) = \kappa_n \left[x - 0.5\left(|x + x_{ni}| - |x - x_{ni}|\right)\right]^n . \tag{1.2}$$

where x_{ni} is the free gap. This modeling enables one to represent adequately the free gap in which the constraints are zero and to smoothen the sharp discontinuity at $|x| = |x_i|$.

The damping effects during impact are spatially localized around the region, $x = \pm x_i$. The localized dissipative force may also be phenomenologically represented for two-sided barrier by the expression

$$F_d = c\left(\frac{x}{c_i}\right)^{2p} \dot{x}, \tag{1.3}$$

where c is a constant coefficient, which is determined experimentally, $p >> 1$ is a positive integer, and a dot denotes differentiation with respect to time. Païdoussis et al ([769], [770]), and Païdoussis and Semler [772] employed similar phenomenological modeling to represent the nonlinear force on the restraint due to impact on a cantilevered pipe conveying fluid. Kim et al [507] presented several smoothening functions and demonstrated their influence on the nonlinear frequency response characteristics of a single degree-of-freedom system. Andrianov and Awrejcewicz [13] developed an asymptotical behavior of a system with damping and high power-form nonlinearity.

Consider the vibro-impact oscillators described by the differential equation

$$\ddot{x} + d\Pi(x)/dx = 0. \tag{1.4}$$

where the potential $\Pi(x)$ possesses the general power form

$$\Pi(x) = \kappa_0 x^{(2m+2)/(2n+1)}. \tag{1.5}$$

$m \geq n, m, n = 0, 1, 2, ...$ and κ_0 is a positive stiffness constant. Equation (1.4) has been addressed by many researchers (see, e.g., [672], [673], [674], [675], [428], [849]). In order to apply the harmonic balance method both numerator and denominator of the exponent have to be odd. Nonlinear models for approximate description of elastic impacts with the help of smooth functions were developed using even-power potential for two-sided impact and inverse-power potential, $1/x^{\alpha} : x > 0$ *and* $\alpha > 0$, for one-sided impact [359]. It was demonstrated that both models may be generalized to describe the case of inelastic impact with velocity-independent recovery coefficient less than unity. The modification was achieved by adding another strongly nonlinear dissipative term. Pilipchuk [850] considered a family of strongly nonlinear oscillators with a generalized power form elastic force and viscous damping. An explicit analytical solution was obtained as a combination of smooth and non-smooth functions.

1.3 Zhuravlev Non-smooth Coordinate Transformation

This transformation was originally introduced by Zhuravlev [1137] who assumed purely elastic barriers as well as inelastic ones. The main rationale of such coordinate transformation is to convert the vibro-impact oscillator into an oscillator without barriers such that the corresponding equation of motion does not contain any impact terms. In the transformed form one can utilize any standard asymptotic approximate technique to solve the equation of motion. This technique has been widely used by researchers in the former Soviet Union [242]. Ezovskikh [304] proposed a change of variables that eliminate infinite discontinuities on the right sides of the equations of motion of vibro-impact systems with moving barriers. A number of typical problems were solved for a system impact with one-sided barrier located at $x = -x_i$, see Fig. 1.2(a), using the transformation

$$x = zsgn(z) - x_i. \tag{1.6}$$

This transformation shifts the barrier to the axis $z = 0$ and maps the domain $x > -x_i$ of the phase plane trajectories on the original plane $(x, \dot{x})$ to the new phase plane $(z, \dot{z})$. The first and second derivatives with respect to time are $\dot{x} = \dot{z}sgn(z)$, and $\ddot{x} = \ddot{z}sgn(z)$. Note that $z\left[d(sgn(z))/dt\right] = 0$, because $sgn(z)$ changes its sign at $z = 0$.

For the case of inelastic impact, the condition $\dot{x}_+ = -e\dot{x}_-$ must be introduced, where e is the coefficient of restitution, and $\dot{x}_+$ and $\dot{x}_-$are the system velocities just after and before impact, respectively. Note that the additional damping associated with inelastic impact may be significant than the inherent system linear and nonlinear damping terms. The coefficient e is assumed

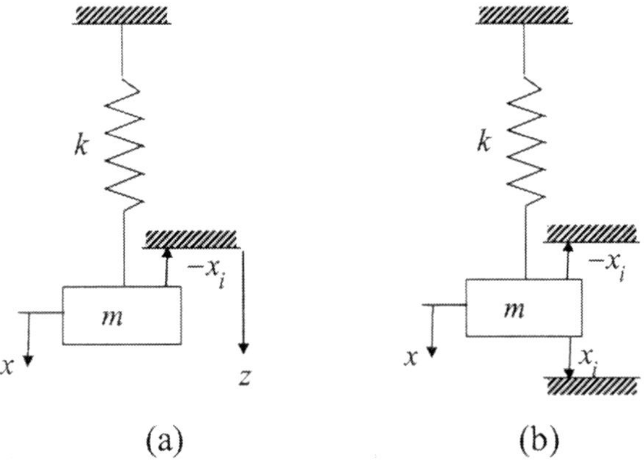

Fig. 1.2. Spring-mass systems with (a) one-sided barrier, (b) two-sided barrier.

to be close to unity, such that $(1-e)$ is considered a small parameter. According to the coordinate transformation given by equation (1.6), the impact is one-sided barrier at $x=-x_i$, and thus the impact condition $\dot{x}_+ = -e\dot{x}_-$, specified at $x=-x_i$, is transformed to

$$\dot{z}_+ = e\dot{z}_- \ at \quad z=0. \tag{1.7}$$

The transformed velocity jump is reduced by an amount proportional to $(1-e)$. It is possible to introduce this jump into the equation of motion using the Dirac delta-function, and thus one avoids using condition (1.7). The additional damping term due to impact may be written in the form

$$\left(\dot{z}_+ - e\dot{z}_-\right)\delta(t-t_i) = (1-e)\dot{z}\delta(t-t_i), \quad provided \quad |\dot{z}_+| < |\dot{z}| < |\dot{z}_-|, \tag{1.8}$$

where t_i is the time instant of impact. Zhuravlev (1976) introduced transformation of variables from the time domain to the space domain. This was done by setting $(t-t_i) = z(t)/\dot{z}(t_i)$, such that $\delta(t-t_i) = \delta(z(t)/\dot{z}(t_i))$, and one can write $\delta(t-t_i) = |\dot{z}|\delta(z)$[1]. In this case, equation (1.8) can be written in the form

$$(1-e)\dot{z}\delta(t-t_i) = (1-e)\dot{z}|\dot{z}|\delta(z). \tag{1.9}$$

Dimentberg [242] provided a systematic description of Zhuravlev coordinate transformation and demonstrated its application to vibro-impact systems under random excitation. A modified form transformation was proposed by Privalov ([867], [868]). The transformation was used to investigate a typical example of a vibro-impact system with unilateral motion constraint.

For the case of a two-sided barrier with a distance $2x_i$ between the two barriers, see Fig. 1.2(b), one may consider a mass spring system where the mass is located in the middle of the gap between the barriers. In this case, the following coordinate transformation may be introduced

$$x = S(z) \tag{1.10}$$

where $S(z)$ is a piecewise saw-tooth piecewise linear function defined as

$$S(z) = \begin{cases} z & if \ -x_i \le z \le x_i \\ 2x_i - z & if \ x_i \le z \le 3x_i \end{cases} \qquad S(z+4nx_i) = S(z), \qquad n=1,2,... \tag{1.11}$$

[1] This is obtained by writing $z(t) = z(t_i) + \dot{z}(t_i)(t-t_i)$, since $z(t_i)=0$, one can write $(t-t_i) = z(t)/\dot{z}(t_i)$. Thus $\delta(t-t_i) = \delta\left(z(t)/\dot{z}(t_i)\right)$.

Now consider $\int\limits_{-\infty}^{\infty} f(z)\delta(z)dz = f(0)$ and $\int\limits_{-\infty}^{\infty} f(z)\delta(\lambda z)dz = \left[\int\limits_{-\infty}^{\infty} f(u/\lambda)\delta(u)du\right]/\lambda = f(0)/\lambda$, one can write $\delta(\lambda z) = \delta(z)/\lambda$, provided $\lambda > 0$.

The Lagrangian of a unit mass-spring system, $L(x, \dot{x}) = (\dot{x}^2/2) - \Pi(x)$, where $\Pi(x)$ is the system potential energy and a dot denotes differentiation with respect to time, t. In terms of the transformed coordinate, the Lagrangian takes the form

$$L = \frac{1}{2}\left[\frac{dS(z)}{dz}\dot{z}\right]^2 - \Pi\left[S(z)\right] = \frac{1}{2}\dot{z}^2 - \Pi\left[S(z)\right]. \tag{1.12}$$

Note that the condition of constraints is automatically satisfied for any z due to $-x_i \leq S(z) \leq x_i$, and the Hamilton principle gives the differential equation of motion with no constraints

$$\ddot{z} + \frac{d\Pi\left[S(z)\right]}{dS(z)} S'(z) = 0. \tag{1.13}$$

where a prime denotes differentiation with respect to z. The function $S(z)$ belongs to a class of continuous but non-smooth functions. Its first derivative, $dS(z)/dz$, has bounded jumps at those points z for which $S(z) = \pm x_i$. Physically equation (1.13) describes a particle moving in the periodic non-smooth potential field as shown in Fig. 1.3. It is seen that the transformed system (1.13) is written in terms of a smooth coordinate without barriers. Note that the velocity, $\dot{x}(t)$, changes its signs at the borders of the interval $-x_i \leq x \leq x_i$, whereas the velocity of the transformed system, $\dot{z}(t)$, remains continuous. Furthermore, one will not deal any more with the problem of matching different pieces of the solution at the points $x = \pm x_i$. The

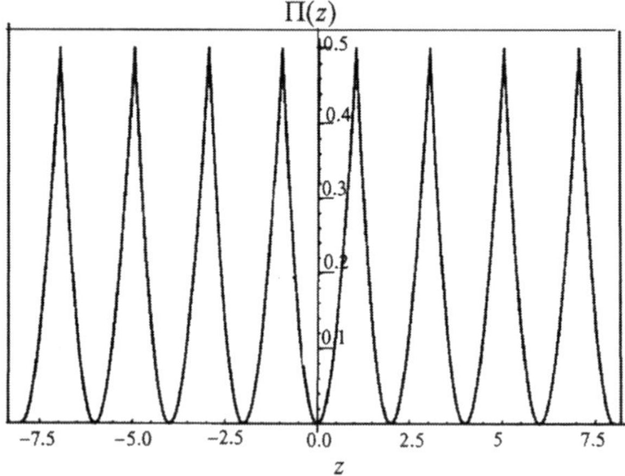

Fig. 1.3. Potential energy of the harmonic oscillator between two absolutely rigid barriers in terms of Zhuravlev coordinates transformation. Each crossing of the peaks corresponds to an impact, [854].

transformation (1.10) provides such matching automatically and gives a uniform expression for the solution that is more convenient for further manipulations. The most important advantage of the transformation is that it brings a special geometrical treatment of the vibro-impact motion in terms of the transformed system. Each impact event corresponds to a transmission (jump) into a new cell of the periodic potential field shown in Fig. 1.3. Such visualization may reveal interesting physical properties when dealing with more complicated systems of many degrees of freedom as will be illustrated in chapter 4.

1.4 Ivanov Transformation

A modified non-smooth coordinate transformation was developed by Ivanov [455] for the case of inelastic impact. In terms of the generalized coordinates, $q_0, q_1, ..., q_n$, subject to the unilateral constraint $q_0 \geq 0$, these systems may be described by the following set of equations in the vector form

$$\dot{\mathbf{X}} = \mathbf{F}(t, \mathbf{X}), \quad \mathbf{F}(t + T, \mathbf{X}) = \mathbf{F}(t, \mathbf{X}), \tag{1.14}$$

where $\mathbf{X} = \left\{ q_0, \dot{q}_0, q_1, ..., q_n, \ p_1, ..., p_n \right\}^{\overline{T}}$, p_i are the generalized momenta. The functions $\mathbf{F}(t, \mathbf{X})$ are periodic with period T, and the superscript $\overline{T}$ denotes transpose. At the instant of impact, $t = t^*$, we have

$$q_0(t^*) = 0, \quad \text{and} \quad \dot{q}_0(t^*) < 0. \tag{1.15}$$

The impact reaction force R^* involves energy dissipation [534] and can be expressed by the viscoelastic model

$$R^* = \begin{cases} -\varpi^2 q_0 - 2\eta\varpi\dot{q}_0 & if \quad q_0 < 0 \\ 0 & if \quad q_0 \geq 0 \end{cases}, \tag{1.16}$$

where ϖ is large and $\eta \in [0, 1)$. The equation of motion of the impact trajectory in the region $q_0 < 0$ is

$$\ddot{q}_0 + 2\eta\varpi\dot{q}_0 + \varpi^2 q_0 = F_2. \tag{1.17}$$

The solution of this equation subject to the initial conditions (1.15) is

$$\begin{aligned} q_0 = & \frac{\mathbf{F}_2^0(t^*, \mathbf{X}(\mathbf{t}^*))}{\varpi^2} \left[1 - \frac{\mathrm{e}^{-\eta\varpi}}{\sqrt{1-\eta^2}} \sin\left(\varpi\sqrt{1-\eta^2}\tau + \arcsin\sqrt{1-\eta^2} \right) \right] \\ & + \frac{\dot{q}_0(t^*)\mathrm{e}^{-\eta\varpi}}{\varpi\sqrt{1-\eta^2}} \sin\left(\varpi\sqrt{1-\eta^2}\tau \right) + O(\tau), \end{aligned} \tag{1.18}$$

where $\tau = t - t^*$, and $X_j(t) = X_j(t^*) + O(\tau)$, $j = 3, ..., 2n+2$. Note that $\tau = O(\varpi^{-1})$, which implies that the duration of impact is very small, and thus the variables X_j do not change during impacts. On the other hand, the displacement q_0 and velocity $\dot{q}_0$, experience two distinct limiting regimes.

The first regime is the impact regime in which $\varpi|\dot{q}_0(t^*)| \gg |\mathbf{F}_2^0|$ and the duration of the impact τ is close to $\pi/(\varpi\sqrt{1-\eta^2})$. The restitution law is

$$\dot{q}_0(t^* + 0) = -e\dot{q}_0(t^* - 0), \quad e = \mathrm{e}^{-\pi\eta/\sqrt{1-\eta^2}} \in [0, 1]. \tag{1.19}$$

Ivanov [455] introduced the new coordinate transformation

$$\begin{aligned} \mathbf{Y} &= \{z, v, q_1, ..., q_n, p_1, ..., p_n\}^T, \\ q_0 &= z sgn(z), \qquad \dot{q}_0 = Rv\ sgn(z), \end{aligned} \tag{1.20}$$

where $R = 1 - ksgn(zv)$, $k = [(1-e)/(1+e)] \in [0, 1)$. The values of z and v are not restricted. The substitution $\mathbf{Y} \to \mathbf{X}$ is irreversible since each vector $\mathbf{X}$ has two inverses for which $z_1 = -z_2$, and $v_1 = -v_2$. However, for every $\mathbf{Y}$ there exists a unique image $\mathbf{X}$ in accordance with the transformation

$$\begin{aligned} \mathbf{X} &= \mathbf{SY} \\ \mathbf{S} &= diag\{sgn(z),\ R\ sgn(z),\ 1, ..., 1\} \end{aligned} \tag{1.21}$$

where the matrix $\mathbf{S}$ is diagonal of dimension $2n+2$. The equations of motion (1.14) will take the transformed form

$$\dot{\mathbf{Y}} = \mathbf{S}^{-1}\mathbf{F}(t, \mathbf{SY}) = \mathbf{G}(t, \mathbf{Y}). \tag{1.22}$$

Equations (1.22) possess a discontinuous right-hand side due to the presence of terms such as $sgn(z)$ and $sgn(zv)$. However, their solutions are continuous vector functions in the time domain, and differentiable provided $zv \neq 0$. Note the first equation in the system (1.22) is

$$\dot{z} = Rv. \tag{1.23}$$

Since $R > 0$, the product zv changes its sign from negative to positive. Before intersection $R = 1 + k$, while after intersection $R = 1 - k$. For the case of purely elastic impact, i.e., $e = 1$, the transformation is reduced to the one developed by Zhuravlev.

The second regime belongs to very small velocity in which $\varpi|\dot{q}_0(t^*)| \ll |\mathbf{F}_2^0|$. According to equation (1.18) the duration of impact is $\tau \approx 2|\dot{q}_0(t^*)|/\mathbf{F}_2^0$, and from equation (1.19) $e = 1$.

According to the impact rule given by equations (1.19), Ivanov [455] introduced an auxiliary phase space such that all trajectories would be continuous. This can be demonstrated by the linear oscillator with one-sided barrier:

$$\ddot{q} + 2\zeta\omega_n\dot{q} + \omega_n^2 q = 0, \tag{1.24}$$

such that $q \geq 0$, and $\zeta < 1$. System (1.24) allows a continuous representation in the form

$$\dot{z} = Rv, \qquad \dot{v} = -2\zeta\omega_n v - \frac{\omega_n^2}{R} z \tag{1.25}$$

The solution of this system is

$$z = \exp(\zeta\omega_n)\left(A_j \cos\omega_n\sqrt{1-\zeta^2} + B_j \sin\omega_n\sqrt{1-\zeta^2}\right) \tag{1.26}$$

where the constants A_j and B_j are obtained at each intersection with the the velocity $v-$axis to preserve the continuity of $z(t)$ and $v(t)$. For the initial conditions $q(0) = 0$ and $\dot{q}(0) = V > 0$, one obtains $A_1 = 0$, and $B_1 = (V/\omega_n\sqrt{1-\zeta^2})$. The first impact occurs at time $t = \pi/(\omega_n\sqrt{1-\zeta^2})$, and during the second period, one obtains $B_2 = e[\exp(-2\pi\zeta/\sqrt{1-\zeta^2})]B_1$. After the second impact at $t = 2\pi/(\omega_n\sqrt{1-\zeta^2})$ the value of B_2 changes to $B_3 = e^2[\exp(-2\pi\zeta/\sqrt{1-\zeta^2})]^2 B_1$. This sequence continues and depending on the damping ratio ζ the phase portrait could be periodic, for $\zeta = 0$, decaying for $\zeta > 0$, or unstable for $\zeta < 0$.

Dimentberg et al [246] employed the path integral (PI) method together with the Zhuravlev-Ivanov coordinate transformation to examine the response probability density function of stochastic vibro-impact systems with high energy losses at impacts. It was found that the response energy probability density function yields relatively larger response energy as the coefficient of restitution increases and the results were confirmed using Monte Carlo simulation. For a detailed description of the PI-method and its implementation the reader may consult references ([704], [532], [705], [259], [260]). A general procedure for analyzing near-elastic vibro-impact problems using a discontinuity-reducing transformation of coordinates together with an extended averaging scheme was proposed introduced in references ([322], [323], [992], [993]). This combined procedure was demonstrated through several applications ranging from mass–spring–dashpot systems with inelastic one-sided barrier to self–excited friction–oscillators with one or two stops.

1.5 Hertzian Contact

1.5.1 Modeling

Hertzian contact refers to the localized stresses that develop as two curved surfaces come in contact and deform slightly under an applied load ([406], [407]). This deformation is dependent on the elasticity of the material in contact, i.e., its modulus of elasticity. The contact stress is usually given as a function of the normal contact force, radii of curvature and the modulus of elasticity of both bodies. The Hertzian contact law expresses the contact

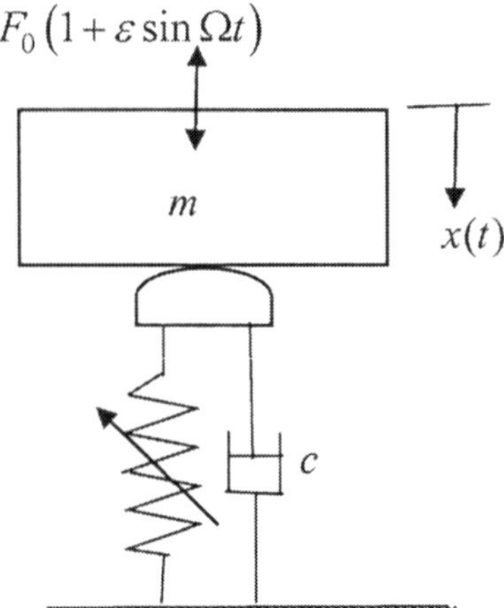

Fig. 1.4. Schematic diagram of an oscillator with Hertzian contact.

normal force, F_c, in terms of the penetration, x, i.e.,

$$F_c = \kappa x^n. \tag{1.27}$$

where κ is the contact stiffness coefficient, which is a function of the elastic properties and geometries of the contact bodies. This coefficient is derived in details by Antoine et al [14]. The value of n for Hertzian elastic point contact is $3/2$ and for extended planar contact (non-Hertzian) is in the range $3/2 - 7/2$. The contact stiffness can be obtained from analytical contact models such as the Hertzian contact model for spherical contacts [481]. Equation (1.27) is valid as long as the contact area is small compared to the geometry of the colliding bodies and the contact areas are perfectly smooth such that the impact is not associated with friction. Furthermore, the materials of the impact bodies are assumed isotropic and linearly elastic.

For the case of rough surfaces in contact, the statistical model developed by Greenwood and Williamson (GW) [377] for rough surfaces is usually used to estimate the contact stiffness. For the case of a sphere-plane Hertzian contact, shown in Fig. 1.4, Perret-Liaudet ([805], [806]) studied second-order subharmonic and superharmonic resonances of the oscillator described by the equation of motion

$$m\ddot{x} + c\dot{x} + kx^{3/2} = F_0\left(1 + \gamma\sin\Omega t\right), \qquad for\ x \geq 0. \tag{1.28}$$

$$m\ddot{x} = F_0\left(1 + \gamma\sin\Omega t\right), \qquad for\ x < 0. \tag{1.29}$$

where F_0 is the static external load component, Ω is the frequency of the harmonic load component, and γF_0 is the amplitude of the harmonic component. The static contact compression and linearized natural frequency were given by the expressions $x_s = (F_0/\kappa)^{2/3}$ and $\omega_n^2 = (3\kappa/2m)\sqrt{x_s}$, respectively. Equations (1.28) and (1.29) can be written in the non-dimensional form

$$q'' + 2\zeta q' + \left(1 + \frac{2}{3}q\right)^{3/2} = 1 + \gamma \sin \nu\tau \qquad for\ q \geq -\frac{3}{2}, \tag{1.30}$$

$$q'' = 1 + \gamma \sin \nu\tau \qquad for\ q < -\frac{3}{2}, \tag{1.31}$$

where the non-dimensional parameter $q = \frac{3}{2}\left(\frac{x}{x_s} - 1\right)$, a prime denotes differentiation with respect to the non–dimensional time parameter $\tau = \omega_n t$, $\nu = \frac{\Omega}{\omega_n}$, and $\zeta = \frac{c}{2}\frac{\omega_n}{m}$.

Perret-Liaudet ([805], [806]) expanded the third expression on the left-hand side of equation (1.30) in a Taylor series to give

$$\left(1 + \frac{2}{3}q\right)^{3/2} \approx 1 + q + \frac{1}{6}q^2 - \frac{1}{54}q^3. \tag{1.32}$$

Fig. 1.5 shows a comparison between the exact and approximate curves of the normal restoring force, $F(q)$. In this case, equation (1.30) takes the form

$$q'' + 2\zeta q' + q + \frac{1}{6}q^2 - \frac{1}{54}q^3 = \gamma \sin \nu\tau. \tag{1.33}$$

Perret-Liaudet ([805], [806]) reported three frequency ranges characterized by the number of coexisting responses. As the frequency decreases, a single stable harmonic response loses its stability while, at the bifurcation point, another stable 2–subharmonic response is created (flip bifurcation). At the

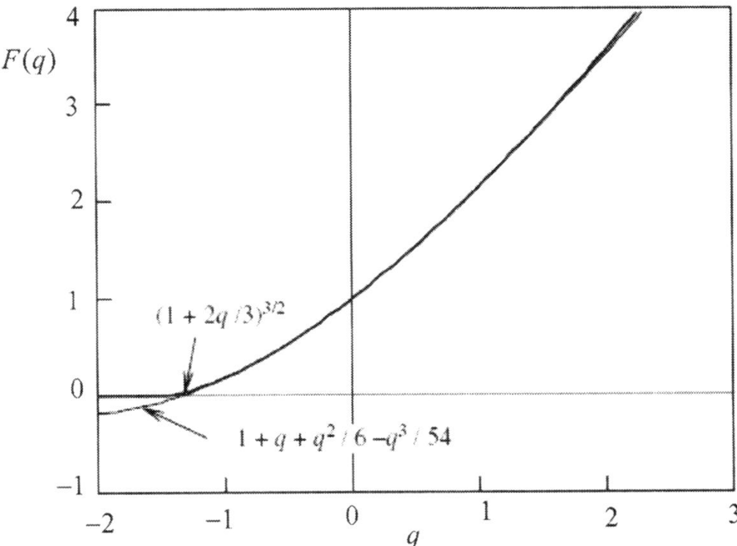

Fig. 1.5. Dependence of restoring contact force model (thick curve) and its approximate form (thin curve), [805].

second bifurcation point, the unstable harmonic response recovers its stability, while another unstable 2−subharmonic response is formed. It was noted that the stable subharmonic response leads to contact loss. Thus, this resonance can initiate vibro-impact phenomena. The occurrence of vibro-impact gives rise to a sudden extension of the superharmonic resonance with high response levels. It was found that both subharmonic and superharmonic resonances precede vibro-impact conditions.

If the contacting surfaces are not spherical, the contact force will differ from the Hertzian law. In real structures, each impact is associated with energy loss and thus the damping has to be included in the analytical modeling. This damping is in general nonlinear function of deformation and velocity. Půst [871] and Půst and Peterka [872] adopted the following modeling of the contact force

$$F_c\left(x, \dot{x}\right) = f(x)\left[1 + g(\dot{x})\right]. \tag{1.34}$$

where the function $f(x) > 0$ is analytic for $x > 0$, and $f(x) = 0$ for $x \leq 0$. The damping function $g(\dot{x})$ is analytic and satisfies the conditions: $\dot{x}g(\dot{x}) \geq 0$, and $\left[1 + g(\dot{x})\right] > 0$. Different forms were developed by Hunt and Crossley [434], Yang and Lin [1095], and Půst and Peterka [872]. These include:

i. The behavior of damped stop with piecewise linear characteristics encountered in leaf springs where damping is caused by dry friction between leaves and the contact force takes the form

$$F_c = \kappa x\left(1 + \alpha\frac{\dot{x}}{|\dot{x}|}\right). \tag{1.35}$$

where α is a constant coefficient.

ii. The simple contact force model [434]

$$F_c = \kappa x\left(1 + \alpha\dot{x}\right). \tag{1.36}$$

iii. Contact force model with quadratic damping function

$$F_c = \kappa x\left(1 + \alpha_1\dot{x}|\dot{x}|\right). \tag{1.37}$$

iv. Hertz contact force with damping

$$F_c = \kappa x^{3/2}\left(1 + \dot{x}\right). \tag{1.38}$$

v. Another type of Hertzian contact force associated with hysteretic damping was developed by Van de Wouw et al ([1024], [1025]). It takes the form

$$F_c = \kappa x^{3/2}\left(1 + \frac{\mu}{\kappa}\dot{x}\right) = \kappa x^{3/2}\left[1 + \frac{3}{5}\left(1 - e^2\right)\frac{\dot{x}}{\dot{x}^-}\right], \quad for\ x \geq 0. \tag{1.39}$$

where e is the coefficient of restitution, $\dot{x}^-$ is the velocity difference of the two colliding bodies at the beginning of impact. Both κ and e were estimated experimentally where several impacts were observed to measure the indentation x, the indentation velocity $\dot{x}$, and the contact force F_c. The dependence of the contact force between the impact of two half spheres on the indentation was obtained by Van de Wouw et al [1024] and is shown in Fig. 1.6. The coefficient of restitution can be obtained by estimating the energy loss per impact, ΔE, which is equivalent to the area of hysteretic loop. This can be written in the form, after using the second expression of the first form of equation (1.39), i.e.,

$$\Delta E = \oint \mu x^{3/2}\dot{x}dx. \tag{1.40}$$

The parameter μ can be estimated from the expression

$$\mu = \frac{\Delta E}{\oint x^{3/2}\dot{x}dx}. \tag{1.41}$$

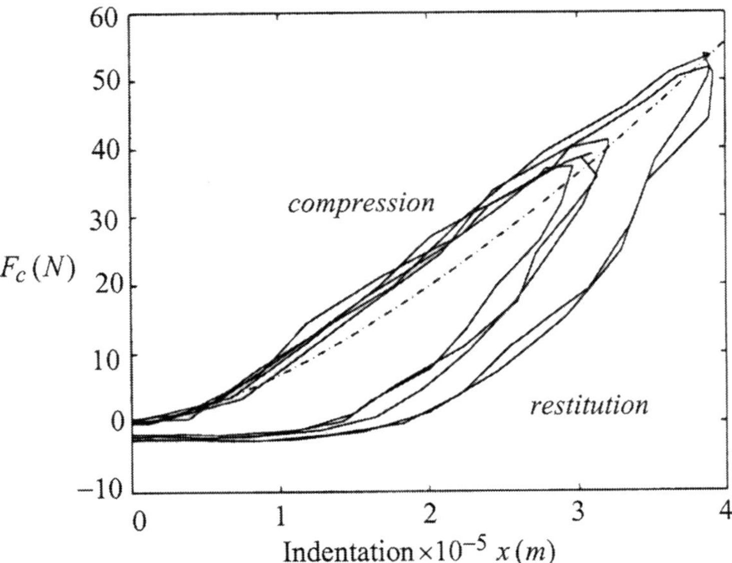

Fig. 1.6. Dependence of contact force on indentation: —— measurement, - - - - analytical $kx^{3/2}$, [1024].

Thus the coefficient of restitution is

$$e = \sqrt{1 - \frac{4\mu\dot{\bar{x}}}{3\kappa}}. \tag{1.42}$$

Note that the Hertzian stiffness κ and the coefficient of restitution can be determined from least-squares estimates from several impact tests.

When the contact stiffness coefficient increases, the response was found to possess a non-zero mean. Furthermore, the clearance was found to be about twice the square root of the mean square response of the corresponding linear system. The stationary random response of single- and two-degree of freedom systems with Hertzian contact law and subjected to a random excitation was analyzed in the literature ([475], [476], [477]). Lin and Bapat [595] introduced a clearance estimation algorithm to study the random response of a single-degree-of-freedom oscillator with two elastic stops. Feng and He [319] derived the mean of impact Poincaré map for a class of random vibro-impact systems. The calculated results revealed complex nonlinear behavior and the bifurcation diagrams exhibited the routes to random chaos.

1.5.2 Contact Stiffness of Braking Systems

In automotive braking systems, contact forces between sliding surfaces are strongly nonlinear and non-smooth; and thus can result in complex dynamic characteristics in the form of chatter and squeal. It is known that contact stiffness, true contact area, and contact forces play an important role in the dynamic behavior of braking systems. These parameters are important in studying contact dynamics and interface modeling. Beside their importance in automotive braking systems, they are important in other applications such as robotic applications, micro-bearings and spindle bearings in magnetic storage hard disk drives, and lightly loaded mechanical joints. The sliding friction measurements revealed the existence of natural normal micro-vibrations whose frequency is determined by the contact stiffness and the slider mass. The contact stiffness was found to be a function of the normal displacement ([1007], [1008]). Tolstoi [1007] found that both the negative friction-velocity slope and the frictional self-excited vibration are closely associated with the freedom of the slider to move in the normal direction.

Carson and Johnson [152] and Gray and Johnson [374] indicated that inelastic deformation, sometimes leading to corrugations, is mainly due to the nature of the Hertzian contact between elastic-plastic solids. Nayak [720] considered the vibration of an elastic ball rolling over an elastic half-space. The contact forces were represented by a viscous damper and a nonlinear contact spring. Accordingly, the equation of motion for contact vibrations is essentially nonlinear. Nayak [720] developed a complicated damping model in which energy loss depends on the frequency of stresses. Sosnovskiy and Sherbakov [953] reported some experimental results dealing with irregular wavy residual damages (troppy phenomenon), which may occur in the contact area in rolling friction as the result of a non-stationary process of cyclic deformation. These damages initiate vibro-impact loading of an active system. It was concluded that vibro-impact emerges as the result of drastic quasi-periodic change of contact pressure accompanied by pulsations of displacements.

Hess and Soom ([411], [412]) examined normal vibrations and friction under harmonic normal loads. They considered Hertzian and rough planar contacts. The Hertzian contact yielded a nonlinear ordinary differential equation with quadratic and cubic stiffness nonlinearities with negative and positive coefficients. They used the method of multiple scales and the results revealed softening amplitude-frequency response characteristics. The softening effect is associated with decreasing contact stiffness. Castravete et al [154] measured the contact stiffness in a pin-disc system and found the normal stiffness is a mixture of soft and hard nonlinearities. It was phenomenologically represented by the following formula

$$\kappa(x) = k_0 - k_1 x + k_2 x^2. \tag{1.43}$$

where k_0, k_1, and k_2 are positive constants and x is the normal penetration. Fig. 1.7 shows the nonlinear dependence of the normal load on the penetration. Equation (1.43) was obtained from the slope of the curve fitting shown in Fig. 1.7.

In order to measure the contact stiffness one should measure the elastic deformation under the influence of contact forces. The onset of friction-vibration instability is sometimes attributed to the inverse variation in the

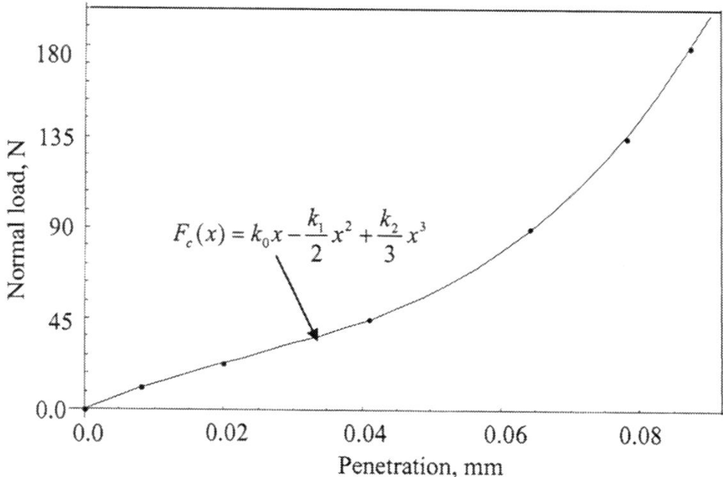

Fig. 1.7. Static test of load-penetration of pin-on-disc system (frictional pin material) points are experimental readings, solid curves are curve fitting, [154].

friction force with the relative velocity of sliding whereby the friction-velocity gradient becomes high compared with the system damping coefficient. Sherif ([936], [937], [938]) studied the influence of the effective contact stiffness on the stability of sliding surfaces. It was found that the instability of frictional vibrations is mainly dependent on the value of the total tangential contact stiffness of sliding surfaces as well as the system stiffness. Furthermore, continuous modification of the equivalent contact stiffness by wear mechanism was found to be the main cause of squeal initiation. The analysis revealed that stability can be attained at all possible squeal frequencies of the friction pad by the good selection of its geometrical configuration (thick, wide and short brake pad) and its material loss factor within certain limits.

1.6 Point-Wise Mapping

The point-wise mapping may be demonstrated by considering the vibro-impact system described by the equation of motion

$$\ddot{x} + 2\zeta\omega_n\dot{x} + \omega_n^2 x = f(t) \qquad for\ x < x_0, \tag{1.44}$$

subject to the velocity constraint

$$\dot{x}_+ = -e\dot{x}_- \qquad for\ x = x_0, \quad and\ 0 < e < 1. \tag{1.45}$$

The solution may be obtained by a step-wise integration ([53], [519], [925], [930], [931], [932]). This is known as the point-wise mapping method, where one has to match the solutions at points of impact. In the former Soviet Union, it is known as the "*stitching*" method. In this method the equations of motion are integrated between impacts. Kinematic impact conditions are then used to switch between time intervals of solution and solutions over times involving several impacts are obtained by gluing together a suitable number of such partial solutions. Equation (1.44) is usually solved for a given set of initial conditions during the free flight trajectory until the mass hits the barrier at $x = x_0$. The impact condition is then imposed to obtain the velocity just after impact, $\dot{x}_+$. This is then taken as the initial condition for the trajectory after impact. Note that the time duration during impact must be known a priori.

Shaw and Holmes ([930], [931], [932]) analyzed a simple model of a periodically forced oscillator with a constraint that leads to motions with impacts. For perfectly plastic impacts, Shaw and Holmes [932] represented the system dynamics by a discontinuous map defined on a circle. They considered the system shown in Fig. 1.8 whose equation of motion may be written in the form

$$x'' + 2\zeta x' + h(x) = f\cos\nu\tau. \tag{1.46}$$

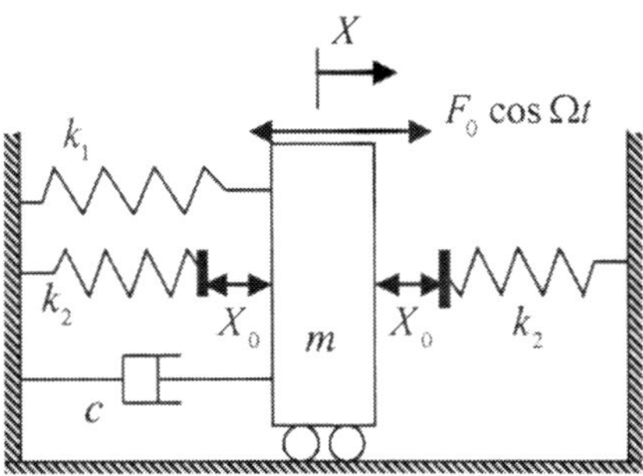

Fig. 1.8. Schematic diagram of two-sided simple degree-of-freedom oscillator with elastic supports.

where a prime denotes differentiation with respect to the non-dimensional time parameter $\tau = \omega_n t$, $\omega_n^2 = k_1/m$, $x = X/X_0$, X_0 is the gap between the mass and spring k_2, $2\zeta = c/(m\omega_n)$, $\nu = \Omega/\omega_n$, $f = F_0/m\omega_n^2 X_0$, $h(x) = x$ for $|x| < 1$ and $h(x) = \varpi^2 x + (1 - \varpi^2)sgn(x)$ for $|x| \geq 1$, $\varpi_i^2 = (k_1 + k_2)/k_1$ for even i (during contact with spring k_2) or $\varpi_i^2 = 1$ for odd i (for free flight motion). Equation (1.46) comprises two equations corresponding to the two cases of $h(x)$. The analytic solution of equation (1.46) for double-contact motion is (modified after Shaw and Holmes [930])

$$x(\tau) = e^{-\zeta(\tau-\tau_1)} \{A_i \cos[\omega_{id}(\tau - \tau_i)] + B_{id}[\omega_{id}(\tau - \tau_i)]\}$$
$$+a_i \cos \nu\tau + b_i \sin \nu\tau + sgn(\tau) sgn\left(x_1^{'}\right)\left(\frac{1-\varpi_i^2}{\varpi_i^2}\right), \quad (1.47)$$

$$i = 0, 1, 2, ..., N.$$

where

$$A_i = -\frac{(\varpi_i^2 - \nu^2)}{(\varpi_i^2-\nu^2)^2 + (2\zeta\nu)^2} f \cos \nu\tau - \frac{2\zeta\nu}{(\varpi_i^2-\nu^2)^2+(2\zeta\nu)^2} f \sin \nu\tau + (-1)^i \frac{sgn(\tau)sign(x_i^{'})}{\varpi_i^2},$$

$$B_i\omega_{id} = x_i^{'} + (-1)^i \frac{sgn(\tau)sign(x_i^{'})}{\varpi_i^2} + \left[\frac{(\varpi_i^2-\nu^2)\nu f}{(\varpi_i^2-\nu^2)^2+(2\zeta\nu)^2} - \frac{2\zeta^2\nu f}{(\varpi_i^2-\nu^2)^2+(2\zeta\nu)^2}\right] \sin \nu\tau$$
$$-\frac{1}{\omega_{id}}\left[\frac{(\varpi_i^2-\nu^2)\zeta f}{(\varpi_i^2-\nu^2)^2+(2\zeta\nu)^2} + \frac{2\zeta^2\nu f}{(\varpi_i^2-\nu^2)^2+(2\zeta\nu)^2}\right] \cos \nu\tau,$$

$$a_i = \frac{(\varpi_i^2-\nu^2) f}{(\varpi_i^2-\nu^2)^2+(2\zeta\nu)^2}, \qquad b_i = \frac{2\zeta\nu f}{(\varpi_i^2-\nu^2)^2+(2\zeta\nu)^2}, \qquad \omega_i = \sqrt{\varpi^2 - \zeta^2}.$$

Equation (1.47) involves contacting and non-contacting regimes. The crossing times, when $x(\tau_i) = 1$ are not known a priori and must be estimated from the roots of the two solutions:

$$x_0(\tau; \tau_0, x_0^{'}) = 1, \qquad and \;\; x_1(\tau; \tau_0, x_1^{'}) = 1. \quad (1.48)$$

The stability of a periodic response of the system may be examined by evaluating the eigenvalues of the Jacobian of the return map at $(\bar{\tau}, \bar{x}^{'})$. The Poincaré section must satisfy the condition

$$\left(\bar{\tau} + \frac{2\pi n}{\nu}, \bar{x}'\right) = \mathbf{P}^k \left(\bar{\tau}, \bar{x}'\right), \tag{1.49}$$

where $\mathbf{P}^k$ is the k-th iterated map and n is the order of subharmonic. Equation (1.49) holds for both symmetric ($2k$ impact motion) and non-symmetric periodic motion after impacts at $x = \pm 1$. The Jacobian of the oscillator is

$$J = \begin{bmatrix} \frac{\partial \tau_{i+1}}{\partial \tau_i} & \frac{\partial \tau_{i+1}}{\partial x_i'} \\ \frac{\partial x_{i+1}'}{\partial \tau_i} & \frac{\partial x_{i+1}'}{\partial x_i'} \end{bmatrix}, \tag{1.50}$$

where $\frac{\partial \tau_{i+1}}{\partial \tau_i} = \frac{e^{-\zeta(\tau_{i+1}-\tau_i)}}{x_{i+1}'} \left\{ x_i' \cos\left[\omega_i\left(\tau_{i+1}-\tau_i\right)\right] - \frac{1}{\omega_i}\left[\zeta x_i' + (-1)^i sign(x_i') - f\cos(\nu\tau_i)\right] + \sin\left[\omega_i\left(\tau_{i+1}-\tau_i\right)\right] \right\},$

$$\frac{\partial \tau_{i+1}}{\partial x_i'} = \frac{e^{-\zeta(\tau_{i+1}-\tau_i)}}{x_{i+1}'\omega_i} \sin\left[\omega_i\left(\tau_{i+1}-\tau_i\right)\right],$$

$$\begin{aligned}\frac{\partial x_{i+1}'}{\partial \tau_i} = {} & \frac{\partial \tau_{i+1}}{\partial \tau_i} a_{i+1} + e^{-\zeta(\tau_{i+1}-\tau_i)} \left\{ \left[2\zeta x_i' + (-1)^i sign(x_i') - f\cos(\nu\tau_i)\right] \times \right. \\ & \cos\left[\omega_i\left(\tau_{i+1}-\tau_i\right)\right] \\ & \left. + \left[x_i' \frac{\omega_i^2-\zeta^2}{\omega_i} + \frac{\zeta}{\omega_i}\left[-(-1)^i sign(x_i') - f\cos(\nu\tau_i)\right]\right] \times \sin\left[\omega_i\left(\tau_{i+1}-\tau_i\right)\right] \right\},\end{aligned}$$

$$\begin{aligned}\frac{\partial x_{i+1}'}{\partial x_i'} = {} & e^{-\zeta(\tau_{i+1}-\tau_i)} \left\{ \cos\left[\omega_i\left(\tau_{i+1}-\tau_i\right)\right] - \frac{\zeta}{\omega_i}\sin\left[\omega_i\left(\tau_{i+1}-\tau_i\right)\right] \right\} \\ & + \frac{\partial \tau_{i+1}}{\partial x_i'} a_{i+1},\end{aligned}$$

$$\begin{aligned}a_{i+1} = {} & e^{-\zeta(\tau_{i+1}-\tau_i)} \left\{ \left(\zeta^2 A_i - 2\zeta\omega_i B_i - \omega_i^2 A_i\right)\cos\left[\omega_i\left(\tau_{i+1}-\tau_i\right)\right] \right. \\ & \left. + \left(\zeta^2 B_i + 2\zeta\omega_i A_i - \omega_i^2 B_i\right)\sin\left[\omega_i\left(\tau_{i+1}-\tau_i\right)\right] \right\} \\ & - a_i\nu^2\cos(\nu\tau_i) - b_i\nu^2\sin(\nu\tau_i).\end{aligned}$$

The map was shown to undergo period-doubling bifurcations followed by complex sequences of transitions, due to discontinuities, in which arbitrarily long super-stable periodic motions occur. Shaw and Holmes [930] considered the case of one-sided spring k_2 and showed that bifurcation occurs between $\nu = 2.4$ and $\nu = 2.42$ for stiffness ratio $\varpi = 4.0$ as shown in the two phase portraits of Fig. 1.9. Other subharmonic orbits were found to exist for different values of stiffness ratio and damping factor ([312], [313]). For two-sided spring k_2, Knudsen and Massih [514] found that the motion can be a stable periodic symmetric attractor, a supercritical breaking bifurcation, and stable and unstable asymmetric periodic solutions depending on excitation frequency ratio ν and initial conditions.

Shaw and Holmes [931] considered the simple harmonic oscillator with harmonic excitation and a constraint that restricts motions to one side of the equilibrium position. Upon reaching a specified displacement, the direction of

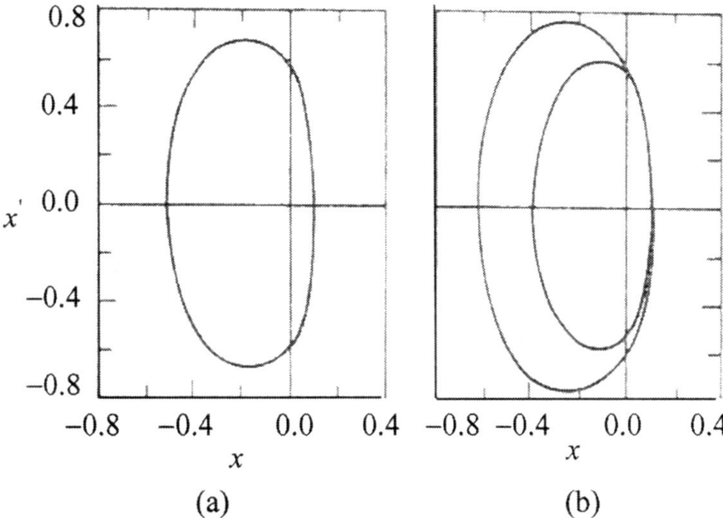

Fig. 1.9. Phase portraits of one-sided second spring k_2 (a) stable period one orbit at $\nu = 2.4$, and (b) stable period two orbit at $\nu = 2.42$. For $f = 1$, $\varpi = 4$, $\zeta = 0.125$, and $x_0 = 0$, [930].

motion is reversed using the simple impact rule. The coefficient of restitution for this impact was taken to be small. For zero coefficient of restitution the motion of the system was studied using a one-dimensional mapping. It was shown that stable periodic orbits exist at almost all forcing frequencies, however, transient non-periodic or chaotic motions can also occur. Moreover, over certain (narrow) frequency windows arbitrarily long stable periodic motions exist.

1.7 Saw-Tooth-Time-Transformation

The saw-tooth-time-transformation (STTT) method is based on a special transformation of time and gives explicit form of analytical solutions for non-linearities of high power. The physical and mathematical principles of the STTT have been formulated in the literature (see, e.g., [843], [852]). The idea of STTT is similar to a great extent to the trigonometric generating functions $\{\sin t,\ \cos t\}$ frequently used in constructing solutions of linear and weakly nonlinear systems. Similarly, one can consider a pair of non-smooth functions, which have relatively simple forms and will be termed as the saw-tooth sine, $\tau(t)$, and the rectangular cosine, $r(t)$, which is the generalized derivative of $\tau(t)$ as shown in Fig. 1.10. The functions $\{\tau(t),\ r(t)\}$ and $\{\sin t,\ \cos t\}$ describe the motions of the two simplest vibrating models, namely, the motion of a particle between two rigid barriers and a mass-spring oscillator, respectively.

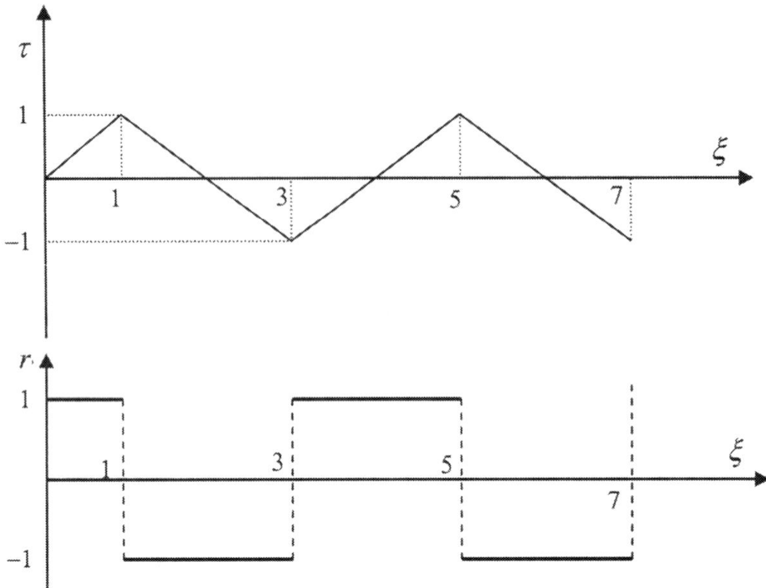

Fig. 1.10. Saw-tooth functions sine and rectangular cosine, $\xi = (t/t_0)$.

The method may be demonstrated by considering the free undamped oscillation of the nonlinear system

$$\mathbf{m}\frac{d^2\mathbf{x}}{dt^2} + \mathbf{kx} + \beta\mathbf{N}(\mathbf{x}) = \mathbf{0}, \tag{1.51}$$

where $\mathbf{m}$ is the system mass matrix, $\mathbf{k}$ is the linear stiffness matrix, $\mathbf{x}$ is the system generalized coordinate vector, $\mathbf{N}(\mathbf{x})$ is the vector of high power nonlinear terms that account for the strong impact nonlinearity, and β is a positive constant parameter. A family of periodic solutions of equation (1.51) may be written in the form [852],

$$\mathbf{x} = \mathbf{X}(\tau), \qquad \tau{=}\tau(t/t_0), \tag{1.52}$$

where t_0 is unknown scaling factor, which is equal to quarter of the period $T = 4t_0$ and must be defined for the autonomous case. Thus, the solution can be constructed as a function of the saw-tooth function τ, which varies in the region $-1 \leq \tau \leq 1$. Note that equation (1.51) admits a group of transformations $\mathbf{x} \to -\mathbf{x}$. As a result the solution can be constructed as an odd function: $\mathbf{X}(-\tau) \equiv -\mathbf{X}(\tau)$. When substituting equation (1.52) into equation (1.51), one should take into account the following differentiation scheme of equation (1.52), and due to the equality $r^2(t/t_0) = 1$, one can write

$$\frac{d\mathbf{x}}{dt} = \frac{1}{t_0}\frac{d\mathbf{X}}{d\tau}r, \tag{1.53}$$

$$\frac{d^2\mathbf{x}}{dt^2} = \frac{1}{t_0^2}\frac{d^2\mathbf{X}}{d\tau^2} + \frac{1}{t_0^2}\frac{d\mathbf{X}}{d\tau}\frac{dr(t/t_0)}{d(t/t_0)}. \tag{1.54}$$

The last term in equation (1.54) contains the series of Dirac delta functions

$$\frac{dr(t/t_0)}{d(t/t_0)} = 2\sum_{j=-\infty}^{\infty}\left[\delta\left(\frac{t}{t_0}+1-4j\right) - \delta\left(\frac{t}{t_0}-1-4j\right)\right]. \tag{1.55}$$

Note that the delta functions in equation (1.55) are 'localized' at points $\{t : \tau(t/t_0) = \pm 1\}$. It means that under the condition $d\mathbf{X}/d\tau = 0$, for $\tau = \pm 1$, all delta functions of the series will be eliminated, and as a result the second derivative in equation (1.54) becomes continuous function. Substituting equations (1.53) and (1.55) into equation (1.51) gives the boundary value problem

$$\mathbf{m}\frac{d^2\mathbf{x}}{d\tau^2} = -a^2\left[\mathbf{kX} + \beta\mathbf{N}(\mathbf{X})\right], \tag{1.56}$$

$$\frac{d\mathbf{X}}{d\tau}|_{\tau=1} = 0, \qquad \mathbf{X}(-\tau) \equiv -\mathbf{X}(\tau). \tag{1.57}$$

The left-hand side of equation (1.56) does not include the linear stiffness term, and by setting the right-hand side to zero, the left-hand side does not represent a harmonic oscillator as in the case of quasi-linear treatment. This means any generating solution for equation (1.56) should be based on the solution of the left-hand side, i.e.,

$$\mathbf{m}\frac{d^2\mathbf{x}}{d\tau^2} = 0. \tag{1.58}$$

By setting the right-hand side of equation (1.56) to zero, the qualitative structure of the periodic motion of the system will be preserved. This property will be destroyed if the same argument is applied to the system equations of motion before performing STTT. The transformed generating equation (1.58) possesses a solution of the form

$$\mathbf{X}(\tau) = \mathbf{X}(0) + \mathbf{X}^{'}(0)\boldsymbol{\tau}. \tag{1.59}$$

A solution of equation (1.58) in the form of a series of successive approximations may be written in the form:

$$X = X_0(\tau) + \epsilon X_1(\tau) + \epsilon^2 X_2(\tau) + ... \tag{1.60}$$

$$a^2 = \epsilon h_0\left(1 + \epsilon\gamma_1 + \epsilon^2\gamma_2 + ...\right). \tag{1.61}$$

where the formal parameter $\epsilon = 1$ is introduced as bookkeeping to identify terms of different orders in the expansion. All terms of the first series are n–component columns, where n is the number the system degrees of freedom. These functions and the constants h_0, γ_1, γ_2, ... are determined using an iterative process. This technique was used to analyze the response of systems involving liquid sloshing impact under support excitation [852]. Pilipchuk [844] studied the dynamics of linear and nonlinear systems under periodic impulsive excitation. The solutions of the differential equations were represented in a special form, which contains a standard pair of non-smooth periodic functions and possesses a convenient structure. This form is also suitable in the case of excitation with a periodic series of discontinuities of the first kind (a step-wise excitation).

For simple oscillators with a non-smooth restoring force described by the equation of motion

$$\ddot{x} + sgn(x)f(x) = 0, \tag{1.62}$$

where $f(x)$ is an even and sufficiently smooth function, Pilipchuk [845] obtained a closed-form solution by using the saw-tooth time transformation. Linear elastic structures oscillating against absolutely rigid constraints were also analyzed by Pilipchuk [847] who obtained exact solutions expressed through a saw-tooth time argument. It was indicated that impact modes exist when their basic frequencies are shifted into the right small neighborhood of any natural frequency of the linearized system in the absence of barriers. The frequencies of the localized impact mode solutions were found to be located at the right of the linear spectrum and have no upper boundary.

Azeez et al [50] employed the non-smooth time transformations to study strongly nonlinear periodic free oscillations of a vibro-impact system with two degrees of freedom. Periodic solutions revealed vibro-impact states with one- and two-sided collisions, including localized states. The non-smooth temporal transformation was adopted by Pilipchuk et al [855] to construct a family of periodic solutions of a weakly nonlinear system under parametric impulsive excitation. The transformation reduced the equation of motion to a standard weakly nonlinear boundary value problem. The numerical simulations showed a principal role of the shifts of the impulses' sequences. In particular, periodic, multi-periodic and stochastic-like regimes were manifested under variations of the shift parameter and small shift was found to lead to branching of curves of the periodic solutions on the parameters plane.

A combination of non-smooth transformation of variables in space and time was proposed by Pilipchuk ([846], [848]) to treat impulsively forced nonlinear oscillator between two absolutely rigid elastic barriers. It was shown that the space component of the transformation eliminates the barriers, whereas the time component removes external δ-pulses. The advantage of this combination was demonstrated in a boundary-value problem with no space- or time-dependent δ-type singular terms.

The Lie group transformation, which can lead to the simplest form of the system equations of motion has been employed in the literature for few cases. It has become a powerful tool for studying differential equations among mathematicians and specialists. The method is essentially based on the work of Zhuravlev [1138] and Zhuravlev and Klimov [1139]. An essential ingredient of the Lie group operators is the Hausdorff formula, which relates the Lie group operators of the original system and the new one, and the operator of coordinate transformation. Zhuravlev [1138] and Zhuravlev and Klimov [1139] made a conjecture that most of averaging techniques reproduce this formula, each time implicitly, during the transformation process. Pilipchuk and Ibrahim [853] employed this technique for an elastic structure carrying a liquid container subjected to parametric support harmonic excitation.

1.8 Closing Remarks

The modeling techniques described in this chapter have been used for different applications. However, none of them has been verified experimentally. Some experimental results have been reported in the literature but were not compared with any analytical results. This will be discussed for the case of signle-degree-of-freedom systems in Chapter 3.

Chapter 2
Mapping of Grazing and C–Bifurcations

2.1 Introduction

In studying vibro-impact systems one may encounter critical orbits that are neither free trajectories nor impact motion. For example, the trajectory of an oscillating mass reaching a barrier at zero velocity separates two regimes of motion, namely, non-impact and impact oscillations. The bifurcation associated with zero velocity just at the barrier is referred to as grazing bifurcation. Of particular interest of grazing impact bifurcation is its mapping. This chapter presents the basic concept of grazing bifurcation and the discontinuity mappings. It also addresses another type in which the fixed point or a periodic orbit may cross or collide one of the boundaries and this type is referred to as border-collision or C-bifurcation. This bifurcation was originally analyzed by Feigin ([313], [314], [315]).

2.2 Grazing Bifurcation

The theory of grazing bifurcation is believed to be motivated by a number of dynamical systems. For example, the bifurcation associated with vibro-impact of mooring towers of off-shore structures was numerically studied by Thompson and Ghaffari [997] and Thompson [994]. Their results revealed an infinite sequence of period-doubling bifurcations resulting in chaos. Shaw and Holmes [930] studied the dynamic behavior of a periodically forced oscillator with one-sided barrier. They modeled the impact dynamics using a discontinuous map, which exhibited period-doubling bifurcations followed by a complex sequence of transitions of long super-stable periodic oscillations.

Grazing bifurcation may be classified into continuous and discontinuous. Continuous grazing bifurcation is characterized by the occurrence of a continuous transition between different system attractors due to a slow variation of the control parameter near its critical value that results in zero contact velocity. The dynamics of this class after grazing contact remains in the vicinity

R.A. Ibrahim: Vibro-Impact Dynamics: Model., Map. & Appl., LNACM 43, pp. 31–54.
springerlink.com

of the original steady state to the grazing trajectory. On the other hand, discontinuous grazing bifurcations are characterized by the occurrence of jumps between different system attractors due to a control parameter variation near its critical value. For this case, there is a significant difference between the system dynamics before and after grazing. Conditions that lead to continuous and discontinuous grazing bifurcations have been developed for the case of pre-grazing periodic attractors in vibro-impact systems ([731], [341]). Grazing bifurcations through a codimension-2 grazing bifurcation were examined by several researchers ([333], [334], [332], [153], [1086], [1087], [1088]). In particular, Foale and Bishop [333] suggested that grazing bifurcations are the limiting cases of typical bifurcations that are encountered in smooth dynamical systems as the impact is hardened. Later, Dankowicz and Jerrelind [205] and Thota et al [1002] provided more rigorous study of grazing bifurcation in a general class of single degree-of-freedom oscillators. There is a borderline, which separates the free flight trajectory and impact motion, known as grazing bifurcation of a periodic orbit ([732], [733], [146], [147], [145], [234], [232]). Associated with grazing contact is the grazing point defined by the intersection of the grazing trajectory and a codimension-one surface in the state space.

Nordmark [734] and Stensson and Nordmark [960] discussed the creation of periodic orbits associated with grazing bifurcations and developed sufficient conditions for their existence. Peterka [819] demonstrated the existence of transition regions, which correspond to grazing bifurcations and by the boundaries of stability corresponding to the period-doubling and saddle-node bifurcations. The transition between neighboring periodic impact motions was found to be non-continuous, with the exception of singular points, where the existence boundaries and stability boundaries intersect. The equation of motion of a grazing impact oscillator was generalized by Hunt and Sarid [433] who included compliant boundaries and impact energy dissipation, yielding the phase diagram, indentation and force. The relationship between phase and set-point amplitude was discussed in terms of energy dissipation, showing that for a parameter space for which the system is highly nonlinear a chaotic response is possible.

Grazing impact displays rich and yet complex nonlinear behavior such as period doubling, multi-periodic orbits, subharmonic resonances, and chaos (see, e.g., [835], [149]). Grazing bifurcation happens when a periodic orbit, with zero or more impacts during the period, is displaced by a parameter change that results in a new impact, which occurs with zero impact velocity. Hu [427] developed a numerical algorithm to predict periodic grazing orbits in a piecewise linear oscillator. The results revealed an abundance of grazing phenomena and incident bifurcations of the oscillator. Nordmark [736] studied the creation of a set of periodic orbits branching off from the grazing bifurcation point. Criteria for the grazing bifurcation of a periodically forced, piecewise linear system together with the initial grazing manifolds were developed ([608], [609], [610]). The initial grazing manifold was found invariant. The dynamics of a piecewise linear oscillator close to grazing has been experimentally examined by Ing et al [445]. The experimental results revealed the occurrence of higher periodic responses after grazing.

The complex dynamics associated with high-dimensional mapping as applied to vibro-impact systems has been considered in several studies. For example, the four-dimensional Poincaré map of the vibro-impact system was reduced to a two-dimensional normal form using a center manifold reduction and a normal form technique [1068]. It was shown that there exists codimension-2 Hopf bifurcation in multi-degree-of-freedom vibro-impact systems. The amplitude surface method was introduced by Avranov [42] to analyze the bifurcation behavior when the excitation frequency and amplitude are changed in a vibro-impact system. Ding et al [255] studied the interaction of Hopf and period doubling bifurcations corresponding to a codimension-2 when a pair of complex conjugate eigenvalues crosses the unit circle and the other eigenvalue crosses -1 simultaneously for the Jacobi matrix. The four-dimensional map was reduced to a three-dimensional normal form using the center manifold theorem and the theory of normal forms. The results revealed curve doubling bifurcation (a torus doubling bifurcation), Hopf bifurcation and period doubling bifurcation. Numerical results indicated that the vibro-impact system presents complicated and interesting curve doubling bifurcation and Hopf bifurcation as the two controlling parameters vary. Xie et al [1088] developed an algorithm to compute Hopf-flip bifurcations of fixed points for high-dimensional maps with multiple parameters.

2.3 Discontinuity Mappings

2.3.1 Nordmark Mapping

The study of vibro-impact systems has been facilitated by the concept of discontinuity mappings introduced by Nordmark [731]. These mappings approximate the near-grazing impacting dynamics using a discrete dynamical system obtained solely from the conditions at the grazing contact. Discontinuity mappings in dynamical systems were defined by piecewise smooth vector fields [737]. The idea of a discontinuity mapping was utilized to establish a methodology for predicting the characteristics of system attractors that occur following a grazing intersection of a two-frequency, quasi-periodic oscillation with a two-dimensional impact surface in a three-dimensional state space by Dankowicz et al [204]. This work includes cases of trajectories that either cross or are quadratically tangent to a codimension-one surface in state space were treated. It was shown that the discontinuity mappings are equal to the identity up to order n.

Near Grazing orbits, the oscillator can be reduced to iterations of a mapping ([731], [735], [341]). The derivation of mappings for special values of the system parameters was considered in the literature ([1073], [1074], [146],

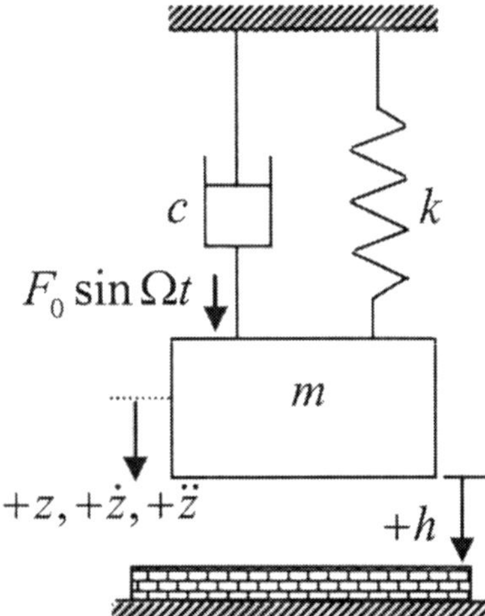

Fig. 2.1. Schematic diagram of a mass-spring-dashpot system with one-sided barrier.

[147], [148], [333]). Bud and Dux [146] and Budd et al [148] studied the dynamic behavior of a vibro-impact oscillator under forced excitation whose frequency is twice the natural frequency of the oscillator. It was shown that the grazing bifurcation can lead to an intermittent chaotic behavior with low velocity impacts followed by an irregular sequence of high velocity impacts. They also analyzed a discontinuous one-dimensional map as a tool to explain the existence of periodic windows in which the period of impacting solutions increases monotonically from one window to another one.

Chin et al ([172], [173]) provided an exhaustive examination of the Nordmark map [731], and predicted the occurrence of a series of transitions from a non-impacting period-1 orbit to period-M orbits, with $M = 2, 3, \ldots$ De Weger et al ([226], [227], [225]) called this series as *period–adding* transitions. They were able to explore these period-adding transitions experimentally. De Weger et al [225] called the M-periodic orbits with one impact per period as "*maximal*" periodic orbits.

Mappings that possess a particular form containing a square-root term occur as local Poincaré mappings of the type considered by Nordmark [731] and Fredriksson and Nordmark [341]. Mechanical systems characterized by these mappings satisfy a condition for the occurrence of impact in the form of a smooth surface of codimension–one in the state space. The equations of motion of these systems are also smooth slightly beyond the impact surface. Furthermore, there is a smooth impact law mapping that takes the system to a new state when reaching the impact surface. This impact law becomes the identity mapping as the impact velocity approaches zero. Consider the linear oscillator, shown in Fig. 2.1, described by the equation of motion

$$\ddot{z} + 2\zeta\omega_n\dot{z} + \omega_n^2 z = \bar{f}_0 \sin \Omega t. \tag{2.1}$$

where z is the position of the oscillator mass m, ζ is the damping ratio, ω_n is the oscillator undamped natural frequency, Ω is the excitation frequency, and $\bar{f}_0 = F_0/m$ is the excitation amplitude. Equation (2.1) may be written in the form

$$z^{''} + \varsigma z^{'} + \varpi^2 z = f_0 \sin(2\pi\tau), \tag{2.2}$$

where $f_0 = 4\pi^2 \bar{f}_0/\Omega^2$, $\varpi^2 = 4\pi^2\omega_0^2/\Omega^2$, $\varsigma = 4\pi\zeta(\omega_n/\Omega)$ and a prime denotes differentiation with respect to $\tau = \Omega t/2\pi$. The mapping from $\tau = n$ to $\tau = n + 1$, where n is an integer, is a Poincaré map on the plane $(z, z^{'})$ with constant phase and thus has the same set of eigenvalues as the Jacobian matrix of the linear Nordmark [731] map

$$\begin{aligned} x_{n+1} &= \alpha x_n + y_n + \Re \\ y_{n+1} &= -\gamma x_n \end{aligned} \qquad for \quad x_n \leq 0, \tag{2.3}$$

$$\begin{aligned} x_{n+1} &= -\sqrt{x_n} + y_n + \Re \\ y_{n+1} &= -\gamma e^2 x_n \end{aligned} \qquad for \quad x_n > 0, \tag{2.4}$$

where x_n and y_n are the transformed coordinates in the position-velocity space evaluated at time t_n, and $\Omega t_n = 2n\pi$, $\Re$ is a parameter related to the amplitude of the external excitation and e is the coefficient of restitution. The parameters α and γ depend on the intrinsic properties of the system equation (2.2) such that the limit $\gamma \to 0$ corresponds to large damping coefficient, and $\gamma = 1$ corresponds to zero dissipation.

Since equations (2.3) and (2.4) are continuously differentiable except at $x_n = 0$, the term $-\sqrt{x_n}$ is referred to as the "*square-root singularity*" because its Jacobian matrix is singular. The relationship between these parameters and the physical system parameters was developed by Chin et al [172].

2.3.1.1 Non-impact Mapping

For the case of non-impacting motion in which the oscillator does not reach the barrier, we consider the general solution of equation (2.2), which may be written in the form

$$z(\tau) = Q(\tau) + \bar{C}_1 e^{s_1\tau} + \bar{C}_2 e^{s_2\tau}, \qquad provided \quad \varsigma^2 - 4\varpi^2 \neq 0, \tag{2.5}$$

where $Q(\tau)$ is the steady-state response, $\bar{C}_1$ and $\bar{C}_2$ are constants determined from initial conditions, and

$$s_{1,2} = \frac{1}{2}\left(-\varsigma \pm \sqrt{\varsigma^2 - 4\varpi^2}\right) \tag{2.6}$$

For $\tau = n$, one can write the response state vector

$$\left\{\begin{array}{c} z(n) \\ z^{'}(n) \end{array}\right\} = \begin{bmatrix} 1 & 1 \\ s_1 & s_2 \end{bmatrix} \left\{\begin{array}{c} \bar{C}_1 e^{s_1 n} \\ \bar{C}_2 e^{s_2 n} \end{array}\right\} + \left\{\begin{array}{c} Q \\ Q^{'} \end{array}\right\}, \tag{2.7}$$

For $\tau = n + 1$, equation (2.7) takes the form

$$\left\{\begin{array}{c} z(n+1) \\ z^{'}(n+1) \end{array}\right\} = \begin{bmatrix} e^{s_1} & e^{s_2} \\ s_1 e^{s_1} & s_2 e^{s_2} \end{bmatrix} \left\{\begin{array}{c} \bar{C}_1 e^{s_1 n} \\ \bar{C}_2 e^{s_2 n} \end{array}\right\} + \left\{\begin{array}{c} Q \\ Q^{'} \end{array}\right\}. \tag{2.8}$$

Pre-multiplying both sides of equation (2.7) by $\begin{bmatrix} 1 & 1 \\ s_1 & s_2 \end{bmatrix}^{-1} = \frac{1}{s_2 - s_1} \times \begin{bmatrix} s_2 & -1 \\ -s_1 & 1 \end{bmatrix}$ and rearranging gives

$$\left\{\begin{array}{c} \bar{C}_1 e^{s_1 n} \\ \bar{C}_2 e^{s_2 n} \end{array}\right\} = \begin{bmatrix} 1 & 1 \\ s_1 & s_2 \end{bmatrix}^{-1} \left\{\begin{array}{c} z(n) \\ z^{'}(n) \end{array}\right\} - \begin{bmatrix} 1 & 1 \\ s_1 & s_2 \end{bmatrix}^{-1} \left\{\begin{array}{c} Q \\ Q^{'} \end{array}\right\}. \tag{2.9}$$

Now setting

$$\begin{bmatrix} e^{s_1} & e^{s_2} \\ s_1 e^{s_1} & s_2 e^{s_2} \end{bmatrix} \begin{bmatrix} 1 & 1 \\ s_1 & s_2 \end{bmatrix}^{-1} =$$

$$\mathbf{B} = \frac{1}{s_2 - s_1} \begin{bmatrix} s_2 e^{s_1} - s_1 e^{s_2} & e^{s_2} - e^{s_1} \\ s_1 s_2 \left(e^{s_1} - e^{s_2}\right) & s_2 e^{s_2} - s_1 e^{s_1} \end{bmatrix}, \tag{2.10}$$

and substituting equation (2.9) into equation (2.8) gives

$$\left\{\begin{array}{c} x_{n+1} \\ y_{n+1} \end{array}\right\} = \mathbf{B} \left\{\begin{array}{c} x_n \\ y_n \end{array}\right\} + [\mathbf{I} - \mathbf{B}] \left\{\begin{array}{c} Q \\ Q^{'} \end{array}\right\} \tag{2.11}$$

The matrix $\mathbf{B}$ has the same set of eigenvalues as the Jacobian, $\mathbf{J}$, of the matrix equation (2.3)

$$\mathbf{J} - \lambda \mathbf{I} = \begin{bmatrix} \frac{\partial x_{n+1}}{\partial x_n} - \lambda & \frac{\partial x_{n+1}}{\partial y_n} \\ \frac{\partial y_{n+1}}{\partial x_n} & \frac{\partial y_{n+1}}{\partial y_n} - \lambda \end{bmatrix} = \begin{bmatrix} \alpha - \lambda & 1 \\ -\gamma & -\lambda \end{bmatrix}. \tag{2.12}$$

This gives the two roots

$$\lambda_{1,2} = \frac{1}{2}\left(\alpha \pm \sqrt{\alpha^2 - 4\gamma}\right). \tag{2.13}$$

The eigenvalues of the matrix $\mathbf{B}$ are evaluated as follows

$$[\mathbf{B} - \lambda \mathbf{I}] = \frac{1}{s_2 - s_1} \begin{bmatrix} \left(s_2 e^{s_1} - s_1 e^{s_2}\right) - \lambda & e^{s_2} - e^{s_1} \\ s_1 s_2 \left(e^{s_1} - e^{s_2}\right) & \left(s_2 e^{s_2} - s_1 e^{s_1}\right) - \lambda \end{bmatrix} = 0 \tag{2.14}$$

Solving the determinant for λ, gives

$$\lambda_{1,2} = e^{s_{1,2}}. \tag{2.15}$$

Equating (2.13) and (2.15) gives

$$\lambda_{1,2} = e^{s_{1,2}} = \frac{1}{2}\left(\alpha \pm \sqrt{\alpha^2 - 4\gamma}\right), \tag{2.16}$$

$$\gamma = \lambda_1\lambda_2 = e^{s_1+s_2} = e^{-\varsigma}, \tag{2.17}$$

$$\alpha = \lambda_1 + \lambda_2 = 2e^{-\varsigma/2}\cosh\left(\frac{1}{2}\sqrt{\varsigma^2 - 4\varpi^2}\right), \tag{2.18}$$

where relation (2.6) has been used. For positive values of $\varsigma > 0$, we have from equations (2.17) and (2.18) the bounds $0 < \gamma < 1$, and $0 < \alpha < 1 + \gamma$.

For the non-impact regime the map is linear. However, the impact orbit possesses a square-root singularity. Close to grazing, the acceleration near impact can be considered constant and the square-root is simply the relationship between elapsed time and traveled distance in systems with constant acceleration. The square-root singularity and the associated extreme stretching of phase space near the point of grazing impact lead to highly non-trivial dynamics. The radical changes in the dynamic behavior of a vibro-impact oscillator owe their origin to the discontinuity in the derivative of the system's Poincaré map as indicated by Whiston [1076]. Janin and Lamarque [465] found that for some values of system parameters a non-differentiable fixed point of the Poincaré map exists. A local expansion of the Poincaré map around such a point was found to be a square-root term on the impact side. It was shown that the periodic solution is stable when the Floquet multipliers are real.

2.3.1.2 Grazing Impact Case

With reference to Fig. 2.1, the oscillator mass hits the barrier when its displacement reaches the value $z = h$. We introduce the coordinate shift $\xi = z - h$, such that the impact occurs when $\xi = 0$. We also introduce the new coordinate $\theta = \Omega t$. There exists a periodic orbit that grazes the rigid barrier such that the orbit reaches the barrier with zero velocity. Let the displacement, velocity, acceleration, and phase at the grazing contact point be ξ_g, v_g, a_g, and θ_g, respectively. At grazing point we have $\xi_g = 0$, $v_g = 0$, and $a_g < 0$. Without loss of generality, we set $\theta_g = 0$. It is possible to choose a Poincaré section to be the surface defined by $v = 0$ in the direction of decreasing v. We define the Poincaré mapping $\mathbf{P}$ as the flow of dynamics that starts from a point on the Poincaré section and ends at the Poincaré section in one period. Under zero forcing excitation, the homogeneous solution of the oscillator in the phase space is

$$\begin{Bmatrix} \xi(t) \\ \dot{\xi}(t) \end{Bmatrix} = \frac{1}{s_2 - s_1} \begin{bmatrix} s_2 e^{s_1 t} - s_1 e^{s_2 t} & e^{s_2 t} - e^{s_1 t} \\ s_1 s_2 \left(e^{s_1 t} - e^{s_2 t}\right) & s_2 e^{s_2 t} - s_1 e^{s_1 t} \end{bmatrix} \begin{Bmatrix} \xi(0) \\ \dot{\xi}(0) \end{Bmatrix} \tag{2.19}$$

For convenience we write

$$\mathbf{P(t)} = \frac{1}{s_2 - s_1} \begin{bmatrix} s_2 e^{s_1 t} - s_1 e^{s_2 t} & e^{s_2 t} - e^{s_1 t} \\ s_1 s_2 \left(e^{s_1 t} - e^{s_2 t}\right) & s_2 e^{s_2 t} - s_1 e^{s_1 t} \end{bmatrix} = \begin{bmatrix} P_{11}(t) & P_{12}(t) \\ P_{21}(t) & P_{22}(t) \end{bmatrix}. \tag{2.20}$$

If the wall did not exist, local stability of the grazing periodic orbit can be determined by the eigenvalues of the Jacobian of $\mathbf{P}$. If all eigenvalues are outside the unit circle, the corresponding fixed point of $\mathbf{P}$ is unstable and the associated periodic orbit is unstable limit cycle. If some, but not all of the eigenvalues of the Jacobian are outside the unit circle, the corresponding solution is unstable limit cycle of the saddle type. These cases belong to what is known as hyperbolic fixed point [722] and the corresponding solutions are referred to as hyperbolic. The solution is non-hyperbolic if one or more of the eigenvalues lie on the unit circle. Under these states a linearization of the Poincaré map may not be sufficient for determining the stability of the trajectory. In other words, in the presence of the barrier, stability of the grazing periodic orbit is not solely determined by the Jacobian because the Poincaré mapping can not be directly applied to all the nearby trajectories.

An impacting trajectory dose not intersect the Poincaré section but jumps across the section. This is demonstrated by Fig. 2.2, which shows a trajectory starts at point A and impacts the barrier with velocity $v_a = v_-$, at point a, then rebounds with velocity $v_b = v_+$, at point b and finally ends at point B. To utilize the Poincaré section, Xiaopeng Zhao[1] proposed to temporally ignore the existence of the wall and extend the trajectory forward in time until it intersects the Poincaré section at point N. Apparently, this virtual intersection point (ξ_n, θ_n) is at the right of the grazing contact point. The next virtual intersection on the Poincaré section is $(\xi_{n+1}, \theta_{n+1})$. Note that if $\xi_n > 0$, we cannot directly apply the Poincaré section. However, imagine the barrier is removed and extrapolate the trajectory backward from point B in time until it intersects the Poincaré section at point C whose coordinate is (ξ_c, θ_c). Obviously, point C is correction to point N and thus can be a function of the coordinate (ξ_n, θ_n). The subsequent virtual intersection on the Poincaré section can be obtained by applying $\mathbf{P}$ to (ξ_c, θ_c) as

$$\xi_{n+1} = P_{11} x_c(\xi_n, \theta_n) + P_{12} \theta_c(\xi_n, \theta_n) \tag{2.21a}$$

$$\theta_{n+1} = P_{21} x_c(\xi_n, \theta_n) + P_{22} \theta_c(\xi_n, \theta_n) \tag{2.21b}$$

For trajectories near grazing, points a, b, N, and C are very close to the grazing point G. The flow from N to C can be solved using local dynamic

[1] Private communication.

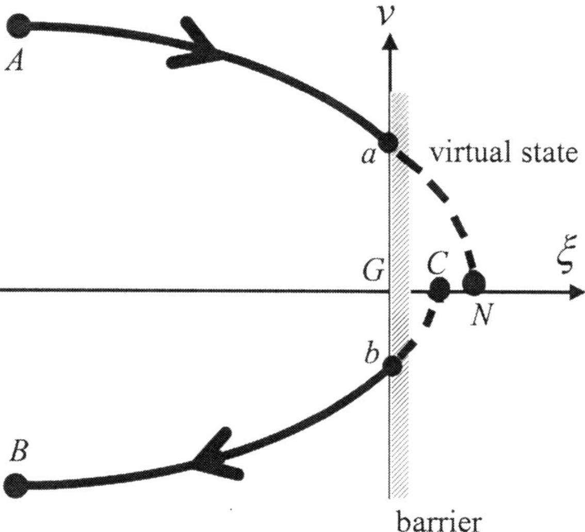

Fig. 2.2. Trajectory iteration near grazing point.

analysis. To this end, consider a trajectory starting from a point whose state coordinates are $\xi, v, \theta,$ and continues for a short time interval Δt. The new state variables at the end of this time interval are $x+v\ \Delta t+\frac{1}{2}a(\Delta t)^2$, $v+a\Delta t$, and $\theta+\Delta t$. Since the motion is studied in the neighborhood of the grazing point, the acceleration at G will be denoted as a_g. Consider the flow in the following three segments:

- Flow from N to a: Let the time interval of this segment be $-\Delta t_1$ and note that $v_n = 0$. The state variables at point a can be written in terms of those of point N as

$$\xi_a = \xi_n + \frac{1}{2}a_g(\Delta t_1)^2 \tag{2.22a}$$

$$v_a = -a_g\Delta t_1 \tag{2.22b}$$

$$\theta_a = \theta_n - \Delta t_1 \tag{2.22c}$$

  Since $\xi_a = 0$, one can write

$$\Delta t_1 = \sqrt{-2\frac{\xi_n}{a_g}} \tag{2.23}$$

- Jump from a to b: Applying the coefficient of restitution law, one can write the state variables at point b as

$$\xi_b = \xi_a = 0 \tag{2.24a}$$

$$v_b = -ev_a = ea_g\Delta t_1 \tag{2.24b}$$

$$\theta_b = \theta_a = \theta_n - \Delta t_1 \tag{2.24c}$$

- Flow from b to C: Let the time interval of this segment be $-\Delta t_2$. The state variables at point C can be written in terms of those of point b as

$$\xi_C = -v_b\Delta t_2 + \frac{1}{2}a_g(\Delta t_2)^2 = -ea_g\Delta t_1\Delta t_2 + \frac{1}{2}a_g(\Delta t_2)^2 \tag{2.25a}$$

$$v_C = v_b - a_g\Delta t_2 = ea_g\Delta t_1 - a_g\Delta t_2 \tag{2.25b}$$

$$\theta_C = \theta_b - \Delta t_2 = \theta_n - \Delta t_1 - \Delta t_2 \tag{2.25c}$$

Since $v_C = 0$, one can show that $\Delta t_2 = e\Delta t_1 = e\sqrt{-2\frac{\xi_n}{a_g}}$. It then follows that

$$\xi_C = -\frac{1}{2}e^2a_g(\Delta t_1)^2 = e^2x_n \tag{2.26a}$$

$$\theta_C = \theta_n - (1+e)\Delta t_1 \tag{2.26b}$$

Now substituting equations (2.26) into equations (2.21) yields the mapping equations for the grazing impact case, $\xi_n > 0$

$$\xi_{n+1} = P_{11}e^2\xi_n + P_{12}[\theta_n - (1+e)\Delta t_1] \tag{2.27}$$

$$\theta_{n+1} = P_{21}e^2\xi_n + P_{22}[\theta_n - (1+e)\Delta t_1] \tag{2.28}$$

Equations (2.27) and (2.28) describe the mapping in the physical coordinates. In order to express them in the most simplest form similar to those derived by Nordmark [731] the following coordinate transformation is introduced

$$\xi_n = l_1x_n \tag{2.29a}$$

$$\theta_n = l_2x_n + l_3y_n \tag{2.29b}$$

where the constants l_1, l_2, and l_3 are evaluated such that upon substituting equations (2.29) into the mapping equations (2.27) and (2.28) the coefficients of the square root $\sqrt{x_n}$ and the coordinate y_n in equation (2.27) each becomes unity, and at the same time the coefficient of y_n in equation (2.28) vanishes upon substitution. These constraints result in the following mapping equations in the transformed coordinates

$$\xi_{n+1} = -sgn(P_{12})\sqrt{\xi_n} + \left[\alpha - (1-e^2)P_{11}\right]\xi_n + \theta_n \tag{2.30}$$

$$\theta_{n+1} = -\gamma e^2 \xi_n \tag{2.31}$$

where $\alpha = P_{11} + P_{22}$, and $\gamma = P_{11}P_{22} - P_{12}P_{21}$. Note that the constraints imposed in reaching equations (2.30) and (2.31) require the following values for the coefficients l_1, l_2, and l_3

$$l_1 = -\frac{2}{a_g}(1+e)^2 P_{12}^2, \tag{2.32a}$$

$$l_2 = -\frac{2}{a_g}(1+e)^2 P_{12}P_{22}, \tag{2.32b}$$

$$l_3 = -\frac{2}{a_g}(1+e)^2 P_{12}. \tag{2.32c}$$

Chin et al [172] numerically generated the bifurcation phenomena for the Nordmark map given by equations (2.3) and (2.4) as the bifurcation parameter, $\Re$, increases through $\Re = 0$ (grazing impact). Figs. 2.3(a) and 2.3(b) show the bifurcation diagrams for $e = 1$, and two sets of the system parameters $(\gamma, \alpha) = (0.05,\ 0.65)$ and $(0.01,\ 0.25)$, respectively. Three basic bifurcation scenarios were reported. The first exhibits bifurcation from a stable period–1 orbit for $\Re < 0$ to reversed infinite period adding cascade as increases through zero. For example, Fig. 2.3(a) shows such cascade where chaos appears in bands between successive windows of periodic behavior. On the other hand, Fig. 2.3(b) reveals a cascade with hysteresis. The bifurcation diagrams of the other two cases are not shown here, however a brief description is provided. The second case is characterized by bifurcation from a stable

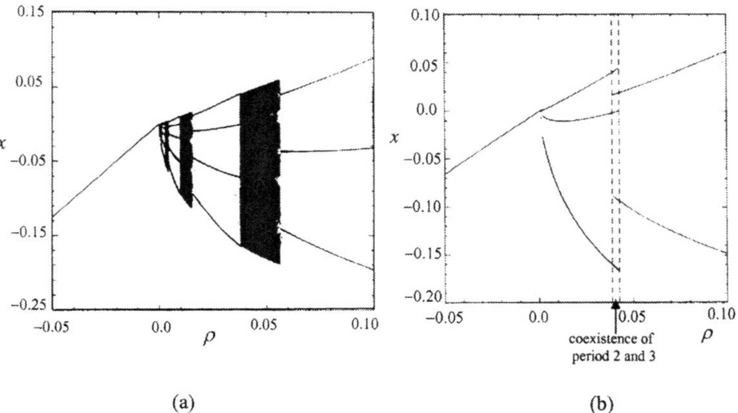

Fig. 2.3. Bifurcation diagrams for $e = 1$: (a) $(\gamma, \alpha) = (0.05,\ 0.65)$ and (b) $(\gamma, \alpha) = (0.01,\ 0.25)$, [172].

period–1 orbit in $\Re < 0$ to chaotic attractor as $\Re$ increases through zero. The third scenario belongs to collision of an unstable period–M maximal orbit (which is a regular saddle, and is created, together with a stable period–M maximal orbit in a saddle-node bifurcation in $\Re < 0$) and the period–1 orbit at $\Re = 0$.

2.3.2 *Molenaar et al Mapping*

Molenaar et al [686] and De Weger et al [225] introduced some modifications to the Nordmark mapping. They obtained the value of the bifurcation parameter, $\Re$, using nonlinear and lengthy analysis, in the form

$$\Re = \frac{1-\alpha+\gamma}{2|A|\left(1+e\right)^2\left[\left(-e^{s_1}+e^{s_2}\right)/\left(s_2-s_1\right)\right]^2}\left(\frac{f_0-f_g}{f_g}\right) \tag{2.33}$$

where A is the acceleration of the particular solution of the oscillator when the excitation amplitude, f_0, approaches its value, f_g, at grazing impact. Molenaar et al [686] observed that the fixed negative sign, which precedes the square-root in equation (2.4), prohibits period-one maximal periodic orbits for the underdamped oscillator. Their nonlinear analysis yielded the modified form of the transformed map

$$\begin{aligned} x_{n+1} &= \alpha x_n + y_n + \Re \\ y_{n+1} &= -\gamma x_n \end{aligned} \qquad for \quad x_n \leq 0, \tag{2.34}$$

$$\begin{aligned} x_{n+1} &= -C_1\sqrt{x_n} + C_2 x_n + y_n + \Re \\ y_{n+1} &= C_3 x_n \end{aligned} \qquad for \quad x_n > 0, \tag{2.35}$$

where

$$\begin{aligned} C_1 &= sgn\left(\frac{-e^{s_1}+e^{s_2}}{s_2-s_1}\right), \\ C_2 &= \alpha - 2(1+e)\left(\frac{-s_1e^{s_1}+s_2e^{s_2}}{s_2-s_1}\right) + (1+e)^2\left(\frac{-s_1e^{s_1}+s_2e^{s_2}}{s_2-s_1}\right)^2, \\ C_3 &= (1+2e)\gamma - (1+e)^2\left(\frac{-s_1e^{s_1}+s_2e^{s_2}}{s_2-s_1}\right)^2. \end{aligned} \tag{2.36}$$

One may observe a significant difference between the two maps given by equations (2.3) and (2.4), on the one hand, and equations (2.34) and (2.35) on the other hand. For example, the negative sign preceding the factor C_1 guarantees the presence of maximal period-one orbits. The extra term, C_2x_n, in the first equation of (2.35), provides a quantitative understanding of the loss of stability of maximal periodic orbits due to an additional impact. The coefficient C_3 is different from that of Nordmark [731].

For the case of flexible barrier the impact mass will penetrate the barrier in a form of elastic deformation. In this case, the coefficient of restitution must be within the range, $0 < e < 1$. For low-velocity impact, the barrier will influence the system dynamics when it absorbs energy, which is measured by the value $e < 1$. For this case, Molenaar et al [686] obtained the same mapping equations (2.35) but with the following coefficients:

$$C_1 = sgn\left(\frac{-e^{s_1}+e^{s_2}}{s_2-s_1}\right),$$
$$C_2 = \alpha - 2(1-e)\left(\frac{-s_1e^{s_1}+s_2e^{s_2}}{s_2-s_1}\right) + (1+e)^2\left(\frac{-s_1e^{s_1}+s_2e^{s_2}}{s_2-s_1}\right)^2,$$
$$C_3 = (1-2e)\gamma - (1-e)^2\left(\frac{-s_1e^{s_1}+s_2e^{s_2}}{s_2-s_1}\right)^2,$$
$$\Re = \frac{1-\alpha+\gamma}{2|A|\,(1-e)^2\left[\left(-e^{s_1}+e^{s_2}\right)/\left(s_2-s_1\right)\right]^2}\left(\frac{f_0-f_g}{f_g}\right) \tag{2.37}$$

This situation was considered in studying a linear oscillator with flexible barrier of stiffness k_2 as shown in Fig. 2.4(a) a linear oscillator. The equation of motion of this system may be written in the form

$$\ddot{z} + \varsigma\dot{z} + K(z) = F\cos(2\pi t + \phi), \tag{2.38}$$

where $K(z) = \begin{cases} \nu^2(1+z) & z \le 0 \\ \nu^2(1+z) + \kappa^2 z & z > 0 \end{cases}$,

$z/Z/\ell, \nu^2 = 4\pi^2 k_1/\left(m\Omega^2\right), F = 4\pi^2\bar{f}_0/\left(m\ell\Omega^2\right)$, ℓ is the gap length between the mass and the flexible barrier, $\varsigma = 2\pi c/(m\Omega)$, $\kappa^2 = 4\pi^2 k_2/\left(m\Omega^2\right)$, the stiffness of the barrier is measured by the ratio $\Gamma = \kappa^2/\nu^2$, such that a rigid barrier possesses $\Gamma = \infty$.

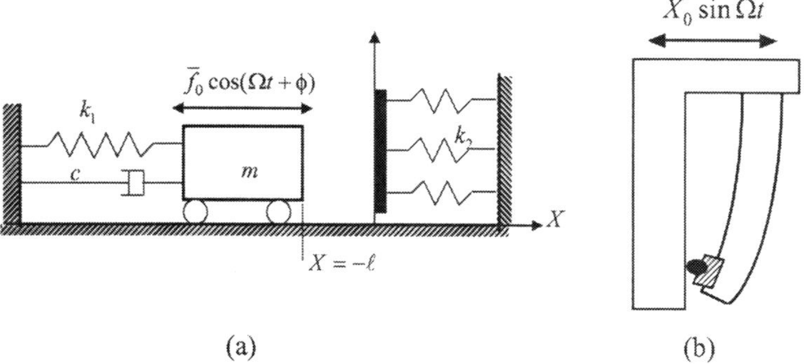

Fig. 2.4. Schematic diagram of (a) mass-spring-dashpot system with flexible barrier and (b) experimental model used by de Weger et al, [225].

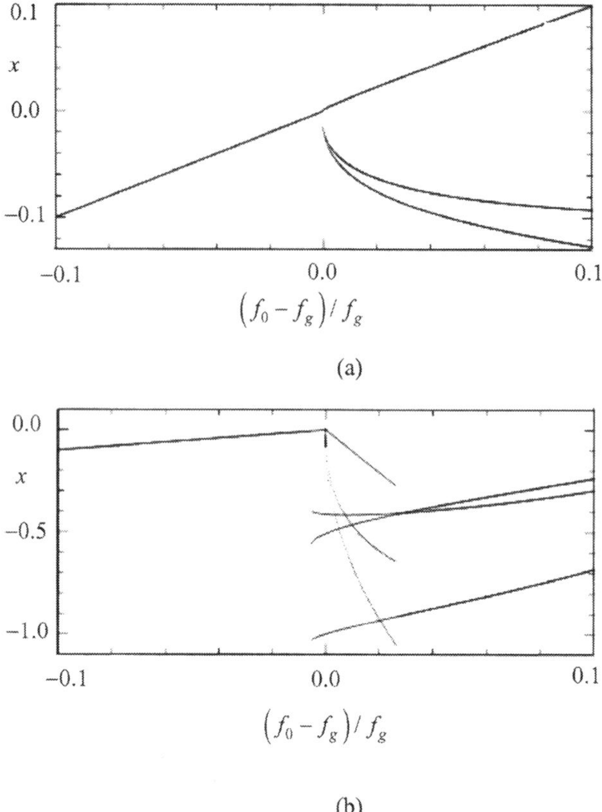

Fig. 2.5. Bifurcation diagrams of the system of Fig. 2.4(a) for stiffness ratio (a) $\Gamma = k_2/k_1 = 10$, and (b) $\Gamma = 1000$, [686].

Equation (2.38) was numerically solved in the presence of grazing impacts by using the analytical solutions between impacts [686]. Figs. 2.5(a,b) show the bifurcation diagrams for two different values of stiffness ratio $\Gamma = 10$ and 1000, respectively. In both cases the transition from period–3 to period–1 is observed and is typical for a square–root map. It was concluded that the map for impacts with compliant barrier is very similar to the one generated for a rigid barrier. In another work, De Weger et al [225] conducted an experimental investigation on a beam model excited by a sinusoidal excitation $X_0 \sin \Omega t$, as shown in Fig. 2.4(b). They measured period-adding transitions $p_1 \longleftrightarrow p_M$ from the non-impacting state to the maximal periodic orbit with period $M = 3$ to 10 as shown in Fig. 2.6. The measured period of free oscillations of the beam was $T = 40.8 \pm 0.02\ ms$, and damping parameter $\zeta\omega_n = 2.1 \pm 0.05\ s^{-1}$. It is seen that the apparent hysteresis may irregularly vary. The transition $p_M \longleftrightarrow p_{M/2}$ was found to give rise to the jump characteristics

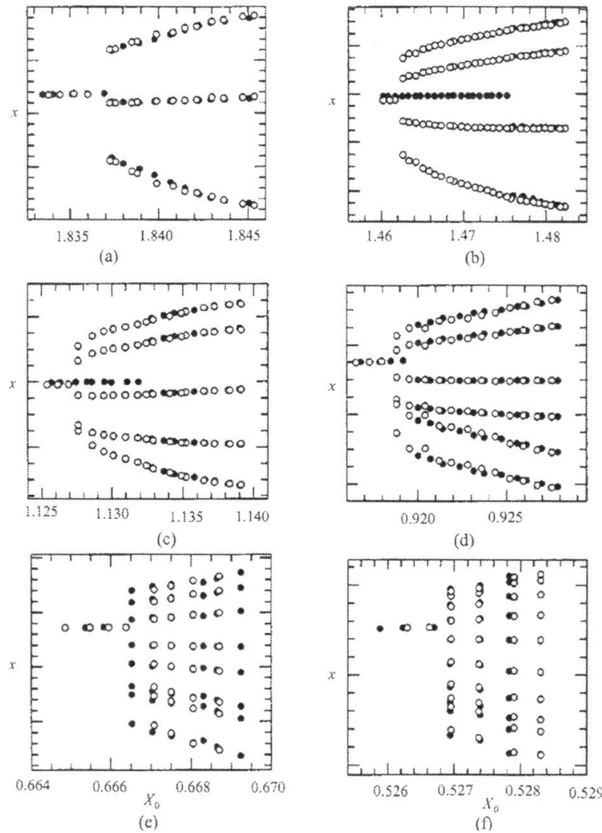

Fig. 2.6. Experimental measurements of bifurcation diagrams of Maximal period M and excitation frequency f: (a) $M = 3$, $f = 20.90$ Hz, (b) $M = 4$, $f = 21.50$ Hz, (c) $M = 5$, $f = 22.20$ Hz, (d) $M = 6$, $f = 22.60$ Hz, (e) $M = 8$, $f = 23.05$ Hz, (f) $M = 10$, $f = 23.40$ Hz. Closed circles (•)belong to the upward scan of the excitation amplitude and open circles (○) for the downward scan, [225].

in the bifurcation diagram shown in Fig. 2.7 for the case of $p_4 \longleftrightarrow p_{4/2}$. It was reported that the additional impact occurred at the turning point that immediately preceded the primary impact. This turning point was the closest to the barrier. At bifurcation, the motion between these two impacts experienced significant changes.

In terms of discontinuity mapping, De Weger et al [225] explicitly showed that the square-root also survives soft impacts with a soft barrier. It was essential to allow for a discontinuity in terms of a coefficient of restitution $e < 1$. For perfectly elastic soft collisions with a soft barrier, however, the square-root vanishes and collisions need to become hard again to restore it. Bifurcation and chaos in piecewise-smooth dynamical systems with impacts were experimentally demonstrated using a cam-follower system characterized by a radial cam

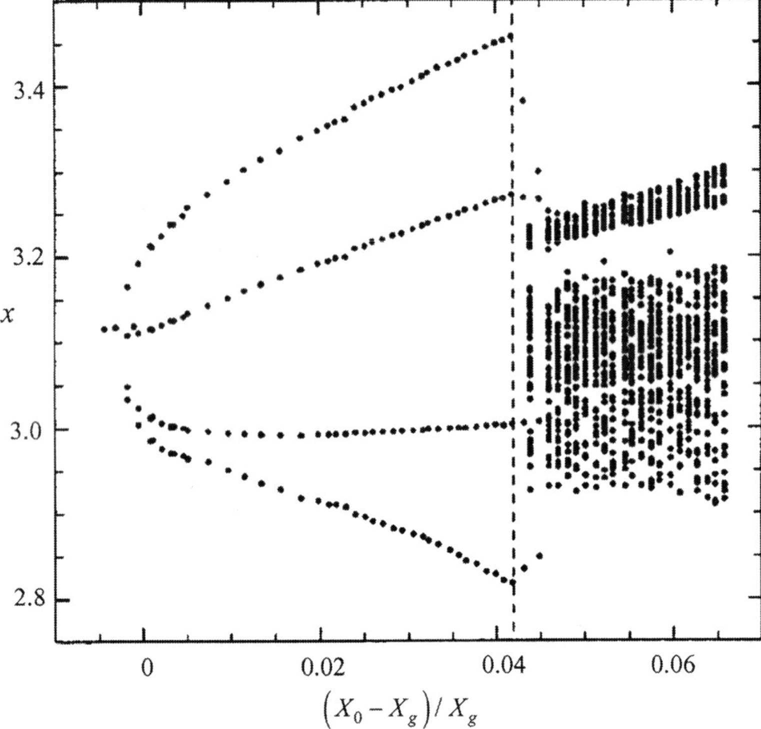

Fig. 2.7. Grazing bifurcation of a maximal period$-M$ orbit to a period$-M/2$ orbit (two impacts per period) at the vertical dashed line, $(X_0 - X_g)/X_g \approx 0.042$, [225].

and a flat–faced follower [10]. Under variations of the cam rotational speed, the follower was observed to detach from the cam and then exhibits periodic impacting behavior characterized by many impacts and chattering[2].

2.3.3 Further Developments

Mathematicians, physicists and applied dynamicists developed other discontinuity mappings for non-smooth dynamics. For example, Thota and Dankowicz ([1000], [1001]) and Thota [999] formulated analytical conditions for the persistence of a local attractor in the immediate vicinity of periodic and quasi-periodic grazing trajectories. They employed a local analysis based on the discontinuity-mapping approach to derive a normal-form description of the dynamics near the grazing trajectory. It was found that the catastrophic

[2] Chattering is a special type of oscillation characterized by very small amplitudes that are decreasing with time, see more on chattering in Section 3.2.

loss of a local attractor and strong instability characteristics of grazing bifurcations are directly associated with the repeated application of a square-root term that appears to lowest order in the normal-form expansion. Furthermore, it was found that the square-root term is absent in the description of the dynamics normal to a quasi-periodic trajectory covering a codimension-one invariant torus resulting in a piecewise linear description of the normal dynamics. In contrast, for codimension-2 or higher, the square-root term is generically present in the normal dynamics. Qian and Sun [874] proved the existence of invariant tori and quasi-periodic solutions for asymptotically linear impact oscillators by using the successor map and some generalized versions of the Moser's twist theorem.

Di Bernardo et al [233] presented a nice review of the one-parameter, non-smooth bifurcations that take place in a variety of continuous time piecewise-smooth dynamical systems. In particular, they defined discontinuity-induced bifurcations as non-trivial interactions of a limit set with respect to a codimension-one discontinuity boundary in phase space. A variety of dynamic phenomena such as the sudden creation or disappearance of attractors, jumps to chaos, bifurcation diagrams with sharp corners, and cascades of period adding were discussed.

The occurrence of chaos in vibro-impact systems under different values of system and excitation parameters was addressed by many researchers (see, e.g., [996], [452], [453], [928], [934], [883], [51], [123], [960], [961], [563], [829], [667], [1054], [826], [229], [601], [617], [619], [77], [830], [553], [255], [256], [257], [624]). Through a corner bifurcation, the system can experience complex dynamics, which can include periodic orbits of arbitrary period and period-adding cascades ([150], [841], [230]).

Di Bernardo et al [241] treated specific transitions when the boundary equilibrium, lying within a discontinuity manifold, is perturbed. They showed that such equilibria can either persist under parameter variations or disappear, and thus yield different bifurcation scenarios. They discussed the occurrence of boundary-equilibrium bifurcations in three classes of piecewise-smooth dynamical systems, namely, piecewise-smooth continuous [325] and impacting systems. Conditions for an equilibrium position of these systems to persist or undergo a non-smooth fold scenario at the discontinuity-induced bifurcation point were given. It was shown that, under certain conditions, limit cycles can branch off the boundary equilibrium bifurcation point. Conditions for boundary-equilibrium bifurcation were illustrated based on the work of Freire et al [344] and Leine [577].

In mechanical systems the inclusion of friction and impacts would lead to discontinuities in the corresponding mathematical models. Oestreich et al ([745], [746]) proposed different methods to treat vibro-impact models and friction oscillators. These methods were also applied to experimental measurements or simulated time signals. Furthermore, these methods include one-dimensional maps, bifurcation and stability analysis, the determination of Lyapunov exponents, and the phase-space reconstruction with the aid of general mutual information and false nearest neighbors.

Svahn and Dankowicz [973] examined the conditions under which a continuous/discontinuous transition occurs as a result of the grazing contact of an initially stationary part of a mechanism with relatively more massive oscillating element in the presence of dry friction. They derived the conditions for the persistence of a local attractor in the near-grazing dynamics of a two-degree-of-freedom vibro-impact oscillator with friction. They also considered a harmonic oscillator with Coulomb friction whose motion is limited by an oscillatory unilateral constraint corresponding to the case of infinite mass ratio. Collisions between the harmonic oscillator and the unilateral constraint were governed by a simple coefficient of restitution impact law. It was argued that the loss of a local attractor and the associated large–amplitude oscillations of the less massive part afford a means for energy transfer through the mechanism by a means of energy damping. Nordmark et al [738] considered a series of studies dealing with the non-smooth dynamics of planar mechanical systems with isolated contact in the presence of Coulomb friction. Their analysis identified boundaries between open regions of initial conditions and parameter values corresponding to distinct forms of the impact law. According to Stronge [964], they defined what is known as the *energetic coefficient of restitution*, e^*, which is obtained by equating the total work done by the normal force, F_N, during restitution, $\int_{v<0} F_n \nu dt$, to the work done by the normal force during compression times, $-e^{*2}\int_{v>0} F_n \nu dt$, where v is the velocity.

2.4 Border–Collision or C–Bifurcation

Border bifurcation occurs when a fixed point or a periodic orbit of a piecewise smooth system crosses or collides one of the boundaries between different phase spaces as a system parameter varies in quasi–statically manner [742]. It is termed as C–bifurcation by Feigin ([313], [314], [315], [316], [318]). Di Bernardo et al [235] described a number of different dynamical scenarios that can be observed after a C–bifurcation including (i) a transition from the orbit involved in the C–bifurcation to an orbit of a different type, (ii) joining two different solutions, and (iii) sudden transition to a chaotic attractor.

Di Bernardo et al [235] presented a systematic description of a local map for C–bifurcations for dynamical systems described by the following set of ordinary differential equations

$$\dot{\mathbf{z}} = \mathbf{f}(\mathbf{z},t,\boldsymbol{\mu}), \tag{2.39}$$

where $\mathbf{f} : \mathbb{R}^{n+1} \to \mathbb{R}^n$ is a piecewise smooth function, $\mathbf{z} \in \mathbb{R}^n$, and $\boldsymbol{\mu} \in \mathbb{R}^p$ is a parameter vector. Let L_0 be a periodic orbit of system (2.39) which is tangent to one of the phase space boundaries, Φ_0, for $\boldsymbol{\mu} = \boldsymbol{\mu}_0$ as shown in Fig. 2.8. The

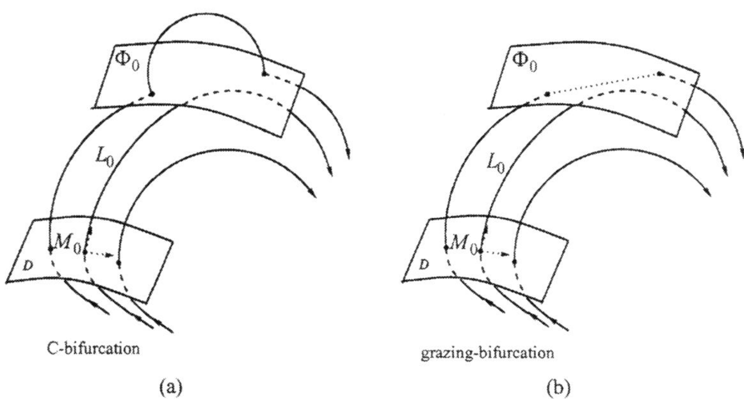

Fig. 2.8. Illustration of (a) C-bifurcation and (b) grazing bifurcation where D is a generic Poincaré section, [235].

system flow intersects transversally the Poincaré plane D at point M_0, which corresponds to the grazing limit cycle, L_0, as shown in Fig. 2.9. All points on the line S passing through M_0 belong to all trajectories that are tangent to the boundary Φ_0 on the line Γ. In this case, the line S splits the plane D into two planes D^- and D^+ such that trajectories emanating from D^- cross the boundary Φ_0 while those crossing D^+ stay away from Φ_0. All points x_i, $i = 1, 2, ..., n$ on plane D, with origin located at point M_0, are designated such that $x_n^* < 0 \Longleftrightarrow x^* \in D^-$, $x_n^* > 0 \Longleftrightarrow x^* \in D^+$, and $x_n^* = 0 \Longleftrightarrow x^* \in S$. If the system is linearized with respect to x_i and the control parameter μ in the neighborhood of M_0, then the motion is governed by the equations of the Poincaré map Π, Di Bernardo et al [235],

$$x^{(k+1)} = \begin{cases} A_1 x^{(k)} + c\mu & for \quad x^{(k)} > 0 \\ A_2 x^{(k)} + c\mu & for \quad x^{(k)} < 0 \end{cases} \tag{2.40}$$

where $A_1 = \frac{\partial \Pi^+}{\partial x}|_{x=0}$, $A_2 = \frac{\partial \Pi^-}{\partial x}|_{x=0}$, and $c = \frac{\partial \Pi^-}{\partial \mu}|_{\mu=0} = \frac{\partial \Pi^+}{\partial \mu}|_{\mu=0}$.

It was assumed that the mapping is continuous when $x_n^{(k)} = 0$ and $\mu = 0$ for all values of k, and smooth for $k = 1, ..., n-1$. The elements of matrices $A_1 = \left[a_{ij}^{(1)}\right]$ and $A_2 = \left[a_{ij}^{(2)}\right]$ are related such that $a_{ij}^{(1)} = a_{ij}^{(2)}$, $j \neq n$. Di Bernardo et al [235] derived a set of elementary conditions describing the possible consequences of the C–bifurcation. These include the existence of a periodic orbit on one side of the boundary and is smoothly converted at the C–bifurcation point ($\mu = 0$) into a solution existing on the other side of the boundary. Another possibility is the presence of two periodic orbits corresponding to two fixed points, M^* in the sub–mapping Π^- and M^{**} in Π^+, both exist on one side of the boundary, collide on the border at a C–bifurcation, $\mu = 0$, and then vanish on the other side. The third case involves a new two-periodic solution originates at the C–bifurcation. The possibility of

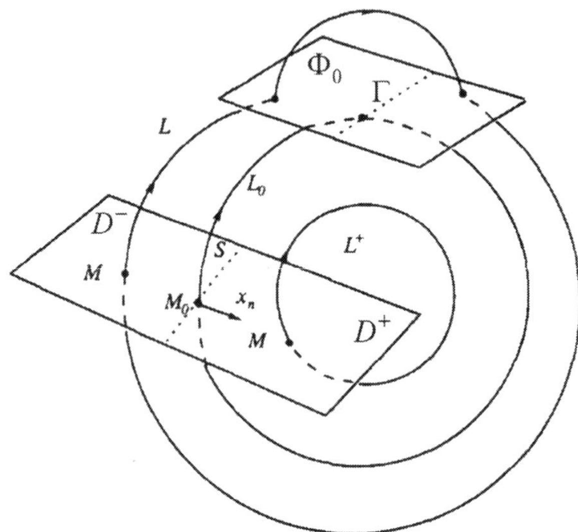

Fig. 2.9. Demonstration of C-bifurcation of a periodic orbit and Poincaré section D, [235].

a sudden jump to a chaotic attractor at the C–bifurcation was demonstrated for the case of a codimension-one map.

Halse et al [385] examined the behavior of piecewise-smooth, continuous, one-dimensional maps that are linear for negative values of the control parameter and nonlinear for positive values. These maps display both period-adding and period-doubling behavior and are described by the general form

$$x_{k+1} = \begin{cases} \alpha x_k - \mu & for \quad x_k \le 0 \\ \beta x_k^{\gamma} - \mu & for \quad x_k > 0 \end{cases}, \tag{2.41}$$

$$x_{k+1} = \begin{cases} \alpha x_k - \mu & for \quad x_k \le 0 \\ \alpha x_k + \beta x_k^{\gamma} - \mu & for \quad x_k > 0 \end{cases}, \tag{2.42}$$

where $\alpha \in \mathbb{R}$, $\beta \in \{1, -1\}$ and $\gamma > 1$. The sign of β is important. For maps with $\gamma = 2$ or $3/2$, the stability and existence conditions of fixed points and period–2 orbits in the vicinity of the border–collision were found analytically. The period–adding behavior was examined in these maps, where analytical solutions for the boundaries of periodic solutions do exist. It was shown that these maps can exhibit different types of C–bifurcations. Conditions for the existence of finite and infinite period adding sequences were shown to be dependent on the map parameters. Closed form conditions for the existence and stability of periodic orbits of increasing periodicity were derived for two specific cases of $\gamma = 2$ and $\gamma = 3/2$, which are associated to grazing and sliding bifurcations. The sliding or Filippov [325] bifurcation is associated with the solution of piecewise systems. Sliding is only possible if the direction of the system vector field on both sides of a discontinuity set points towards the set itself so that nearby trajectories are constrained to evolve on it [240]. *Sliding* motion may be understood as repeated switching between two different system configurations as reported in references [239] and [231].

Non-smooth bifurcations are encountered in a number of physical and engineering systems described by two or more sets of differential equations ([83], [1013], [1140]). Examples of such systems include the Colpitts oscillator [633], thyrister controlled reactor circuits [879], dc-dc converters [81], sigma-delta modulators [310], digitally controlled systems [384], and other electronic circuits [923]. These systems give rise to piecewise smooth maps ([80], [1115], [1116]).

Nordmark and Kowalczyk [739] developed an analytical unfolding of codimension-2 bifurcation involving grazing sliding using discontinuity mapping techniques. It was shown that the fold curve is one–sided and cubically tangent to the grazing curve locally to the codimension-2 point. Kowalczyk et al [531] proposed a strategy for the classification of codimension-2 discontinuity-induced C–bifurcations of limit cycles in piecewise smooth systems of ordinary differential equations. Specifically, codimension-2 grazing bifurcations that are local in the sense that the dynamics can be fully described by an appropriate Poincaré map from a neighborhood of the grazing point (or points) of the critical cycle to itself. Codimension-2 grazing bifurcations were divided into three distinct types: 1) the grazing point is degenerate, 2) the grazing cycle is itself degenerate (e.g., non–hyperbolic) or 3) simultaneous occurrence of two grazing events.

The bifurcation theory of one– and two–dimensional continuous piecewise smooth maps was developed by Banerjee and Grebogi [76] and Banerjee et al [82]. Routroy et al [898], Hogan et al [420], and Dutta et al [281] extended Feigin's approach to the case of discontinuous maps. They obtained the conditions of existence of period–1 and period–2 fixed points in general multi–dimensional maps. Banerjee and Grebogi [76] presented some experimental results to establish a theoretical framework and classification of bifurcations resulting from border collision. It was found that there are eleven qualitatively different types of border collision bifurcations depending on the parameters of the normal form. Parui and Banerjee [791] developed the theory of border collision bifurcation for the special case where the state space is piecewise smooth, but two–dimensional in one–side of the borderline, and one–dimensional in the other side. Jain and Banerjee [462] presented a classification of border–collision bifurcations in one–dimensional discontinuous maps in the neighborhood of the point of discontinuity. For each range of parameter values, the condition of existence and stability of various periodic orbits and of chaos were defined. The dynamics of a mechanical switching system in which the state variables are continuous was studied by Banerjee et al [79]. The first derivative of the vector field was found to change discontinuously across the switching boundary at the switching events.

Zhusubalyev et al ([1143], [1144], [1142]) and Zhusubalyev and Mosekilde [1141] described the main bifurcations and routes to chaos exhibited by piecewise–smooth dynamical systems. In particular, they examined the complex behavior of multidimensional piecewise–smooth systems with border–collision bifurcations. A new route to quasi–periodicity in the two–dimensional piecewise–linear normal form map was developed. It was shown that border–collision bifurcations can lead to the birth of a stable closed invariant curve associated with quasi–periodic or periodic dynamics. It was demonstrated that a two–dimensional torus can arise from a periodic orbit through a bifurcation in which two complex–conjugate Poincaré characteristic multipliers jump abruptly from the inside to the outside of the unit circle.

Kousaka et al [525] studied a system interrupted by its own state and a periodic interval. Tanaka and Ushio [978] considered a switching system modeled by a discrete–time flow and exhibited a lot of border–collision bifurcations. Di Bernardo [228] and Di Bernardo et al ([230], [231]) analyzed the corner–collision bifurcation in piecewise–smooth systems of ordinary differential equations whose periodic solution grazes with a corner of the discontinuity set. Their results contrasted with the equivalent results when a periodic orbit grazes with a smooth discontinuity set. Zhusubalyev et al [1145] studied some resonance phenomena in a piecewise–smooth dynamical system with external periodic excitation and examined transitions to chaos via border–collision bifurcations of cycles on a two–dimensional torus. The analysis provided the structure of border–collision bifurcation boundaries of synchronization tongues and transitions to chaos via border–collision bifurcations of cycles on a two–dimensional torus.

2.5 Border Bifurcation in Switching Circuits

The bifurcation theory for piecewise smooth systems has attracted significant research attention in physical systems such as piecewise linear electronic circuits (see, e.g., [762], [269], [741], [236], [229], [1116]). In such systems, as a control parameter varies through a critical value typical transitions exhibit periodic orbits with period–1 bifurcating to a chaotic orbit or a periodic orbit vanishing as it hits the border, resulting in an abrupt change in the Jacobian matrix. These anomalous bifurcation phenomena lead to border collision bifurcations. Recent investigations on bifurcations in switching circuits revealed that many typical bifurcations can occur in piecewise smooth maps that cannot be classified among the generic cases like saddle–node, pitchfork, or Hopf bifurcations occurring in smooth maps. The bifurcation structure associated with grazing impact belongs to a more general class of *border–collision bifurcations* that arise in non–smooth systems as reported by Nusse and Yorke ([742], [743]) and Nusse et al [741].

Ohnishi and Inaba [748] observed a strange bifurcation route to chaos in a piecewise–linear second–order non–autonomous differential equation derived from a simple electronic circuit. When a limit cycle loses its stability, the attractor changes directly to chaos (instant chaos) without undergoing a period doubling bifurcation or intermittency. The width of the attractor band was found continuous at the bifurcation point, and the chaotic band grows larger continuously as the system parameter varies. It was shown that the singular phenomenon arises from the piecewise–linearity of the system and the Lyapunov exponent jumps discontinuously from minus to plus at the bifurcation point.

In the area of control theory, piecewise linear switched systems constitute a special class of hybrid systems ([478], [590]). These systems arise when some nonlinear components such as switching, dead–zone, saturation, relays, and hysteresis are encountered. The local stability of periodic orbits in switched discontinuous piecewise affine periodically driven systems was studied by El Aroudi et al [288]. These systems are described by a set of affine differential equations together with periodic (in time) switching rules to commute between them. The switching rules were described by switching functions that are periodic in time and linear in state. The feedback control of piecewise smooth discrete time systems that undergo border collision bifurcations was studied by Di Bernardo et al [238], Hassouneh et al [399] and Hassouneh and Abed [398]. The goal of the control was to modify the bifurcation so that the bifurcated steady state is locally attracting and locally unique. In such bifurcations, a fixed point remains asymptotically stable at both sides of the critical bifurcation point, but at the critical bifurcation value, all orbits starting from all points other than the fixed point diverge to infinity. Do [261] gave an opposite view and investigated the mechanism causing dangerous border collision bifurcations. At the critical bifurcation value the qualitative type of the fixed point can be induced by invariant manifolds of the periodic saddle orbit.

A model allowing a double impacting regime for a particle undergoing simple harmonic motion was considered by Gutierrez and Arrowsmith ([382], [383]). Control equations were developed followed by strategies for preserving and annihilating resonant periodic orbits. Gutierrez and Arrowsmith [383] adopted a class of dynamic systems characterized by the discontinuous motion of a spring–mass constrained by the motion of a feedback–assisted actuator. It was shown that the combined effects of mechanical restitution coefficient and displacement feedback can be exactly represented by a single equivalent dissipation coefficient. The stabilization of grazing periodic trajectories in a certain class of vibro–impact oscillators in the presence of control was formulated by Dankowicz and Svahn [206]. They proposed a control algorithm that guarantees the local persistence of system attractors with at most low–velocity contact in vibro–impact oscillators. Sufficient conditions were formulated on the linearization of the control strategies along a grazing periodic trajectory to ensure the asymptotic stability of the grazing trajectory.

2.6 Closing Remarks

Theoretical developments of mapping associated with grazing and border bifurcations have uncovered many complex dynamic characteristics of vibro–impact systems. Three major types of grazing bifurcations have been reported [172]. These are bifurcations from stable period–1 to a reversed infinite period adding cascade, bifurcation from a stable period–1 orbit to attracting chaos occupying a full interval of the bifurcation parameter, and collision of an unstable maximal periodic orbit and period–1 orbit. The square–root singularity in the Nordmark mapping is directly linked to the phenomenon of period adding. The Nordmark map [731] was originally developed for absolutely rigid barriers. A modified map developed by Molenaar et al [686] was developed to incorporate the effect of flexible barriers. It was found that the map for impacts with a flexible barrier is very similar to the one developed for rigid barriers. The only difference between the two cases was found in a change of the scale of the bifurcation parameter. One may note that all mapping techniques described in this chapter have been applied to piecewise linear systems. The influence of the system inherent weak nonlinearity on grazing bifurcation may cause a change in the value of the control parameter depending on whether the system nonlinearity possesses soft or hard characteristics.

Chapter 3
Single–Degree–of–Freedom Systems

3.1 Introduction

The theory of vibro–impact dynamics has been applied to classical lumped discrete systems represented by single–, two–, and multi–degree of freedom against one– or two–sided barriers. One freedom systems in the form of mass–spring–dashpot with one–sided barrier have been extensively studied in the literature. Other systems such as a ball bouncing on an oscillating table, a simple pendulum with one– or two–sided barriers, and ship roll dynamics interacting with icebergs will be considered. The study of these systems has revealed different and complex response characteristics such as periodic and quasi–periodic oscillations, grazing and period doubling motions, chattering and chaotic oscillations. A simple idealization of a vibratory plow impacting against an immovable relatively rigid obstruction was analyzed to determine possible periodic motions and the stability of these motions [921]. It should be mentioned that these systems may exhibit some peculiar periodic or chaotic response regimes (see, e.g., [811], [813], [417], [4], [418], [419], [124], [1077], [1078], [945], [12], [127], [910], [23], [199], [552]).

Peterka and Kotera [828] described the behavior of impact motion of single freedom systems experiencing more than one impact of the body against the rigid stop during one excitation period. They presented some domains of attraction of initial conditions from which one of several possible impact motions becomes stabilized after cessation of the transient response. The steady state vibro–impacting responses of harmonically excited linear oscillators were studied by Whiston [1075]. By using the modern theory of dynamical systems together with numerical simulation, it was found that the steady state motions are attracting sets in the system phase space and capture initial conditions in their domains of attraction. The phase space of a vibro–impact system was found to be inhabited by many attracting sets. For example, there are sub–harmonic, multi–impact, periodic orbits and chaotic, steady state responses. An attempt was made to build generic topological models of their phase spaces for physically significant parameter ranges.

R.A. Ibrahim: Vibro-Impact Dynamics: Model., Map. & Appl., LNACM 43, pp. 55–95.
springerlink.com

3.2 Bouncing Ball on a Vibrating Platform and Pile Drivers

3.2.1 Analysis

The original elastic version of the problem of ball–platform collisions was introduced by Fermi [321]. Earlier theoretical treatments, based on the Chirikov [174] "high–bounce" approximation, chaos was observed to occur for the partially elastic bouncing ball. The study of ball bouncing on an oscillating surface was then introduced as an attempt to understand the impulsive noise generated from metal–to–metal collision. Impulsive noise is a daily problem for those who are operating pneumatic hammers, riveting machines, punch presses and peening guns. This problem is believed to be considered by Wood and Byrne ([1082], [1083]) who studied impact process under random vibrating surface. It was then studied by Holmes [423] and Guckenheimer and Holmes [380] under sinusoidal surface excitation. With reference to Fig. 3.1, the table of mass m_t is moving sinusoidally, $Z_0 \sin \Omega t$, where Z_0 and Ω are the table amplitude and frequency. The recurrence relationship between the state of the system at the $(j+1)$–impact and j–impact was given by the nonlinear mapping

$$t_{j+1} = t_j + 2\frac{V_j}{g}, \tag{3.1a}$$

$$V_{j+1} = eV_j + (1+e)W(t_j + 2V_j/g), \tag{3.1b}$$

where V and W are the absolute velocities the ball rebound and the table, e is the coefficient of restitution, t_j is the time of the j–th impact, and g is the gravitational acceleration. Under sinusoidal excitation of the table, Holmes [423]

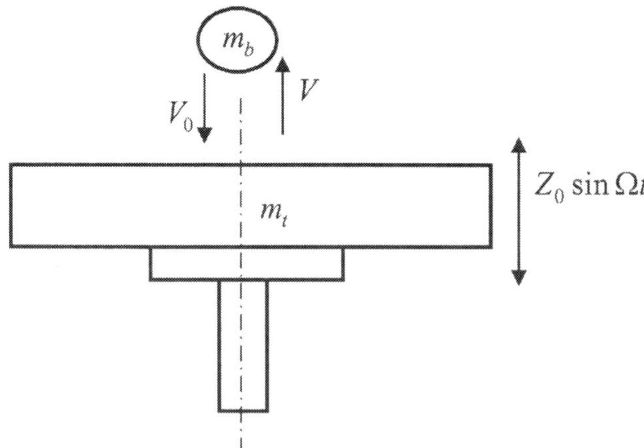

Fig. 3.1. Ball bouncing on an oscillating table.

found that the fixed points with $\Omega t_n < \pi \sin(\Omega t_n + 2V/g) > 0$ are all saddle points of the first kind. Fixed points with $\Omega t_n > \pi \sin(\Omega t_n + 2V/g) < 0$, were found to be sinks (centers) if

$$n\pi\left(\frac{1-e}{1+e}\right) < \frac{\Omega^2 Z_0}{g} < \sqrt{n^2\pi^2\left(\frac{1-e}{1+e}\right)^2 + 1}. \tag{3.2}$$

On the other hand, saddle point of second kind, i.e., $\pi \sin(\Omega t_n + 2V/g) < 0$ occurs if

$$\frac{\Omega^2 Z_0}{g} > \sqrt{n^2\pi^2\left(\frac{1-e}{1+e}\right)^2 + 1}. \tag{3.3}$$

From relation (3.2) the following local bifurcation points are obtained

$$\frac{\Omega^2 Z_0}{g} = n\pi\left(\frac{1-e}{1+e}\right), \tag{3.4}$$

$$\frac{\Omega^2 Z_0}{g} = \sqrt{n^2\pi^2\left(\frac{1-e}{1+e}\right)^2 + 1}. \tag{3.5}$$

The value given by equation (3.4) establishes fixed points in a saddle–node bifurcation. Equation (3.5) establishes the occurrence of stability change known as a flip.

Luo and Han (1996) revisited Holmes work [423] and obtained the following stability condition, for $\pi \sin(\Omega t_n + 2V/g) > 0$,

$$n\pi\left(\frac{1-e}{1+e}\right) < \frac{\Omega^2 Z_0}{g} < \sqrt{n^2\pi^2\left(\frac{1-e}{1+e}\right)^2 + 4\frac{(1+e^2)}{(1+e)^4}} \tag{3.6}$$

Stability and bifurcation boundaries were generated by Luo and Han [605] for period–1 motion, coefficient of restitution $e = 0.5$, and $n = 1$. The dependence of the type of motion on the initial impact velocity and excitation frequency is shown in Fig. 3.2, while the stability boundaries are shown in Fig. 3.3. Luo and Han [605] compared their results with Holmes solution [423] and showed that the range of stable motion is wider. The Poincaré mapping sections of the unstable period–1 motion indicated the existence of identical Smale horseshoe structures and fractals.

For the case of perfect elastic impacts, the equation is reduced to the 'standard mapping'. For sufficiently large excitation velocities and a coefficient of restitution close to one, the bouncing ball–table system was found to exhibit large families of irregular non–periodic solutions in addition to the expected harmonic and subharmonic motions. Veluswami and Crossley [1028] and Veluswami et al [1029] experimentally and analytically studied the motion of a steel ball while impinging against two–end plates excited by an

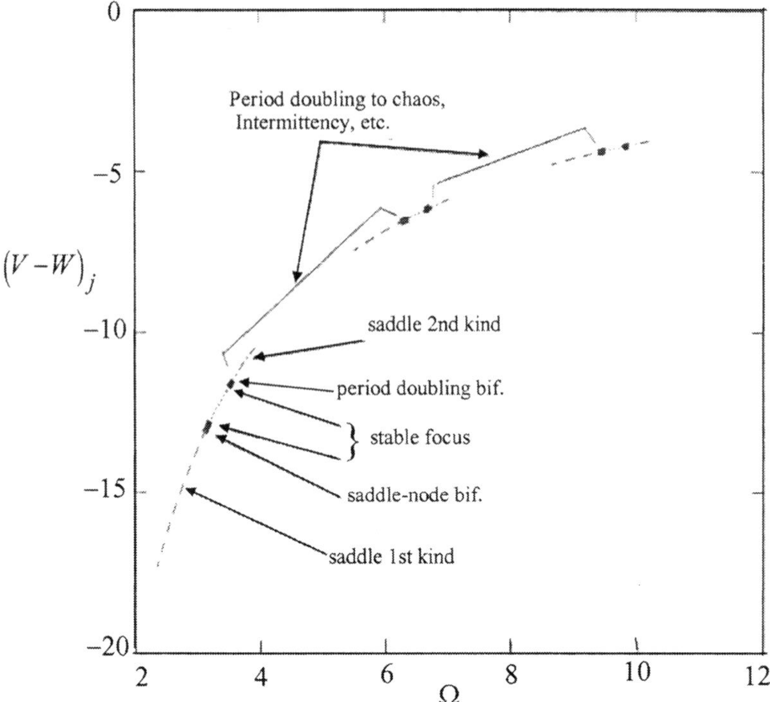

Fig. 3.2. Different regimes of motion of a bouncing ball, [605].

electromagnetic shaker. Over the frequency range ($0-60\ Hz$) and amplitudes ($0-5\ mm$) the ball was found to hit several times on one boundary before passing over to strike the other, per cycle of the shaker. Later, Veluswami et al [1029] developed a mathematical model to describe the vibro–impact dynamics of the system with contact stiffness modeled by the Hertzian spring stiffness. During each impact the motion of the ball was taken to be a brief half wave, due to the highly nonlinear forces of surface compliance and surface damping. It was found that linearization of the surface stiffness does not reproduce the observed phenomena.

Everson [303] presented a detailed study of a two–dimensional manifold mapping describing an imperfectly elastic ball bouncing on a vibrating platform. Quasi–periodic motion on one–dimensional manifold was predicted numerically at low forcing. At high forcing Smale horseshoes were reported. The evolution of the attracting set with changing parameters was studied. A new type of chaos, in which a trajectory alternates between two distinct chaotic regions, was described and explained in terms of manifold collisions. Everson [303] was able to obtain an analytic expression for Lyapunov exponents in the quasi–periodic regime under the influence of high values of forcing excitation. Marudachalam and Bursal (1995) developed a numerical algorithm

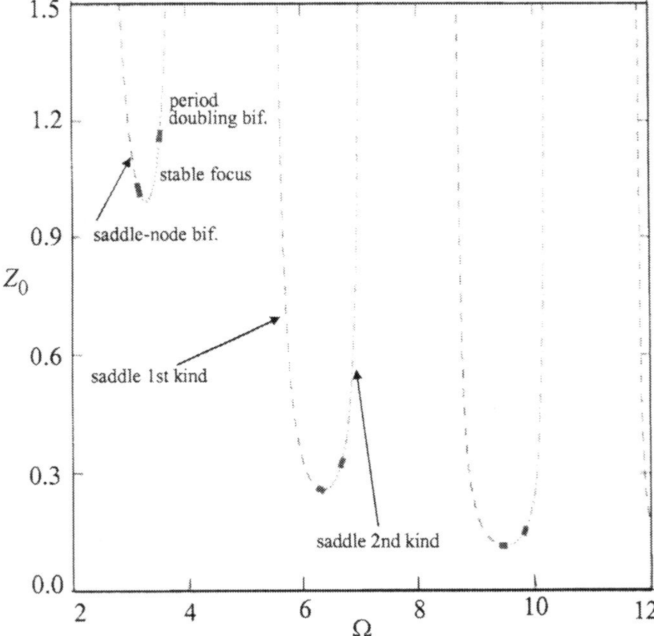

Fig. 3.3. Stability boundaries of different motion regimes, [605].

for studying the global dynamics of an impact oscillator with two–sided rigid barriers. The study included a peculiar type of solution in which the trajectories on phase space from a certain set of initial conditions merge in finite time, making the dynamics non–invertible. The effect of double prime grazing on the dynamics of the system was considered.

Ivanov [455] introduced a unified approach to the analysis of bouncing ball whose center of mass does not coincide with its geometric center. The mathematical essence of this approach was the continuous representation of the impulsive motion in some auxiliary variables. He derived an explicit formula for the fundamental matrix that allows the construction of the characteristic equation. Zimmerman et al [1146] developed the electronic analog of the bouncing ball to demonstrate the inherent nonlinear chaotic features of this system. They used an operational amplifier circuit to simulate the free fall of a ball in constant gravity. Current feedback through a precision diode rectifier was used to model free fall when the rectifier is backward biased and the bounce when the rectifier is conducting.

The motion of a mass sliding on a rod with two end–barriers was experimentally investigated by Blazejczyk–Okolewske et al [123] when the rod excited axially. The response of the mass was characterized by long intervals with regular behavior (two impacts per excitation period). This motion was occasionally interrupted with multi–periods with zero, one or three impacts.

By increasing the excitation frequency the length of regular two–impact intervals was found to decrease and periods with zero, one or three impacts become more frequent.

3.2.2 Experimental and Simulation Results

Experimental studies of a bouncing ball on a vibrating table confirmed the results of the exact model. For example, Tufillaro et al ([1017], [1015]) found that all the major bifurcations predicted by the exact model were observed experimentally within 2% accuracy. The experimental tests were conducted by slowly increasing the amplitude of the oscillating table. The motion of the ball was monitored through an experimental impact map, which is similar to a next return map. Experimental bifurcation diagrams were generated for different values of coefficient of restitution (obtained by changing the material of the ball). It was observed that a chaotic invariant set was established at the end of the period doubling cascade, but for a further increase in the driving amplitude, the strange attractor was destroyed in a crisis. The dynamics of the ball after this crisis was found to result in motion quickly approaching a period sticking solution (for small coefficient of restitution) or to exhibit long transients known as "transient chaos" following the "shadow of the strange attractor" (for larger coefficient of restitution).

Fig. 3.4 shows a bifurcation diagram obtained by direct simulation of the exact model as reported by Tufillaro [1014]. This figure reveals a period doubling route to chaos for coefficient of restitution $e = 0.5$ similar to the same behavior observed in the experimental tests reported by Tufillaro et al ([1017], [1015]). At $Z_0 = 0.012$, the strange attractor was found to be stable for over 10^6 impacts. Over the excitation amplitude range, $0.0118 < Z_0 < 0.019$, a first crisis was predicted that expands the size of the strange attractor. Over the narrow excitation amplitude range, $0.0121 < Z_0 < 0.0122$, another crisis occurred, which destroyed this strange attractor. Above the excitation amplitude $Z_0 > 0.0122$ the orbit was found to follow the shadow of the strange attractor for a number of impacts but eventually converge to a sticking solution (typically after 100 to $1,000$ impacts). In both experiments and numerical simulations, the pre–crisis (chaotic) dynamics and post–crisis (eventually periodic) dynamics can be distinguished.

The effect of a near–resonant perturbation in a bouncing ball was examined to suppress the onset of the first period doubling bifurcation by Tufillaro and Albano [1016] and Wiesenfeld and Tufillaro [1079]. Near the bifurcation point, the full dynamical equations were reduced to a discrete–time map governing the dynamics on a slowly oscillating center manifold. The geometry of the phase–space dynamics was utilized to clarify several points of the effects of strong near–resonant perturbations. Topological parameters were estimated by Tufillaro [1014] from a chaotic time series generated by a

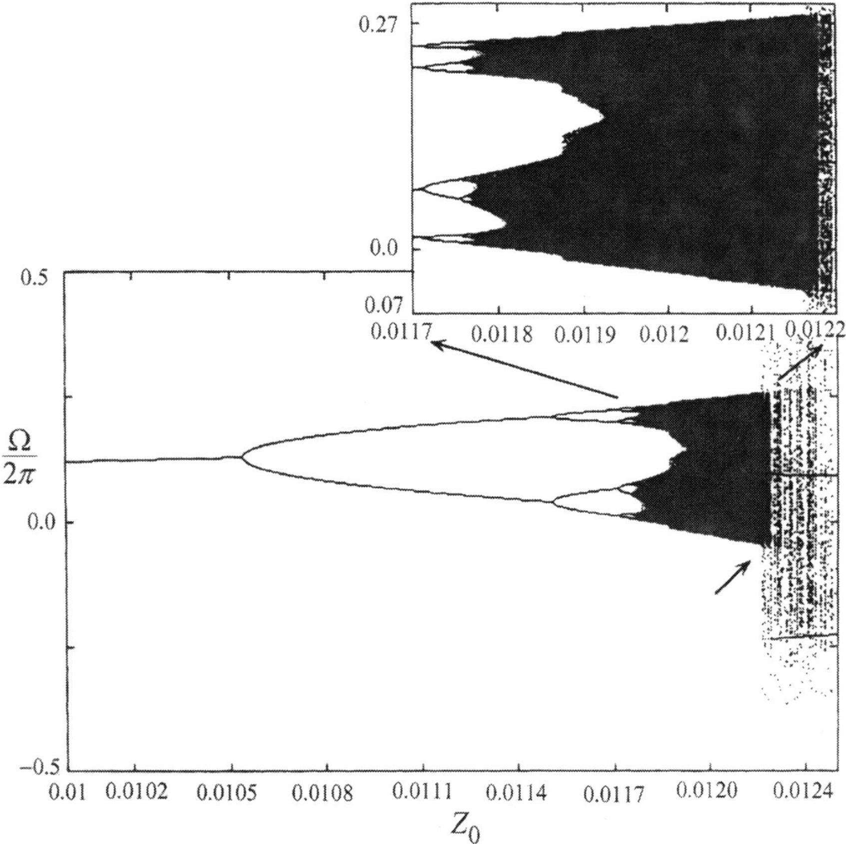

Fig. 3.4. Bifurcation diagram for the bouncing ball system, $e = 0.5$, [1014].

dissipative bouncing ball. Tufillaro [1014] used the *braid analysis* and *pruning front approach*. Braids arise as periodic orbits in dynamical systems modeled by three–dimensional flows. On the other hand, the pruning procedure deals with chaotic two–dimensional diffeomorphisms under certain rules to prune certain orbits of higher period. Kanso and Papadopoulos [490] examined the dynamics of a pseudo–rigid ball impacting on an oscillating rigid foundation.

3.2.3 Chattering Phenomenon

The dynamics of the completely inelastic ball, $e = 0$, exhibited complexity in its temporal behavior [666]. Mehta and Luck [666] developed an approximate map and showed that its period–doubling sequence terminated abruptly in a locking region. It was demonstrated that in this regime, the intervals in the space of the table acceleration parameter, $Z_0\Omega^2/\pi g$, were characterized by different rational winding numbers. For any non–zero value of the coefficient of restitution, $e \neq 0$, the ball was found to perform a large number of smaller and smaller bounces, referred to as *chattering*, provided it lands deep enough in the absorbing region of the phase space ([603], [147]). Chattering is one of the most interesting properties of an impacting system, which is characterized by an infinite number of impacts occurring in a finite time. The ball was shown to wait until it rebounds at the beginning of the next platform cycle with the initial conditions:

$$\tau_0 = \frac{\Omega t_0}{2\pi} = \frac{1}{2\pi} \arcsin\left(\frac{g}{Z_0 \Omega^2}\right), \qquad and \quad V_{0rel} = 0 \tag{3.7}$$

where V_{0rel} is the relative "take–off" velocity of the ball.

Giuseppponi et al [363] presented a quantitative description of the chattering dynamics of an ideal inelastic ball bouncing on a vibrating platform. The velocity of the bouncing ball was sampled at each impact with the plate (asynchronous sampling). Its random nature revealed that the chattering mechanism, through which the ball gets locked on the plate, is accomplished within a limited interval of the plate oscillation phase. Furthermore, chattering trajectories and strange attractors were found to coexist for appropriate ranges of some parameter values. Structure and substructure of the chattering bands were explained in terms of a simple impact map rule. A ball undergoing inelastic collisions with its walls on an oscillating cart was numerically studied by De Souza et al [219]. A multistable regime was found to take place characterized by the coexistence of different attractors with a complicated basin boundary structure in the phase space. Time history records of typical chattering will be provided in Section 3.4 for the case of an inverted pendulum oscillating against two rigid barriers.

3.2.4 Applications

It is interesting to note that applications based on the dynamics of a bouncing mass have been reported in the literature. These applications include the vibration hammers, inertial shakers, and pile drivers or vibro–impact moling systems ([799], [801], [800], [796], [797], [798], [1080], [1081], [728], [729], [615]). For example, Pavlovskaia and Wiercigroch [796] developed an analytical transformation for vibro–impact systems with an impacting mass. The transformation converts high–frequency low–amplitude excitation into low–frequency high–amplitude response. It allows the vibrating system to generate an appropriate internal force to overcome resistance forces of the media and move downwards.

The dynamics of a small vibro–impact pile driver was studied by Luo et al [623] using three–dimensional mapping, which is of piecewise property and singularities. The existence and stability of period$-n$ single–impact motions of the pile driver were obtained analytically and numerically. The study included the influence of singularity of the Poincaré mapping, caused by the grazing contact of driver and pile, on global bifurcations and transitions to chaos. It was found that these transitions are not regular bifurcations but emerge from piecewise property and singularity of the impact mapping. Sijin et al [942] obtained the Poincaré map of period$-n$ motions with single–impact events for a linear vibro–impact system and determined grazing bifurcation conditions and bifurcation equations.

Luo and Lv [615] developed an analytical model, which describes oscillatory and progressive motions in the dynamics of a plastic impact oscillator with a frictional slider. Dynamics of the impact oscillator was analyzed using a five–dimensional map, which describes free flight and sticking solutions of two masses of the system, between impacts, supplemented by transition conditions at the instants of impacts. Piecewise property and singularity were found to exist in the Poincaré map. The piecewise property is caused by the transitions of free flight and sticking motions of impacting masses immediately after the impact, and the singularity of the map is generated via the grazing contact of impacting masses immediately before impact. These properties were shown to exhibit particular types of sliding and grazing bifurcations of periodic–impact motions under parameter variation.

The problem of two–dimensional vibratory impact of a sphere bounces on a massive flat horizontal surface was considered by Kozol and Brach [533]. Regions of periodic and chaotic responses were determined. Higher–order periodic motions were obtained through period–doubling bifurcations. Frictionless periodic motions remain periodic and of the same order in the presence of friction. Transition to chaos was also found unaffected by friction. Hill et al [416] demonstrated that under certain conditions the ball can perform a 'big' bounce followed by a 'little' bounce, and then simply repeat the sequence. Clark et al [184] experimentally found that an electronic impact oscillator, simulating a ball bouncing on a vibrating surface, reveals interesting chaotic regimes that are consistent with the numerical simulations. The problem of the normal impact of an elastic sphere or bar on a Timoshenko beam was considered by Rossikhin and Shitikova [897]. The impact process was found to be accompanied by material local deformation and propagation of strong discontinuity wave surfaces in the beam.

The case of a ball oscillating in a container subjected to harmonic excitation was examined by Luo ([606], [607]). The stability and bifurcation of the system revealed period–doubling bifurcation for unsymmetrical period–1 motions instead of symmetrical period–1 motion. Stability, saddle–node and period–doubling bifurcation conditions for the model motion were determined analytically and numerically. The stability and bifurcation conditions for all motions of this impact oscillator were found to depend strongly on the initial impact phase instead of excitation frequency.

3.3 Mass–Spring System

3.3.1 Introduction

This classical system models some mechanical systems such as power hammers and rams. The early study of this system is believed due to Rusakov and Kharkevich [900], then followed by analytical and experimental studies (see, e.g., [593], [94], [109], [901], [33], [930], [931], [932], [927], [928]). Kim and Noah [509] developed a general approach for determining the periodic solutions and their stability of nonlinear oscillators with piecewise–smooth characteristics. The approach was based on a modified harmonic balance/Fourier transform procedure. The method was applied to a forced oscillator interacting with a stop of finite stiffness. Flip and fold bifurcations were manifested in the response characteristics.

The bifurcation problem of a spring–mass system vibrating against an infinitely large plane was studied by Xie [1086]. It was shown that there exist phenomena of codimension–two bifurcations when the ratios of frequencies are in the neighborhood of the same special values and the coefficient of restitution approaches unity. This simple system exhibited flip bifurcations, Hopf bifurcations of fixed points, and those of period of two points. Lazer and McKenna [568] and Bonheure and Fabry [128] considered simple undamped and damped oscillators driven by a periodic excitation, $F(t)$, with a barrier located at its equilibrium position and is free to oscillate on the right side of the barrier. The existence of periodic oscillations for the resonance case, i.e., when the natural frequency is identical to the number $n/2$ where n is an integer number, was found to depend on the number of zeros of the function

$$\Phi_{n,p} = \int F(t)|\cos[n(t+\theta)/2]|dt. \tag{3.8}$$

The basic idea of this result is that the impact oscillator is considered as a limiting case of an asymmetric oscillator.

3.3.2 Unperturbed Motion

With reference to Fig. 3.5(a), the mass–spring under unperturbed oscillations is governed by the second order differential equation

$$\ddot{x} + \omega_n^2 x = 0. \tag{3.9}$$

subject to the initial conditions, $x(t=0) = h$ and $x(t = 2\pi/\omega_i) = h$, where ω_i is the impact natural frequency, i.e., when the mass interacts with the barrier, h is the gap between the mass m in its static equilibrium position and the barrier. It is not difficult to show that the subsequent motion of the mass is given by the following solution

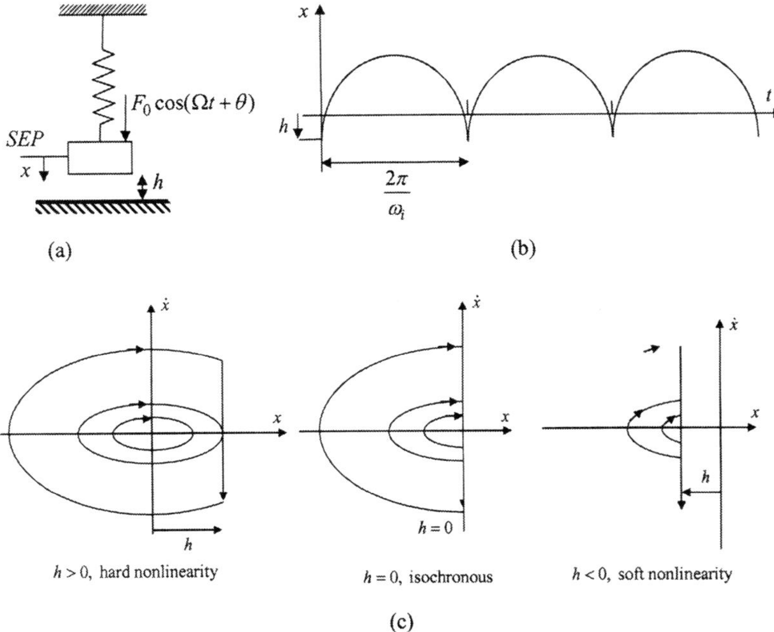

Fig. 3.5. (a) Mass spring system oscillating against one-sided barrier, (b) typical time history record, (c) possible phase portraits for different values of gap.

$$x(t) = h\frac{\cos(\omega_n t - \pi\omega_n/\omega_i)}{\cos(\pi\omega_n/\omega_i)}, \qquad t \in [0,\ 2\pi/\omega_i]. \tag{3.10}$$

Babitsky [53] showed that this motion is also valid over the entire time domain as shown in Fig. 3.5(b). The phase portraits for three different locations of the barrier are shown in Fig. 3.5(c). For the case of $h > 0$ there are three different orbits, namely, the periodic, grazing, and impact. The length of the impact orbit is proportional to the period of oscillations of impact orbit, $2\pi/\omega_i$. This implies that $\omega_n/\omega_i \leq 1$. Thus the three cases corresponding to $h > 0$, $h = 0$, and $h < 0$, correspond to hard nonlinear, isochronous, and soft nonlinear characteristics, respectively.

3.3.3 Perturbed Motion

The perturbed motion of the mass–spring system is considered under sinusoidal excitation. The equation of motion may be written in the form

$$\ddot{x} + \omega_n^2 x = \frac{F_0}{m}\cos(\Omega t + \theta), \tag{3.11}$$

where θ is the initial phase angle. The following initial conditions are based on the assumption of one impact per one period of oscillation

$$\begin{aligned} &x(t=0)=h, \qquad \dot{x}(t=0)=\dot{x}_{+}, \qquad x(t=2\pi j/\Omega)=h \\ &\dot{x}(t=0)=\dot{x}_{-}, \qquad \dot{x}_{j+}=-e\dot{x}_{j-}, \qquad \textit{and} \quad \dot{x}_{-}\geq 0 \end{aligned} \tag{3.12}$$

The solution of equation (3.11) subject to the initial conditions (3.12) may be written in the form, [53],

$$\begin{aligned} x(t)=&\left(h-\frac{F_0/k}{|1-\nu^2|}\right)\left\{\cos\left(\omega_n t\right)+\tan\left(\frac{\pi k}{\nu}\right)\sin\left(\omega_n t\right)\right\}+ \\ &\frac{F_0/k}{|1-\nu^2|}\cos(\Omega t+\theta) \end{aligned} \tag{3.13}$$

$$\sin\theta=-\frac{(1-e)}{2\Omega}\frac{|1-\nu^2|}{F_0/k}\dot{x}_{-}, \tag{3.14a}$$

$$\cos\theta=\left[1+\frac{(1+e)\dot{x}_{-}}{2h\omega_n}\cot(\pi j/\nu)\right]\frac{|1-\nu^2|h}{F_0/k} \tag{3.14b}$$

where $\nu=\Omega/\omega_n$. The value of the velocity just before impact can be obtained from equations (3.13) and (3.14) in the form

$$\dot{x}_{-}=-2h\omega_n\frac{\left\{1\pm\sqrt{1-\left[1-\left(\frac{F_0}{kh|1-\nu^2|}\right)^2\right]\left[1+\frac{1}{\nu^2}\left(\frac{1-e}{1+e}\right)^2\tan^2(\pi j/\nu)\right]}\right\}}{(1+e)\left[1+\frac{1}{\nu^2}\left(\frac{1-e}{1+e}\right)^2\tan^2(\pi j/\nu)\right]} \tag{3.15}$$

Real velocity values occur if the expression under the radical sign is always positive, i.e., if

$$\frac{|h|}{F_0/k}|1-\nu^2|\leq\sqrt{1+\frac{\nu^2}{\tan^2(\pi j/\nu)}\left(\frac{1+e}{1-e}\right)^2} \tag{3.16}$$

Czolczynski [199] considered a mass–spring–dashpot system under sinusoidal excitation with one stop. The equation of motion of the system in the non–dimensional form is

$$X^{''}+2\zeta X^{'}+X=\cos(\nu\tau+\theta) \tag{3.17}$$

where $X=xk/F_0$, a prime denotes derivative with respect to the non–dimensional time parameter $\tau=\omega_n t$, $\nu=\Omega/\omega_n$, and $\zeta=c/\left(2\sqrt{km}\right)$.

The solution of equation (3.17) may be written in the form

$$X(\tau) = A\mathrm{e}^{-\zeta\tau}\cos\left(\sqrt{1-\zeta^2}\tau+\psi\right) + \frac{\left(1-\nu^2\right)}{\left(1-\nu^2\right)^2+(2\zeta\nu)^2}\cos(\nu t+\theta)$$
$$+\frac{2\zeta\nu}{\left(1-\nu^2\right)^2+(2\zeta\nu)^2}\sin(\nu t+\theta) \tag{3.18}$$

Equation (3.18) describes the system periodic motion in the absence of impact with a period equivalent to the excitation period when the homogeneous solution is completely decayed. Czolczynski [199] obtained the conditions for periodic oscillations with impacts. It was assumed that the time interval between two subsequent impacts is equal to a multiple of the forcing period $2\pi/\nu$, i.e., when $X(0) = X(2\pi j/\nu) = -hk/F_0$, $j = 1, 2, ...$, where the displacement x was assumed positive upward. This resulted in the value of the phase angle $\psi = -\pi k/\nu$. Using the impact law, $X^{'}(0) = -eX^{'}(2\pi j/\nu)$, the coefficient A was obtained in the form

$$A = -\frac{(1+e)\nu\left[2\zeta\nu\sin\theta-\left(1-\nu^2\right)\cos\theta\right]}{\left[\left(1-\nu^2\right)^2+(2\zeta\nu)^2\right]M} \tag{3.19}$$

where

$$M = \zeta\cos\psi+\sqrt{1-\zeta^2}\sin\psi+e(\exp(-2\pi j\zeta/\nu))\times$$
$$\left[\zeta\cos\left(\tfrac{2\pi j}{\nu}\sqrt{1-\zeta^2}+\psi\right)+\sqrt{1-\zeta^2}\sin\left(\tfrac{2\pi j}{\nu}\sqrt{1-\zeta^2}+\psi\right)\right].$$

The phase angle was given by the relationships

$$\sin\theta = \frac{-\left(hk/F_0\right)K_s \pm K_c\sqrt{K_s^2+K_c^2\left(hk/F_0\right)^2}}{K_s^2+K_c^2} \tag{3.20a}$$

$$\cos\theta = \frac{-\left(hk/F_0\right)K_s \mp K_c\sqrt{K_s^2+K_c^2\left(hk/F_0\right)^2}}{K_s^2+K_c^2} \tag{3.20b}$$

where $K_s = \frac{1}{\left[(1-\nu^2)^2+(2\zeta\nu)^2\right]}\left\{2\zeta\nu-\frac{(1-\nu^2)}{M}(1+e)\nu\cos\psi\right\}$, and

$$K_c = \frac{1}{\left[(1-\nu^2)^2+(2\zeta\nu)^2\right]}\left\{\left(1-\nu^2\right)-\frac{2\zeta\nu^2}{M}(1+e)\cos\psi\right\}.$$

The lower sign before the square roots in equations (3.20) correspond to unstable solutions. Furthermore, the solution given by equations (3.20) is real if

$$-\sqrt{K_s^2+K_c^2} \le (hk/F_0) \le \sqrt{K_s^2+K_c^2}. \tag{3.21}$$

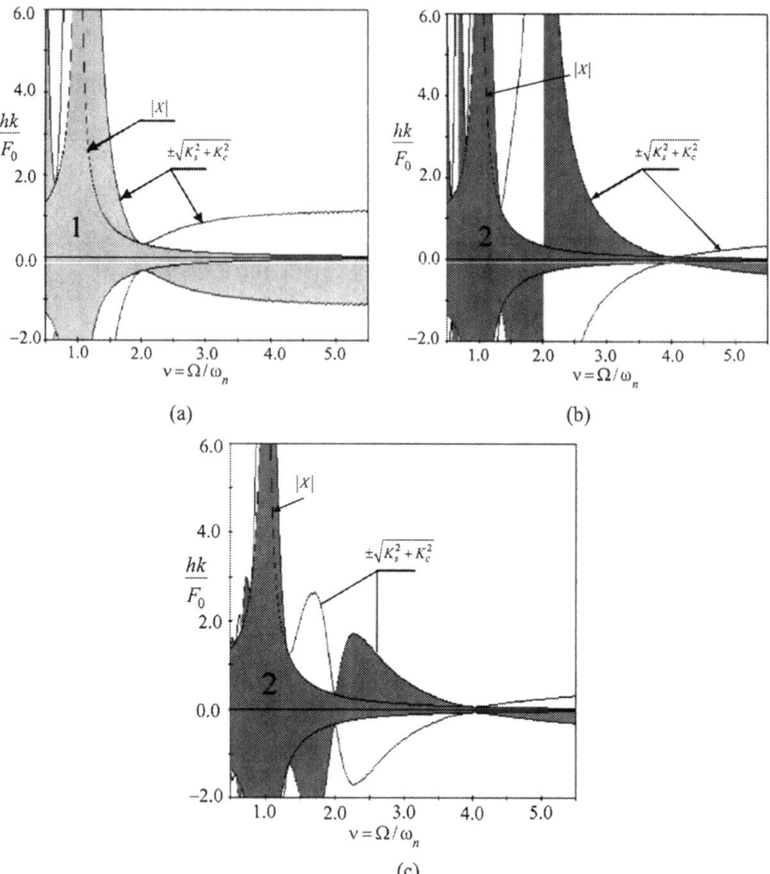

Fig. 3.6. Amplitude frequency response and regions of existence of periodic solutions with impacts for $e = 0.6$: (a) $j = 1$, $\zeta = 0.0$, (b) $j = 2$, $\zeta = 0.0$, (c) $j = 1$, $\zeta = 0.025$, [199].

The existence of periodic solutions with impacts and the stability of these solutions were determined under the condition $\dot{x}(2\pi j/\nu) < 0$. Fig. 3.6 shows the regions of the existence of periodic solutions with impacts for $e = 0.6$, $j = 1$ or 2, and two different values of damping factor $\zeta = 0$ or 0.025 in the domain of gap and excitation frequency parameters hk/F_0. The influence of the coefficient of restitution on the regimes of the oscillator motion with impact was obtained by Czolczynski [199] in the absence and presence of damping. Fig. 3.7 shows typical plots for two different values of coefficient of restitution $e = 0.6$ and 0.8. It is seen that the large coefficient of restitution results in widening the regions of periodic solutions. These regions also exist for negative values of the gap clearance of the barrier.

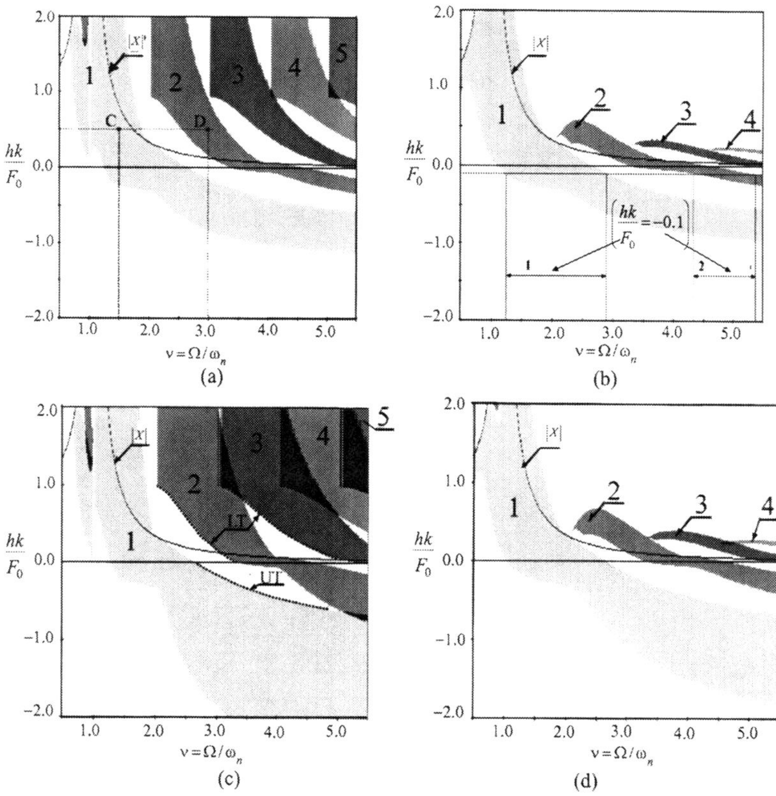

Fig. 3.7. Influence of the coefficient of restitution on the regimes of motion with impact: (a) $e = 0.6$, $\zeta = 0.0$, (b) $e = 0.6$, $\zeta = 0.1$, (c) $e = 0.8$, $\zeta = 0.0$, (d) $e = 0.8$, $\zeta = 0.01$, [199].

The dynamic characteristics of a mass–spring–dashpot system, with two barriers, under support excitation, were studied by Shaw ([926], [927]). The governing equation of motion of this system may be written in the form

$$X^{''} + 2\zeta X^{'} + X = \chi \cos \nu\tau, \qquad |X| < 1, \tag{3.22}$$

$$X_{+}^{'} = -eX_{-}^{'} \tag{3.23}$$

where a prime denotes differentiation with respect to the non–dimensional time parameter $\tau = \omega_n t$, $\omega_n = \sqrt{k/m}$, $X = x/x_0$, x_0 is the gap between the mass and each barrier as shown in Fig. 3.8, $\nu = \Omega/\omega_n$, $\chi = \chi_0/x_0$ and $\zeta = c/\left(2\sqrt{km}\right)$. Shaw [926] conducted stability analysis of the response periodic motion. Stability boundaries were determined by evaluating the modulus of the eigenvalues of the first derivative of the system response Poincaré map. At a saddle–node bifurcation two orbits of the same period and number of

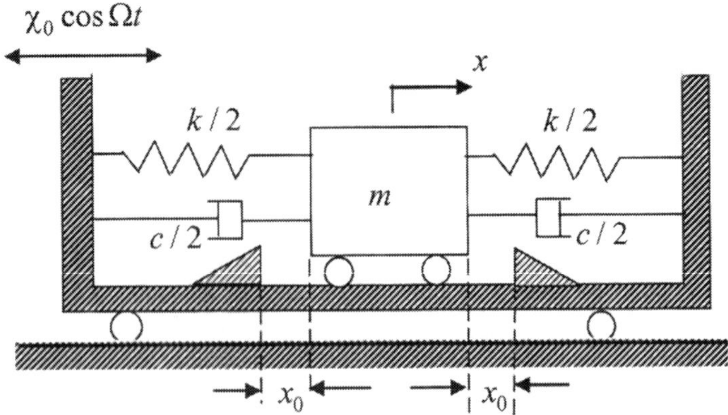

Fig. 3.8. Schematic diagram of mass-spring-dashpot system with two barriers.

impacts, one a stable node and the other a saddle, were found to coalesce and annihilate one another. This state was found to occur under the following condition

$$\chi = \frac{|1-\nu^2|(1-e)\left[1+\cos(\pi n/\nu)\right]}{\sqrt{(1-e)^2\left[1+\cos(\pi n/\nu)\right]^2+\nu^2(1+e)^2\sin^2(\pi n/\nu)}} \tag{3.24}$$

Fig. 3.9 shows the saddle–node bifurcation curves for different values of coefficient of restitution and subharmonic order n as estimated from equation (3.24). These curves define the excitation frequency at which the jump phenomenon takes place. As one system parameter varies, say the excitation frequency parameter, ν, and an eigenvalues passes through +1, Shaw [926] found three possible bifurcations. For $\nu = 1.6$, the simulations revealed a supercritical bifurcation characterized by a stable period–1 double–impact symmetric orbit. For $\nu = 1.45$ the orbit becomes unstable in the form of two stable antisymmetric orbits.

Natsiavas [714] and Natsiavas and Gonzalez [719] obtained an exact solution for the periodic, symmetric, double–crossing response of a single–degree–of–freedom strongly nonlinear oscillator, subjected to harmonic excitation. The nonlinearity of the system was described by a symmetric tri–linear function of the system displacement. Both period–one and subharmonic motions were analyzed. For some combinations of the system parameters, no stable periodic motion was found to exist and the system exhibited chaotic behavior. These studies examined the effect of asymmetries in the response due to unequal gaps as well as unequal stiffness and damping coefficients. It was found that the behavior of the system resembles the response of similar nonlinear systems with continuous characteristics, such as the response of the Duffing and van der Pol oscillators.

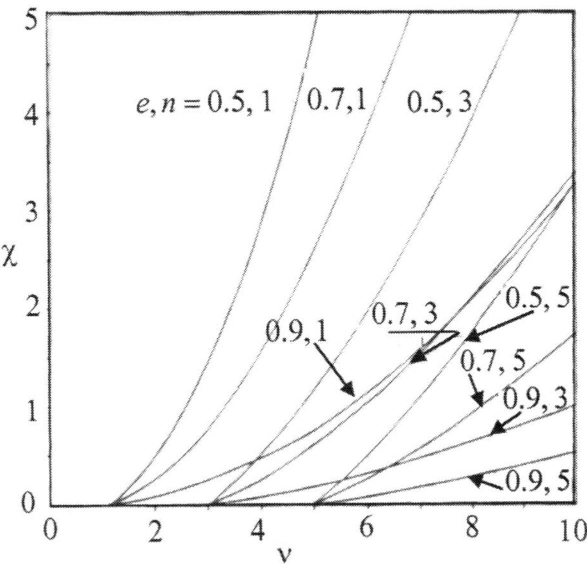

Fig. 3.9. Saddle-node bifurcation boundaries excitation amplitude-frequency domain for different values of coefficient of restitution and subharmonic order−n, [926].

Natsiavas ([715], [716]) examined the long time response of a class of harmonically excited, single degree–of–freedom, strongly nonlinear oscillators. The nonlinearity appears in both the damping and restoring forces, which are bi–linear functions of the system velocity and displacement, respectively. It was found that the system exhibits a regular behavior for some combinations of its parameters. However, there are sets of parameters for which the system undergoes bifurcations which may lead to loss of stability of periodic solutions and the appearance of a chaotic response. The most general n–periodic steady state solutions with an arbitrary number of contacts per response cycle were analyzed. It was shown that, when the damping is positive, the change of stability of the steady state response is not possible through a Hopf bifurcation. Alexander et al [8] developed a finite element model and an equivalent single–degree–of–freedom closed–form solution to predict the dynamic parameters and response of an experimental structure interacting with a gap. The equivalent model was represented by a piecewise linear system. The results suggested that the closed–form solution approximates the response of the experimental structure with accuracy greater than that of the finite element model. Moorthy et al [693] presented a numerical solution algorithm of the chaotic problem of impacting single–degree–of–freedom oscillators, using the Newmark method. The scheme incorporates an equilibrium iteration and variable time–stepping algorithm based on convergence criteria which ensure minimum error at each step.

Chillingworth [171] classified the local geometry of the discontinuity set together with associated local dynamics for a single degree–of–freedom oscillator impacting against a fixed barrier. Generic transitions were found to occur in the discontinuity set as the position of the barrier is smoothly varied. Analytical and numerical solutions of two impacts per n periods of the excitation force and its stability were obtained by Kotera and Peterka ([523], [524]) and Peterka and Vacik [833]. Peterka and Formanek [827] developed analogue and numerical simulation algorithms to predict a wide spectrum of periodic and chaotic impact motions of single-degree-of-freedom systems with strong nonlinearities (impacts and dry friction). The main outcomes of these algorithms are the time and phase trajectories, bifurcations characteristics, Lyapunov exponents, Poincaré maps and autocorrelation functions. Peterka [822] studied the dynamics of a single degree–of–freedom impact oscillator for the case when the stiffness of the barrier changes from zero to infinity. The Kelvin–Voigt and piecewise linear model of soft impact was considered. New phenomena were observed in the dynamics of motion with soft impacts in comparison with known dynamics of motion with rigid impacts.

Kryzhevich [552] considered the absolute elastic impact against a fixed limiter and analyzed the concept of overtaking phenomenon, which is a characteristic feature of impact systems. Gendelman and Meimukhin [360] examined the response regimes of integrable strongly nonlinear damped oscillator under periodic impact loading. For particular model coefficients the system was made integrable. Stable and unstable response regimes corresponding to single–period responses were predicted. For some regions of space of parameters and initial conditions, different response regimes were found to coexist [1070]. Cheng and Xu [167] determined the stable periodic motion, saddle–node, grazing and periodic doubling bifurcation conditions of a single–degree–of–freedom impact oscillator. The grazing bifurcation, period doubling bifurcation and periodic motions were demonstrated on a Poincaré surface defined at a constant excitation phase in terms of excitation amplitude. For fixed values of coefficient of restitution and viscous damping, the hysteretic region, in the amplitude–frequency space, was found to grow with increasing the excitation amplitude and frequency above the resonant frequency.

Lee and Nandi [573] performed signal analysis of the observed impacting signals measured from an experimental model in the chaotic region. Two stages of the signal processing were considered: (a) blind deconvolution and optimization of the observed data, and (b) Lyapunov exponents and noise reduction. The phenomena of the complete, incomplete and chaotic chattering of a mass–spring–dashpot oscillator with end–barrier were examined in terms of p–impact period–n motion using impact map and Poincaré map [572]. Later, Lee and Nadi [574] used blind deconvolution techniques for impact force identification.

Cheng and Xu [167] considered the same system but in the dimensional form, i.e.,

$$x^{''} + 2\zeta x^{'} + x = \chi_0 \cos \nu\tau, \tag{3.25}$$

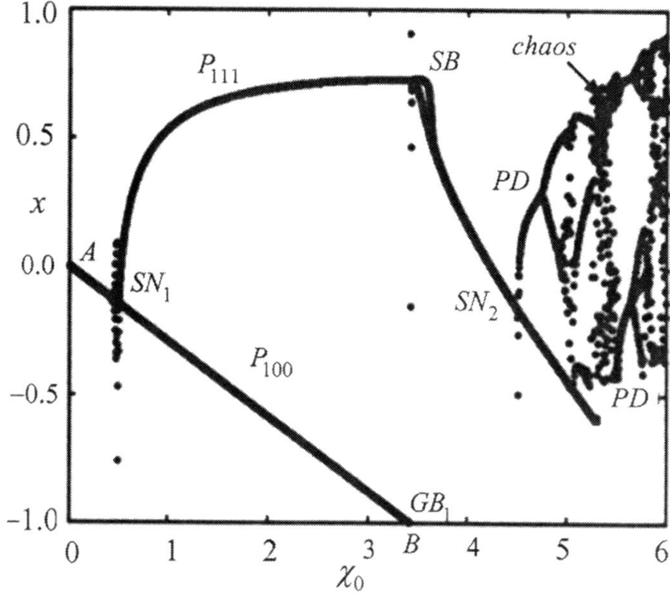

Fig. 3.10. Bifurcation diagram showing the dependence of response amplitude on excitation amplitude showing the phenomenon of hysteresis, for excitation frequency ratio $\nu = 2.1$, $e = 0.7$, and symmetric clearance $\pm x_0$, [167].

with symmetric and asymmetric clearance. They examined the periodic motions of the system for orbits designated by P_{npq}, where n is the number of excitation cycles, p and q are the number of impacts on the first and second barriers, respectively. The hysteretic behavior involving impacting and non–impacting periodic orbits was analyzed. The bifurcation plot using the excitation amplitude as the control parameter was generated by numerical simulation of equation (3.25) for the case of symmetric clearance with $e = 0.7$. The excitation amplitude was initially set to $\chi_0 = 6.0$ and the initial conditions were set to $x(\tau = 0) = 0.9$ and $x^{'}(\tau = 0) = -1.1$, which is a point in the chaotic region. With each decrement of χ_0 the initial conditions were set to be the previous values. Fig. 3.10 shows the bifurcation diagram for damping ratio $\zeta = 0.05$ and excitation frequency ratio $\nu = 2.1$. The only coexisting stable attractors for $\chi_0 < 6.0$ were traced by increasing the excitation amplitude from 0.0 to 6.0. Fig. 3.10 also shows four different types of local transitions for period–1 orbit. The point at $\chi_0 = 3.41$ marks the occurrence of grazing bifurcation (GB_1). Thus over the excitation amplitude range $\chi_0 = 0.0$ to 3.41, the system exhibits non–impacting periodic oscillations of the type P_{100} solution. At $\chi_0 \approx 0.5$ the system possesses a saddle–node bifurcation (SN_1) and the only stable state occupies the region $\chi_0 = 0.0$ to 0.5. At $\chi_0 \approx 0.5$ a pair of symmetric impacting periodic orbits is created as a result of the saddle node bifurcation in which one is a stable node and the

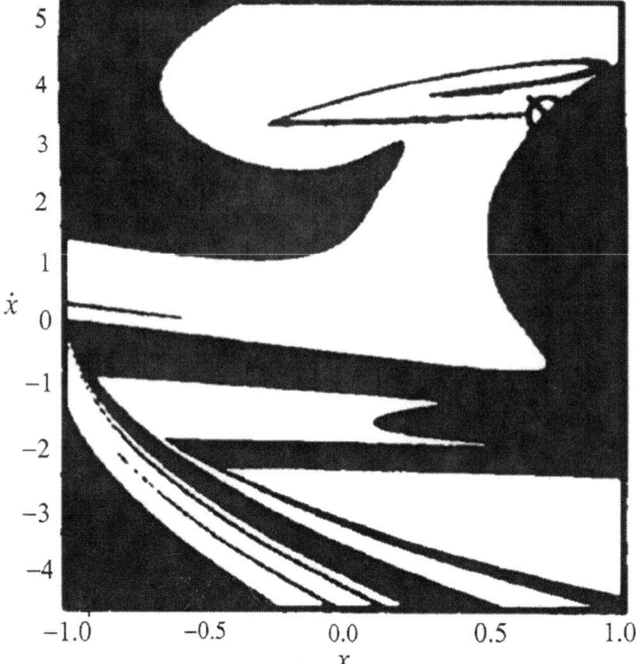

Fig. 3.11. Domains of attraction showing the chaotic attractor by the white region for $\nu = 2.1$ and $\chi_0 = 4.65$, [167].

other is unstable saddle node. Thus, over the region $\chi_0 = 0.0$ to 3.41 both the stable impacting and the non–impacting periodic orbits coexist, and the system experiences hysteresis. At excitation amplitude $\chi_0 = 3.45$ the response experiences symmetry breaking (SB) and two similar stable period–1 asymmetric orbits are created. One of these asymmetric orbits experiences grazing bifurcation (GB_2) at $\chi_0 = 5.215$ and the system settles down to the chaotic attractor. At $\chi_0 = 4.51$, a saddle–node bifurcation (SN_2), there arise stable and unstable asymmetric orbits (P_{112} and P_{121}). These orbits are shown between point SN_2 and the point of period doubling (PD). These orbits lose their stability as a result of period doubling bifurcation at $\chi_0 = 4.735$ and lead to chaotic attractor at $\chi_0 = 5.24$. Thus between $\chi_0 = 4.51$ and $\chi_0 = 4.735$ there are four different coexisting attractors including a pair of asymmetric P_{111} coexisting with asymmetric P_{112} and P_{121} orbits or two disjointed chaotic attractors or a pair of high periodic stable attractor. Fig. 3.11 shows the basins of attraction and two connected pieces of chaotic attractor. The white region belongs to the basin of attraction for the chaotic attractor. The stable manifold of the unstable symmetric periodic orbit is the basin boundary. The same scenario applies to the case of asymmetric clearance.

Experimental investigations of an impact oscillator with a one–sided elastic constraint were reported by Ing et al [446]. The results revealed different bifurcation scenarios under varying the excitation frequency near grazing for different values of excitation amplitude. When a non–impacting periodic orbit bifurcates into an impacting one via grazing mechanism, the resulting orbit was found to be stable. However, in many cases it was observed to lose stability through grazing. The evolution of the attractor is governed by a complex interplay between smooth and non–smooth bifurcations.

3.4 Pendulum Oscillating Against One– or Two–Sided Barrier

The simple pendulum is a classical example extensively used to demonstrate complex dynamic characteristics. It possesses rich dynamics even in the absence of impact. It has also been utilized to simulate liquid sloshing dynamics in partially filled containers both for impact and non–impact liquid motion regimes [436]. The inverted pendulum was also used to model the ship roll dynamics [696]. Blinov [125] analyzed the motion of a gyroscopic pendulum with restricted rotation of the rotor, operating under an ideal unilateral constraint on the angle nutation. Blinov estimated the upper and lower bounds for the period of the pendulum with respect to the angle of nutation. Other applications include gantry crane dynamics and vibration absorbers.

The early work of vibro–impact oscillations of pendulums was reported by Babitsky [52], Shaw ([927], [928]) and Sharif–Bakhtiar and Shaw [922]. These studies dealt with the existence and stability of certain periodic and chaotic motions of the pendulum. Moore and Shaw [691] experimentally investigated the harmonic excitation of a pendulum against rigid barriers. The pendulum was considered in both normal (downward) and inverted positions. The response of the pendulum to sinusoidal excitation revealed non–impacting motions, stable subharmonics, and chaotic motions. These were experimentally found to occur in the parameter regions predicted analytically. The inverted pendulum was found to have 10 distinct possible steady–state responses at a fixed driving amplitude and frequency, each of which was obtained simply by changing the initial conditions. Sharif–Bakhtiar and Shaw [922] studied the effects of motion–limiting stops on the dynamic behavior of a centrifugal pendulum vibration absorber. Their analysis revealed the existence and stability of nonlinear impacting periodic motions. One of the important observations was the coexistence of impacting and non–impacting periodic motions at the desired operating frequency.

Bayly and Virgin [99] and Slade et al [949] conducted numerical simulations and experimental measurements to study the forced excitation of nonlinear impacting pendulum. They reported periodic and chaotic regimes. Garza and Ertas [357] and Todd and Virgin [1005] presented experimental studies of pendulum impacts including inverted and normal pendulums. They

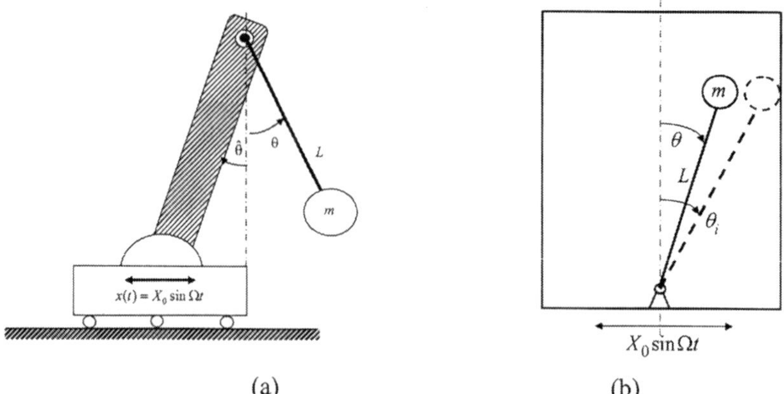

Fig. 3.12. Schematic diagrams of (a) pendulum on an inclined support and (b) inverted pendulum under sinusoidal excitation.

demonstrated the chaotic responses between windows of periodic orbits. It was indicated that the chaotic regime is characterized by finger–like attractors whose number is governed by the periodicity of the previous periodic orbit. A "double–impulse" model of a clock with one counter impulse and one pushing impulse in the period was constructed by Amelkin and Kalitin [11].

The problem of an oblique frictional impact was found to introduce complicated dynamics such as stick–slip motions. Based on a hybrid analysis of vibro–impact dynamics, kinematics and complementary conditions, Dongping and Haiyan [264] developed a piecewise analysis method to describe the sliding motion during an oblique impact. They adopted a parametrically excited planar pendulum between two parallel rigid walls as an illustrative example. It was shown that the sliding impacts occur in such a system with a set of properly selected parameters.

The dynamic behavior of a parametrically excited planar pendulum subjected to a motion–dependent discontinuity was studied by Mann et al [637]. The contact force was modeled by nonlinear elastic and viscoelastic forces. The model incorporated Hertzian contact law for elastic conformal contact. This modeling was similar to the one adopted by Půst and Peterka [872] who described the nonlinearity of the restoring contact force between solid bodies as function of deformation and velocity. Such modeling was found convenient for viscoelastic material barriers in which the material has a velocity–dependent impact force. Experimental and numerical results revealed the presence of multiple periodic attractors, subharmonic, quasi–periodic, and chaotic oscillations.

Piiroinen et al [842] reported some experimental results of a free swinging pendulum colliding with a rigid stop. The pendulum support was inclined at an angle $\widehat{\theta}$ as shown in Fig. 3.12(a) and subjected to a harmonic excitation, $x(t) = X_0 \sin \Omega t$. The inclination of the support was introduced to reduce the

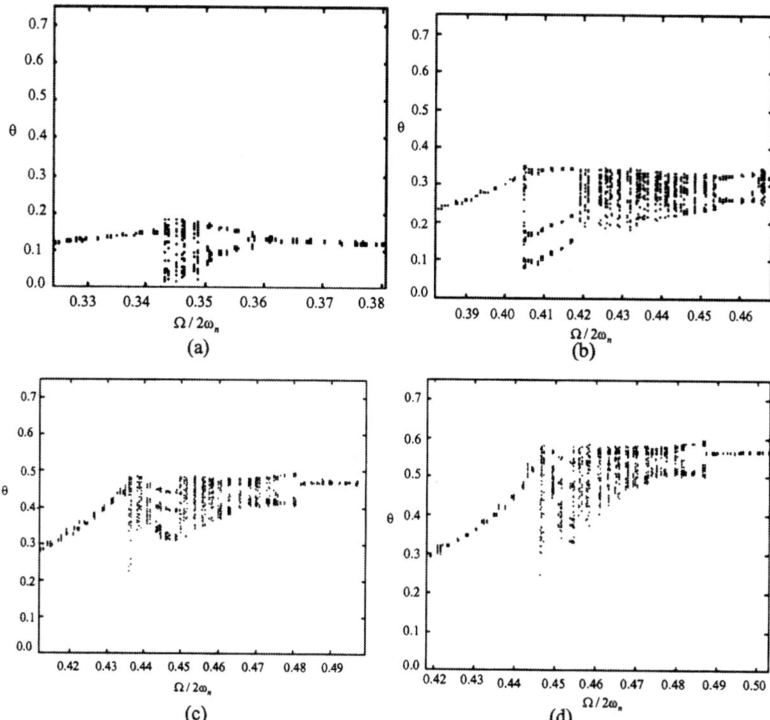

Fig. 3.13. Bifurcation diagrams of experimental measurements in which the response was sampled once every forcing cycle for (a) $\widehat{\theta} = 10^o$, (b) $\widehat{\theta} = 20^o$, (c) $\widehat{\theta} = 30^o$, and (d) $\widehat{\theta} = 40^o$, [842].

influence of gravity, which results in a reduction of the natural frequency of the pendulum, ω_n. The equation of motion of the pendulum may be written in the non–dimensional form

$$\theta^{''} + \frac{2\zeta}{\nu}\theta^{'} + \frac{1}{4\nu^2}\sin\theta = x_0 \cos\theta \sin\tau \tag{3.25}$$

where a prime denotes to differentiation with respect to the non–dimensional time parameter $\tau = \Omega t$, $\nu = \Omega/2\omega_n$, $\omega_n = \sqrt{g/L}$, and $x_0 = X_0/L$. Under impact, equation (3.25) must be augmented with the impact law

$$\theta^{'}_{+} = -e\theta^{'}_{-} \tag{3.26}$$

Note that in the absence of any barrier, the response of the pendulum as governed by equation (3.25) possesses soft nonlinear characteristics. The numerical simulation of equations (3.25) and (3.26) for different values of the barrier angle $\widehat{\theta}$ revealed grazing bifurcation that lead to a rapid change in the pendulum motion, from non–impacting period–one periodic motion to impacting chaotic motion containing windows of stable periodic orbits in a period–adding cascade. The experimental results confirmed the numerical simulation and Fig. 3.13 shows the dependence of the pendulum response, θ, on the excitation frequency ratio, ν, for four different values of inclination angle, $\widehat{\theta} = 10^o$, 20°, 30°, and 40°. There is a critical frequency ratio below which a period–1 limit cycle exists and the pendulum does not experience any impact. At a critical forcing frequency the pendulum was observed to graze with the stop and impacting oscillations took place at frequencies greater than that critical value. The impacting oscillations were characterized by a complex sequence of chaotic and periodic motion. At a large enough excitation frequency the motion settles into a period–1 motion that impacts precisely once per period. The critical forcing frequencies were $\nu_c \approx 0.343$ for $\widehat{\theta} = 10^o$, $\nu_c \approx 0.405$ for $\widehat{\theta} = 20^o$, $\nu_c \approx 0.435$ for $\widehat{\theta} = 30^o$, and $\nu_c \approx 0.446$ for $\widehat{\theta} = 40^o$. It is known that the route to chaos is a period–doubling cascade, however, Fig. 3.13 shows what is known as *period–adding cascade.*

Du and Zhang [265] considered a class of nonlinear impact oscillators based on an inverted pendulum impacting against rigid walls under external periodic excitation. They extended Melnikov method to non–smooth systems. The absence of closed form solutions was found to create some difficulties in estimating the gap between the stable manifold and unstable manifold. Du and Zhang [265] and Du et al [266] introduced an algorithm to compute the Melnikov functions up to the nth–order to obtain conditions of parameters yielding homoclinic cycles and subharmonic bifurcation of the pendulum.

Under horizontal sinusoidal excitation and with reference to Fig. 3.12(b), the inverted pendulum equation of motion may be written in the form

$$\Theta^{''} + 2\zeta\Theta^{'} - \Theta = \chi \sin \nu\tau, \qquad |\Theta| < 1, \tag{3.27}$$

$$\Theta^{'}_{+} = -e\Theta^{'}_{-}, \qquad \Theta = \pm 1 \tag{3.28}$$

where $\Theta = \theta/\theta_i$, $\nu = \Omega\sqrt{L/g}$, $\chi = (X_0/L)\,\nu^2$, and a prime denotes differentiation to the non–dimensional time parameter $\tau = \omega_n t$. Shaw and Rand [934] presented analytical and numerical solutions to determine bifurcation conditions including the appearance of subharmonics by saddle–node bifurcations, secondary bifurcations, and global bifurcation. The particular solution equation (3.27) for relatively small values of external excitation amplitude may be written in the form

$$\Theta_p(\tau) = \frac{\chi \sin(\nu\tau - \psi)}{\sqrt{(1+\nu^2)^2 + (2\zeta\nu)^2}}, \qquad \tan\psi = \frac{2\zeta\nu}{1+\nu^2}. \tag{3.29}$$

As long as the response amplitude $\Theta_0 = \chi/\sqrt{(1+\nu^2)^2 + (2\zeta\nu)^2} < 1$, the motion is a non–impacting periodic of period–1. There exist other types of periodic motions involving impacts at $\Theta = \pm 1$. In particular, two types of periodic motions were identified. The first is referred to as type–I periodic

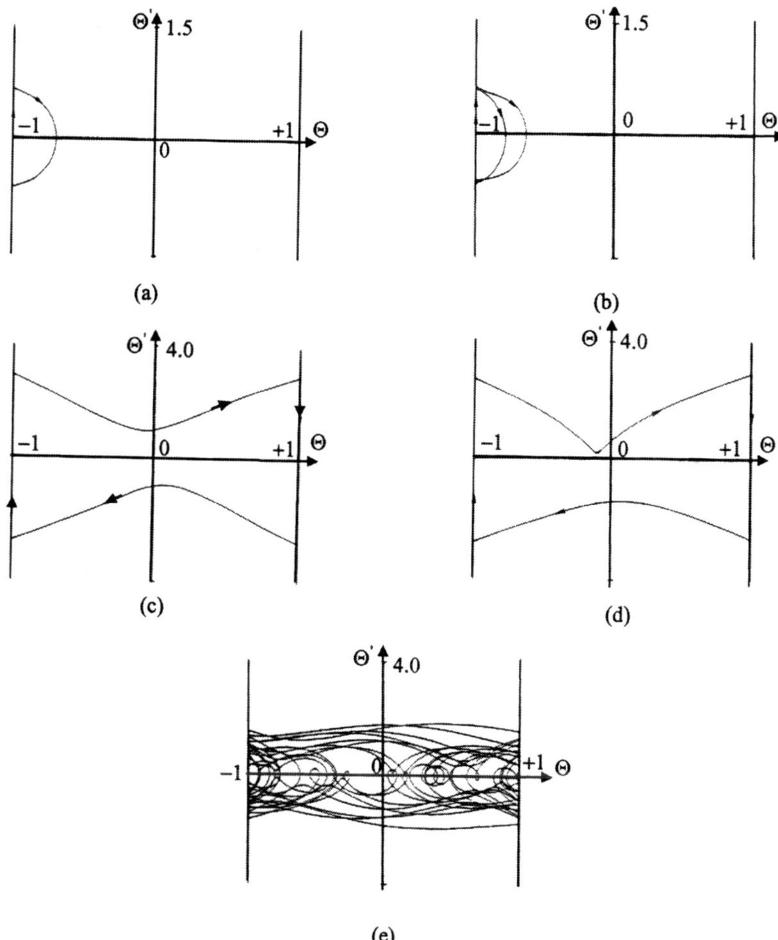

Fig. 3.14. Phase portraits generated for $e = 1$, $\zeta = 0.1$: (a) Type I motion $\chi = 0.6$, $\nu = 4.0$, period T, (b) Type I motion $\chi = 0.65$, $\nu = 4.0$, period $2T$, (c) Type II motion $\chi = 2.3$, $\nu = 1.95$, period T, (d) Type II motion $\chi = 2.3$, $\nu = 1.8$, unsymmetric period T after pitchfork bifurcation, (e) Chaotic type motion $\chi = 2.0$, $\nu = 4.0$, [934].

motions correspond to the pendulum impact at one side of the barrier wall ($\Theta = +1$ or -1) with period–1 or period–2 as shown in the phase portrait of Figs. 3.14(a) and 3.14(b) respectively. Type–II periodic motions were found to be either symmetric or unsymmetric as shown in Figs. 3.14(c) and 3.14(d), respectively. These motions take place in the form of the pendulum impacts on the two barriers. Fig. 3.14(e) exhibits a strange attractor type motion. Rest motion in which the pendulum mass is in contact with one of the barriers was found to coexist with stable subharmonics and/or chaotic motions. Moore and

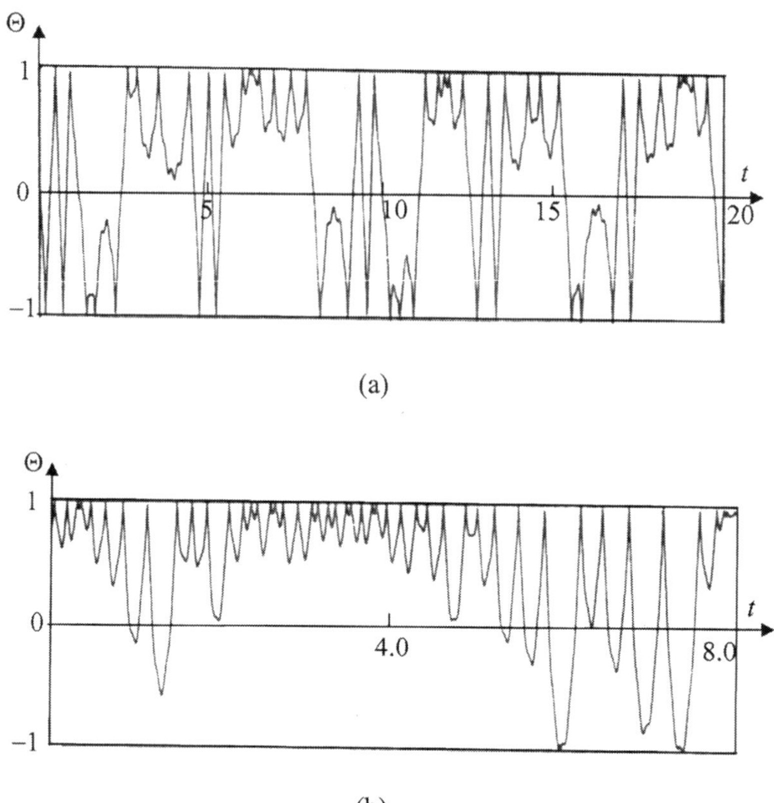

Fig. 3.15. Measured time response records in the chaotic motion regime: (a) impacts on both barriers for $\chi = 1.42$, $\nu = 5.05$, (b) impacts on the right barrier $\chi = 1.39$, $\nu = 10.3$, [691].

Shaw [691] conducted a series of experimental tests on normal and inverted pendulums. Fig. 3.15(a) shows a typical time history record of chaotic motion with impacts on both barriers for $\chi = 1.42$ and $\nu = 5.05$. Fig. 3.15(b) reveals chaotic motion with impacts on the right barrier for $\chi = 1.39$ and $\nu = 10.3$.

Lenci and Rega [583] extended the work of Shaw and Rand [934] in an attempt to reduce the regions of chaotic motions of the inverted pendulum. They derived a closed form solution for the maximum reduction by means of two impulses of amplitude. Later, Lenci and Rega [584] and Rega and Lenci [885] considered a class of periodic motions of an inverted pendulum with rigid lateral constraints. The system was excited by an arbitrary periodic excitation. The periodic solutions were determined as fixed points of the stroboscopic Poincaré map. It was shown that the stability is lost through classical saddle–node or period–doubling bifurcations. The existence paths were determined, both geometrically and analytically, on the basis of a function that can easily be derived from the periodic excitation function.

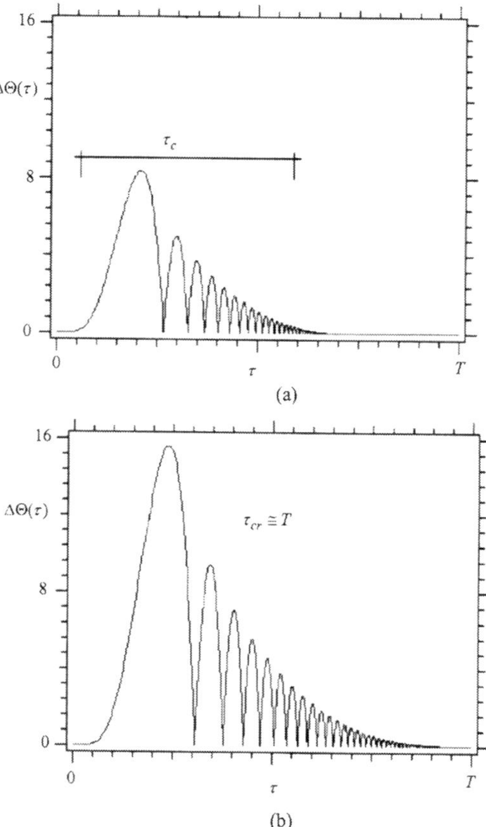

Fig. 3.16. Time history records of chattering oscillations regime where $\Delta\Theta(\tau) = [1 + \Theta(\tau)]10^4$ for excitation frequency ratio $\nu = 5.0$: (a) $\chi = 1.07$, (b) $\chi = 1.0967$, [215].

Furthermore, the dynamics and control of an inverted pendulum with rigid unilateral constraints was considered under optimal excitations that would reduce the chaotic region in the parameter plane. The performance regimes of the optimal excitations were numerically evaluated by comparison with the reference case of harmonic force, and it was shown that it is possible to improve some technical requirements of the dynamics through proper implementations of the optimal excitations.

Demeio and Lenci [215] analyzed the chattering oscillations of impact dampers made up of an inverted pendulum impacting between two lateral walls. An emphasis was given to estimate the time period required by the micro–oscillations to come to rest. When the time period becomes equal to the excitation period the chattering was observed to cease. Chattering oscillations are characterized by very small amplitude, which decreases in time as shown in Fig. 3.16. These oscillations consist of infinite number of impacts occurring in finite time period, called the chattering time, τ_c, which is a fraction of the excitation period. As the excitation amplitude gradually increases the chattering amplitude and time increase until the chattering time reaches its critical value, i.e., $\tau_{cr} = T$. Above that value, the chattering disappears, and after a transient time the response approaches another attractor, which may not coexist with chattering over the range $1 < \chi < \chi_{cr}$. It was shown that such critical period is asymptotically proportional to the square root of the excitation amplitude. Later, Lenci et al [582] performed a numerical investigation of the nonlinear dynamics of an inverted pendulum and identified three different regimes. These are periodic, chaotic, and rest positions with subsequent chattering.

3.5 Ship Impact Interaction with Ice

3.5.1 Introduction and Modeling

Ice impact loading can cause significant damage to offshore structures and ships. The impact arises when drifting ice sheets, ice floe, and icebergs are moving with considerable speed under the action of environmental conditions. When a fast–moving ice feature crushes against a narrow structure, the force of impact is irregular, random and contains repetitive fluctuations. The random fluctuations can be explained by random variations of ice properties as well as ice failure at random locations along the contact area. For a comprehensive account of the ice impact with ships and ocean structures the reader may refer to the review article by Ibrahim et al [438].

Powell et al [866] studied the dynamic impact interaction between ice and structures for a large ice floe striking a tubular steel platform. Mathematical models of ice feature impacts with offshore platforms or vessels include (i) head–on impacts, when an iceberg has a single degree–of–freedom and an absolutely rigid non–compliant structure [151], (ii) more complex models taking into account the structure compliance and its local and global deformations ([194], [974], [975], [280]), and (iii) eccentric impacts ([349], [911], [725]). Ship hull–ice loading has been measured either with actual ships or small–scale model basin tests, or controlled field tests [354]. The impact of a ship hull with an ice cusp was experimentally simulated by Gagnon et al [353]. The impacting ship or offshore structure was represented by a mass–spring–dashpot system having a constant velocity relative to the ice sheet. The nonlinear dynamic response was due to intermittent ice breakage and intermittent contact of the structure with the ice. Periodic motions were found and the periodicity for a particular system was found to be dependent upon initial conditions. A description of some of the effects of random variations in system parameters was also discussed.

Related to ship impact dynamics is the problem of two–dimensional motions of a point mass and a rigid body supported by two cables, which model a floating breakwater. This problem was studied in the upside–down configuration by Plaut and Farmer [858]. Buoyancy and the weight of the body were included, and the wave forces were modeled as harmonic forces, which

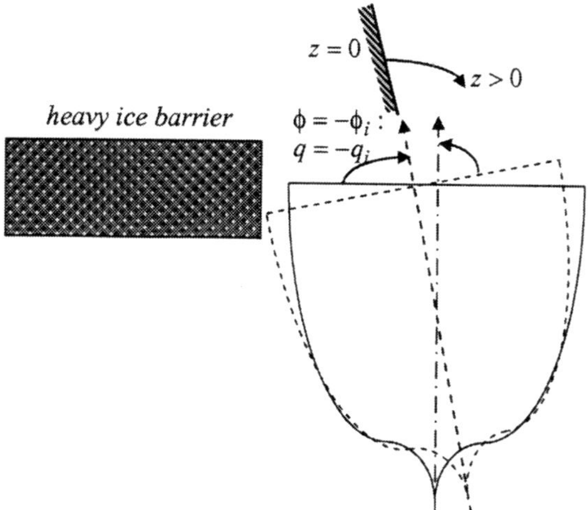

Fig. 3.17. Schematic diagram of one-sided ice barrier impact with ship in roll oscillation showing the coordinates of motion.

follow an elliptical path in a vertical plane. The mooring lines were assumed to have no effect on the breakwater when they are slack and to provide an instantaneous impulsive force when they become taut. In the absence of wave forces, the body 'bounces' on the boundaries and approaches the equilibrium position in which both cables are taut. Under forcing excitation, the body was found to settle down, or for sufficiently high forcing amplitudes may reach the level of the supports.

During ship roll oscillations, ice sheets may impact with the ship side surface. Let the ice forms a barrier with the ship when it hits ice at a roll angle $\phi = -\phi_i$ as shown in Fig. 3.17. The ship equation of motion may be written in the form [373].

$$\ddot{\phi} + 2\zeta\omega_n\dot{\phi} + a\dot{\phi}|\dot{\phi}| + \omega_n^2\phi + C_3\phi^3 + C_5\phi^5 = \xi(t), \qquad for\ \phi \geq -\phi_i \tag{3.30}$$

where ζ and a are the linear and nonlinear hydrodynamic damping factors, respectively. ω_n is the roll natural frequency of small oscillations of the ship, C_3 and C_5 are the ship nonlinear restoring moment coefficients. $\xi(t)$ is ocean wave excitation moment, which can be deterministic or random. When the total restoring moment vanishes, i.e., $\omega_n^2\phi + C_3\phi^3 + C_5\phi^5 = 0$, the ship is either at its equilibrium position, $\phi = 0$ (for unbiased ship equilibrium) or has reached the capsizing roll angle, ϕ_c. Introducing the non–dimensional parameters $\tau = \omega_n t$ and $q = \phi/\phi_c$, equation (3.30) takes the form

$$q^{''} + \overline{\zeta}q^{'} + \gamma q^{'}|q^{'}| + q + \overline{C}_3 q^3 + \overline{C}_5 q^5 = Z(\tau), \tag{3.31}$$

where $\overline{\zeta} = 2\zeta$, $\gamma = a\phi_c^2$, $\overline{C}_3 = C_3\phi_c^2/\omega_n^2$, $\overline{C}_5 = C_5\phi_c^4/\omega_n^2$, and $Z(\tau) = \xi(t)/\left(\omega_n^2\phi_c\right)$. Note that at impact we have $q_i = \phi_i/\phi_c$.

Equation (3.31) is a nonlinear differential equation describing the ship roll dynamics under nonlinear hydrodynamic sea waves. For one–sided barrier representing impact of floating ice on one side of the ship at an impact angle $q = -q_i$, the following Zhuravlev transformation is introduced

$$q = zsgn(z) - q_i. \tag{3.32}$$

This transformation shifts the barrier to the axis $z = 0$ and maps the domain $q > -q_i$ of the phase plane trajectories on the original plane $(q, q^{'})$ to the new phase plane $(z, z^{'})$. In this case the ship equation of motion takes the following from

$$z^{''} + \overline{\zeta} z^{'} + \gamma z^{'2} sgn(z^{'}) + z$$

$$+sgn(z)\left[-q_i + \overline{C}_3 \left(zsgn(z) - q_i\right)^3 + \overline{C}_5 \left(zsgn(z) - q_i\right)^5\right] = Z(\tau) sgn(z). \tag{3.33}$$

Equation (3.33) describes the roll motion of the ship in terms of Zhuravlev non–smooth coordinate z. It is seen that this equation does not explicitly includes any impact terms. The free and forced dynamics will be discussed in the next subsections.

3.5.2 Unperturbed Ship Dynamics

In the absence of damping the unperturbed motion equation takes the form

$$z^{''} + \Gamma(z) = 0 \tag{3.34}$$

where $\Gamma(z) = z + sgn(z)\left[-q_i + \overline{C}_3 \left(zsgn(z) - q_i\right)^3 + \overline{C}_5 \left(zsgn(z) - q_i\right)^5\right]$ is the nonlinear restoring moment of the ship and is shown in Fig. 3.18(a) for the selected impact angle $q_i = -0.4$, and nonlinear coefficients $\overline{C}_3 = -0.3$, and $\overline{C}_5 = 0.1$. It is seen that the restoring moment vanishes at $z = 0.4$. The potential energy, $\Pi(z)$, is obtained by integrating the restoring moment $\Gamma(z)$ over the limits q_i and z, i.e.,

$$\Pi(z) = \int_{q_i}^{z} \Gamma(y) dy = a_6 z^6 + a_5 z^5 + a_4 z^4 + a_3 z^3 + a_2 z^2 + a_1 z + a_0 \tag{3.35}$$

where the coefficients a_i are functions of $\overline{C}_3$, $\overline{C}_5$, and q_i. Note that the choice of the lower limit, q_i, is chosen such that at q_i the potential energy is minimum as shown in Fig. 3.18(b). It is seen that at $z = 0$ (corresponding to ship

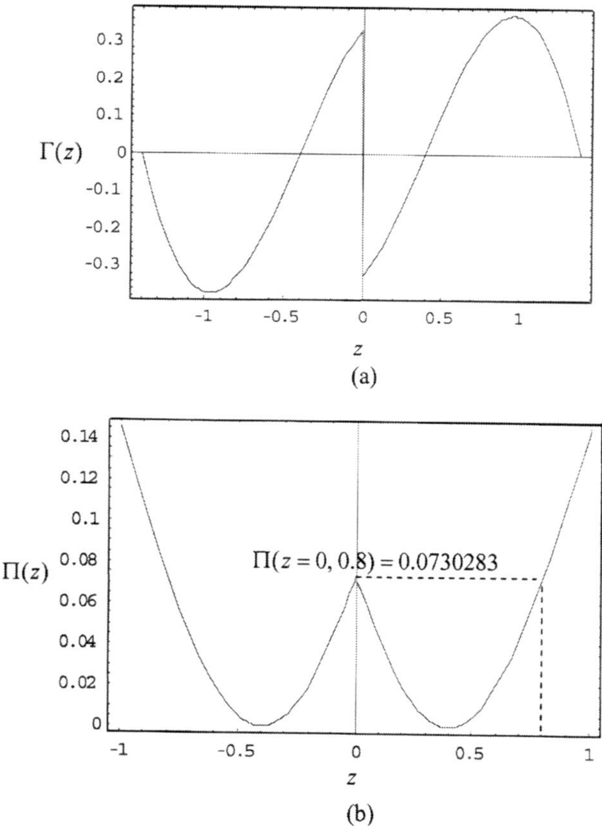

Fig. 3.18. Restoring moment and potential energy of the ship for the case of $q_i > 0$: (a) Restoring moment, and (b) Potential energy in terms of Zhuravlev's non-smooth coordinate z, [373].

angle $q = -q_i$) the potential has a maximum of $\Pi(z = 0) = 0.0730283$. The Hamiltonian of system (3.34), $H = \left(z'^2/2\right) + \Pi(z)$, possesses the first integral of motion

$$z' = \pm\sqrt{2\left[H - \Pi(z)\right]}. \tag{3.36}$$

As long as $H > \Pi(z)$ the phase diagram is periodic closed orbit in the phase space $\{z, z'\}$ as shown in Fig. 3.19. With reference to Fig. 3.18(b), H reaches its maximum value $H_{\max} = \Pi(z = 0) = 0.0730283$. The periodic orbits are only restricted inside the domain $D = \{(z, z')|\ H \leq H_c\}$, where $H_c = H_{\max} - \Delta H$, and ΔH is sufficiently small. H_c is the critical energy level above which impact of the ship will take place, and the trajectories of the motion will be structurally unstable. The motion corresponding to $H_{\max} = 0.0730283$ follows a critical

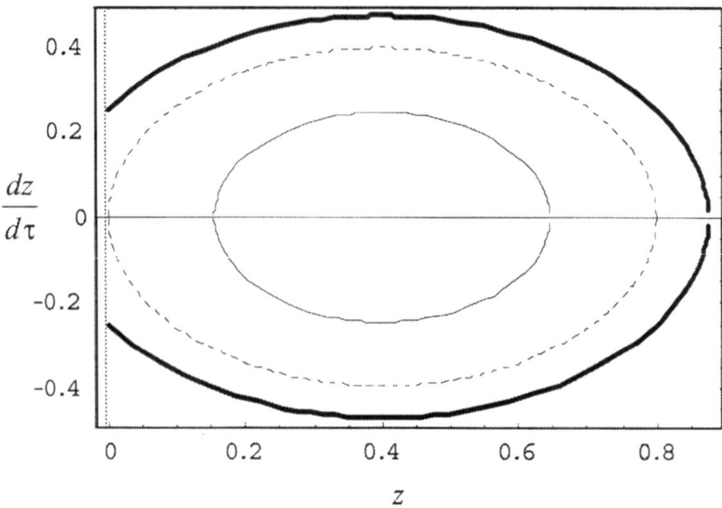

Fig. 3.19. Phase prorait showing three regimes of motion, impact motion (thick solid curve), Grazing impact (dashed curve), $H = 0.03$ periodic non-impacting oscillation (thin curve).

orbit shown by the dashed curve in Fig. 3.19. This orbit describes the grazing impact of the ship with one–sided barrier.

Let the system be given an initial velocity z_0', i.e., $H = z_0'^2/2$. The period of oscillation, T, can be estimated from equation (3.36) as

$$T = \frac{1}{\sqrt{2}} \int_0^z \frac{dz}{\sqrt{(z_0'^2/2) - \Pi(z)}}. \tag{3.37}$$

Note that $\Pi(z) < 0.0730283$ is for the entire range of the ship motion before capsizing as shown in Fig. 3.19. The character of motion depends on the value of initial velocity $\left(z_0'^2/2\right)$. For $z_0'^2/2 < 0.0730283$ the integrand is always real and can assume any value within a range governed by the condition $\left(z_0'^2/2\right) - \Pi(z) = 0$. For a given initial energy, $z_0'^2/2 < 0.0730283$, the ship will oscillate between two values z_1 and z_2 and the corresponding period of oscillation is

$$T = \frac{1}{\sqrt{2}} \int_{z_1}^{z_2} \frac{dz}{\sqrt{(z_0'^2/2) - \Pi(z)}}. \tag{3.38}$$

If $z_0^{'2}/2 < 0.0730283$, the integrand is real and approaches ∞ and the ship is at the verge of capsizing. If $z_0^{'2}/2 > 0.0730283$ the integrand is always real and the value of z increases indefinitely. In this case, the motion is unbounded and the ship will acquire a rotational motion.

3.5.3 Perturbed Ship Dynamics

Under sinusoidal excitation $Z(\tau)$ =a$\sin \nu\tau$, where a is the excitation amplitude, and $\nu = \Omega/\omega_n$, equation (3.33) was solved numerically under different values of excitation amplitude and frequency. The numerical solution was obtained for all initial conditions occupying the grazing orbit shown in Fig. 3.19

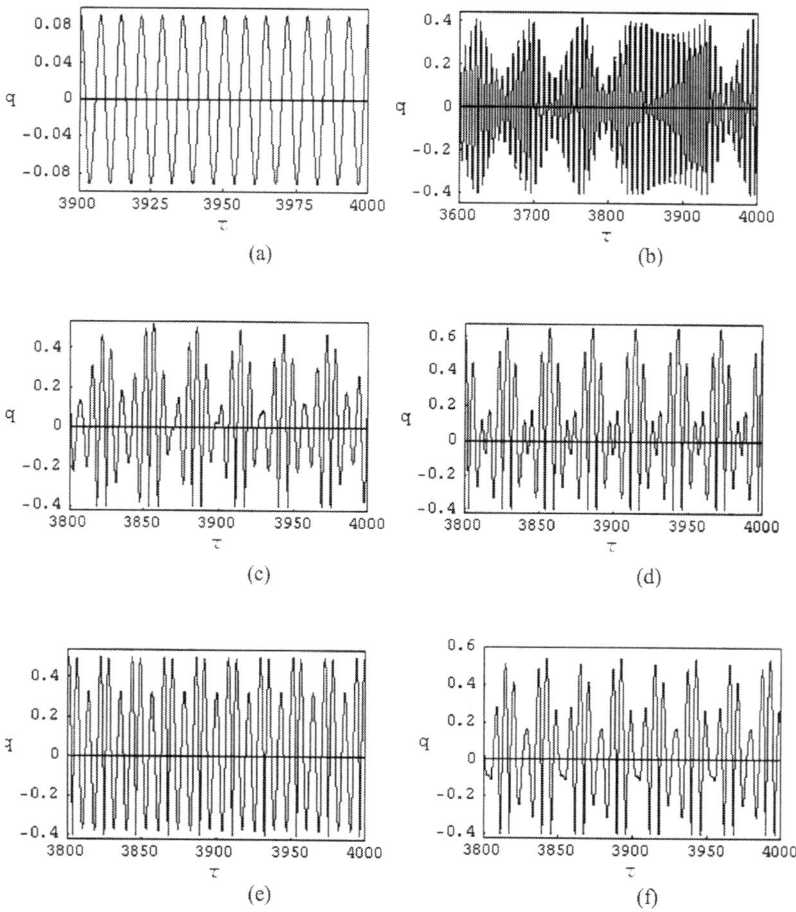

Fig. 3.20. Samples of time history response records for excitation frequency ratio ν = 0.88, and excitation amplitude: (a) a=0.02, (b) a=0.046, (c) a=0.084, (d) a=0.094 (Period four), (e) a=0.106 (Period three), (f) a=0.11, [373].

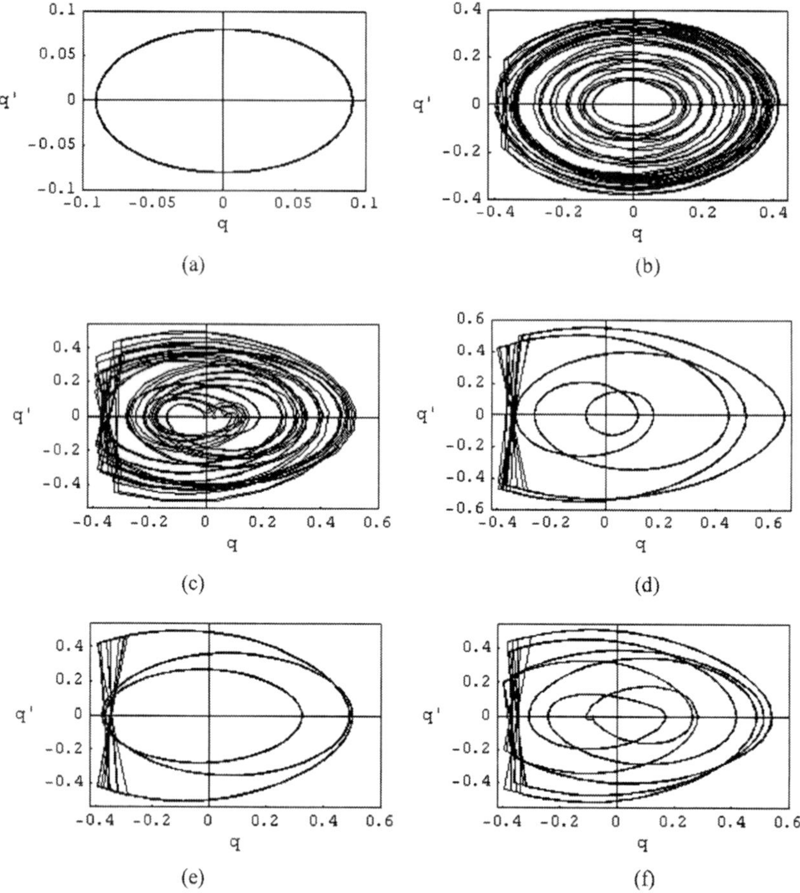

Fig. 3.21. Phase portraits corresponding to the time history records of Fig. 3.20, [373].

by the dashed closed curve. Figs. 3.20(a)–(f) show six selected samples of the response time history records for excitation frequency ratio $\nu = 0.88$ and six different values of excitation amplitude, a $= 0.02, 0.046, 0.084, 0.094, 106,$ and 0.11, respectively. It is seen that for relatively low excitation amplitude, a $= 0.02$, the response is periodic and the ship roll amplitude does not reach the barrier. This is confirmed by the phase portrait shown in Fig. 3.21(a) and the Poincaré map shown in Fig. 3.22(a) reveals period–1 fixed point. As the excitation amplitude increases the response experiences grazing bifurcation and assumes amplitude modulated pattern as shown in Fig. 3.20(b) for a $= 0.046$. Figs. 3.21(b) and 3.22(b) show the corresponding phase diagram and Poincaré map, respectively. The response experiences one impact per 10 excitation periods. At excitation amplitude a $= 0.084$, the response is bounded chaotic with multi–impacts as shown in Figs. 3.20(c), 3.21(c) and 3.22(c). As

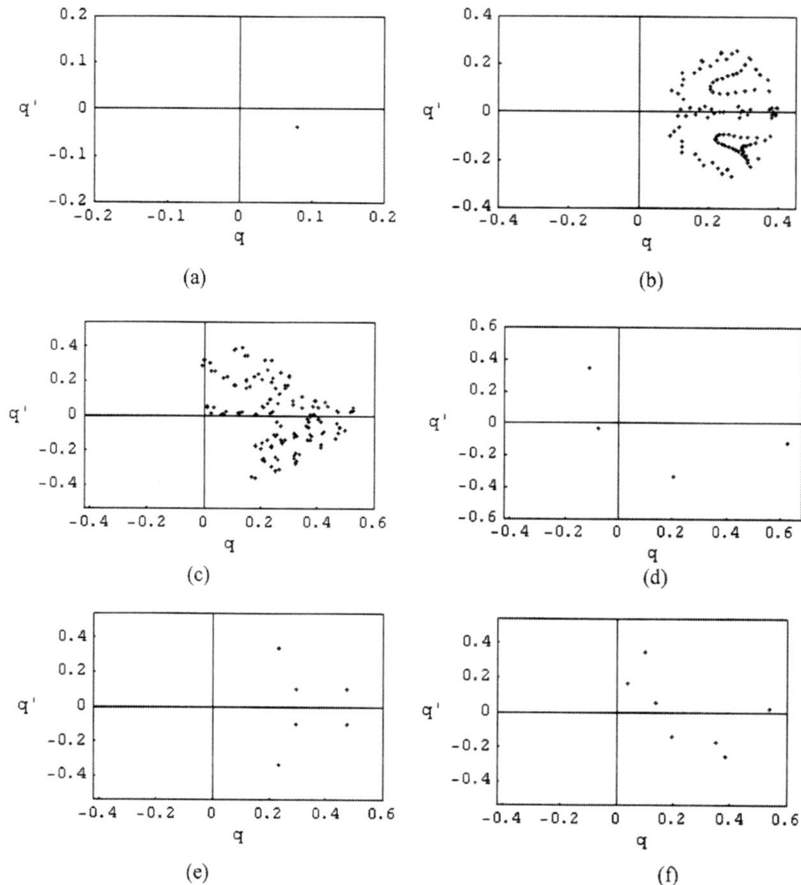

Fig. 3.22. Poincaré maps corresponding to the time history records of Fig. 3.20, [373].

the excitation amplitude increases, e.g., a = 0.094, the response possesses periodic motion with period–4 as shown in Figs. 3.20(d), 3.21(d) and 3.22(d). It then assumes period–3 for excitation amplitude a = 0.106 as shown in Figs. 3.20(e), 3.21(e) and 3.22(d). This is followed by period–7 for excitation amplitude a = 0.11 as revealed in Figs. 3.20(f), 3.21(f) and 3.22(f). For any excitation amplitude, a $\geq$ 0.12, the ship experiences rotational motion indicating the occurrence of capsizing. Note that these scenarios are obtained for given sets of initial conditions. However, for other initial conditions there is a possibility of other attractors that may coexist under the same excitation parameters.

Under forced excitation the ship roll dynamics is governed by the excitation amplitude and frequency and the values of damping factors. The ship response may be non–impacting or impacting bounded or can experience rotational

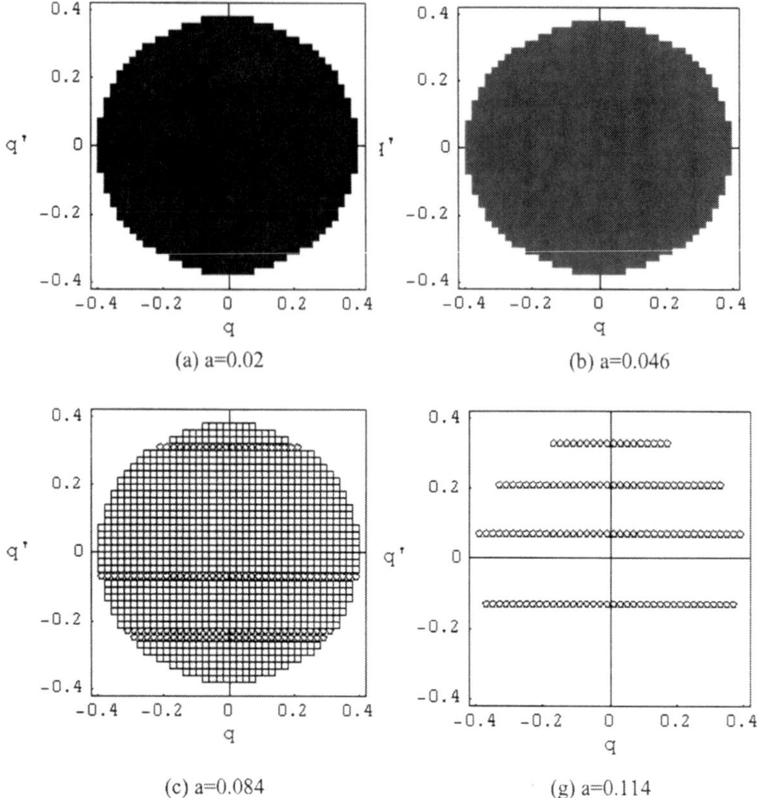

Fig. 3.23. Domains of attraction for different excitation amplitudes for excitation frequency $\nu = 0.88$. ■ Period-1 response, □ Modulated response, (gray squares) Multi - periodic response, (quintet symbol) Chaotic motion, empty space: Rotational Motion, [373].

motion leading to capsizing depending on initial conditions. For different sets of initial conditions covering the entire domain of the phase diagram bounded by the grazing orbit shown in Fig. 3.19, equation (3.33) was numerically solved. Note that an initial condition that leads to a rotational motion is the one that leads to response amplitude that exceeds the exit value ϕ_c. Thus, the safe basin corresponds to the set of initial conditions that lead to bounded response amplitudes smaller than ϕ_c.

Fig. 3.23 shows samples of safe basins of attraction for different values of excitation amplitude and for excitation frequency parameter $\nu = 0.88$. It is seen that for relatively small values of excitation amplitude the entire domain bounded by the grazing orbit experiences non–impact bounded oscillations of period–one as shown in Fig. 3.23(a) by the black region. As the excitation amplitude gradually increases the response assumes modulated

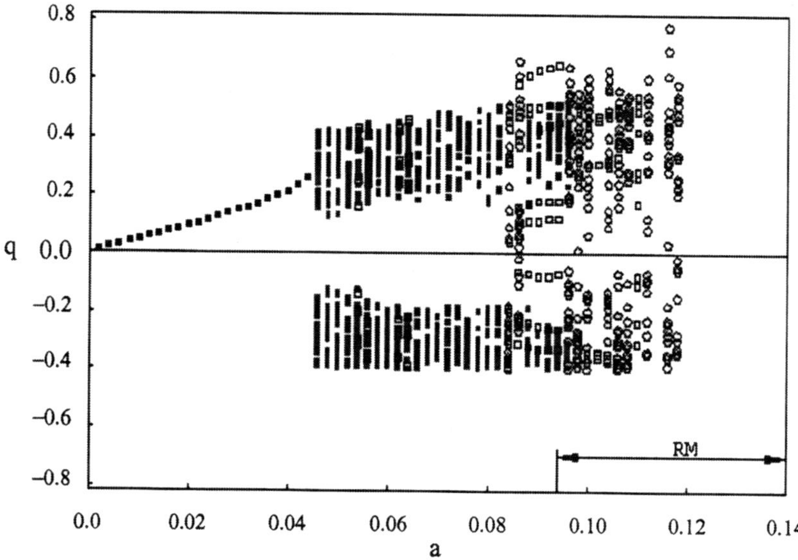

Fig. 3.24. Bifurcation diagram for excitation frequency ratio $\nu = 0.88$. ■ Period-one response, □ Modulated response, (quentet symbol) Multi-periodic response, Chaotic motion, RM = Rotational Motion, [373].

motion shaded by dark gray region as shown in Fig. 3.23(b). This motion is characterized by one impact every ten excitation periods. The modulated motion then coexists with multi–periodic oscillation (shown by empty squares, □) and the chaotic motion (indicated by the quintet symbol) as shown in Fig. 3.23(c) up to excitation amplitude $a < 0.094$. Above that excitation amplitude the region is eroded by regions of rotational motion as shown in Fig. 3.23(d). For excitation amplitude $a \geq 0.12$ the entire region belongs to ship rotational motion or capsizing.

Fig. 3.24 shows the bifurcation diagram on the plane of response–excitation amplitudes for frequency ratio $\nu = 0.88$. This figure summarizes all possible regimes of ship dynamics. Note that for other values of excitation frequency, the bifurcation diagram may be different particularly as the excitation frequency approaches the resonance frequency.

Fig. 3.25 shows the dependence of stability fraction, S_f, on excitation amplitude for three different values of excitation frequency, $\nu = 0.88$, 0.94 and 1.20. The stability fraction is also known in the literature (see, e.g., [565], [995], [998]) as the safety integrity factor (S.I.F.). It is obtained by estimating the ratio of the area of the stable region in the phase plane (area of the safe basin) to the total area encompassed by the grazing orbit, which is the safe basin in the absence of external excitation. For excitation amplitudes less than a critical value, governed by the excitation frequency, there is no erosion at

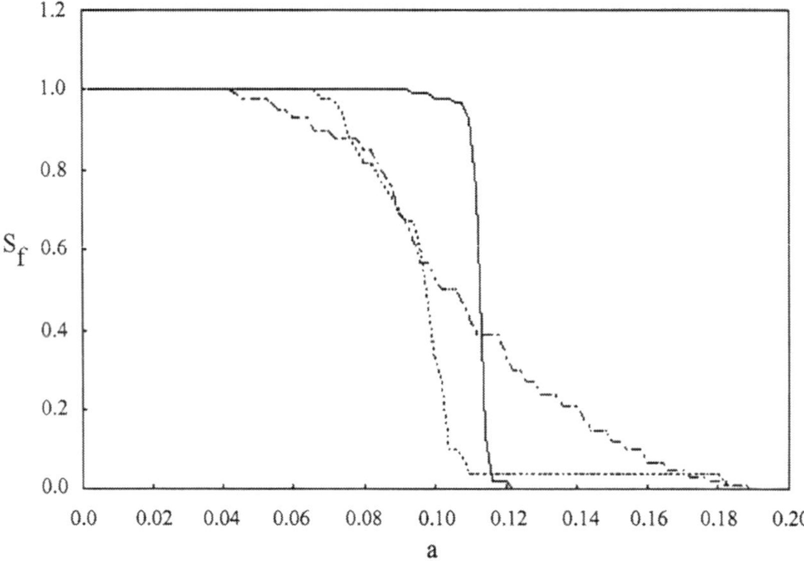

Fig. 3.25. Dependence of stability fraction on excitation amplitude for three different values of excitation frequency: —— $\nu = 0.88.$, ... $\nu = 0.94$, —.—.— $\nu = 1.2$, [373].

all for the safe basin. Above this critical value, the value of the safe basin area shrinks and the stability fraction drops. It is seen that as the excitation frequency increases the upper excitation level increases. This is attributed to the fact that the system is governed by soft nonlinear characteristics and more force is required to cause large response amplitude as the excitation frequency increases. Another important feature of the decreasing curve of the stability fraction is that it becomes progressively less steep when the excitation frequency increases above the resonant frequency.

3.5.4 Inelastic Impact Modeling

For the case of inelastic impact, the impact condition $q'_+ = -eq'_-$ must be introduced, where e is the coefficient of restitution, and q'_+ and q'_- are the ship velocities just before and after impact, respectively. The coefficient e is assumed to be close to unity, such that $(1-e)$ is considered a small parameter. According to the coordinate transformation given by equation (3.32), the impact condition $q'_+ = -eq'_-$ specified at $q = -q_i$, is transformed to

$$z'_+ = -ez'_- \qquad \text{at} \qquad z = 0 \tag{3.39}$$

It is possible to introduce this jump into the equation of motion using the Dirac delta–function, and thus one can avoid using condition (3.39). The additional term due to ice impact with the ship may be written in the form

$$(z'_{+} = -ez'_{-})\delta(t - t_i) = (1 - e) z' \delta(t - t_i) \qquad \text{provided } \left|z'_{+}\right| < \left|z'\right| < \left|z'_{-}\right| \tag{3.40}$$

where t_i is the time instant of impact. Since $\delta(t - t_i) = z' \delta(z)$, equation (3.33) can be written in the following form for case of inelastic impact

$$z'' + \overline{\zeta} z' + \gamma z'^2 sgn\left(z'\right) + z + sgn(z)\left\{-q_i + \overline{C}_3 \left[z sgn(z) - q_i\right]^3 + \overline{C}_5 \left[z sgn(z) - q_i\right]^5\right\} + (1 - e) z' \left|z'\right| \delta(z) = Z(\tau) sgn(z) \tag{3.41}$$

Under sinusoidal excitation and for coefficient of restitution $e = 0.8$, equation (3.41) was solved numerically under different values of excitation amplitude. A comparison between the ship response due to inelastic impact ($e = 0.8$) and its response to elastic impact ($e = 1$) for excitation frequency $\nu = 1$ and excitation amplitude a= 0.08 is shown in Figs. 3.26 and 3.27 for two different sets of initial conditions. For the initial conditions $z_0 = 0.15$, and $z'_0 = 0.05$, the ship experiences periodic impact roll oscillations as shown

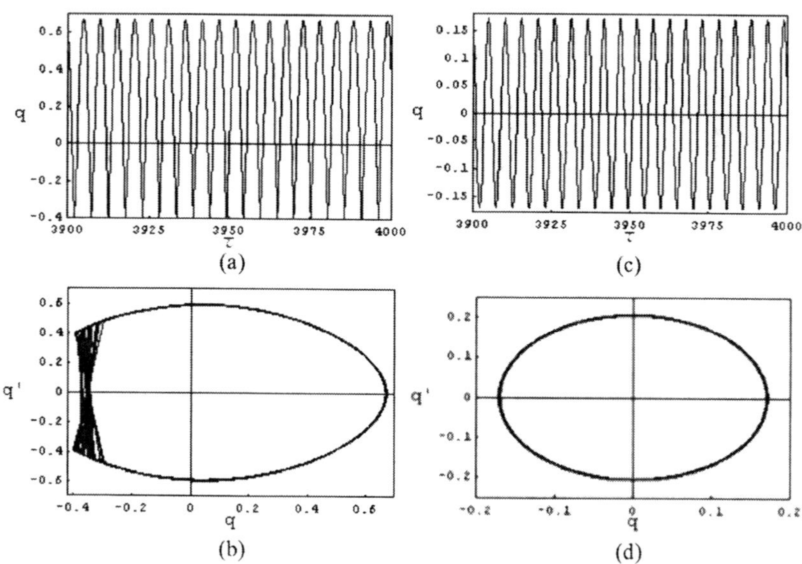

Fig. 3.26. Time history records and phase portraits of ship response under wave excitation frequency $\nu = 1.2$, amplitude a= 0.08, and initial conditions $z_0 = 0.15$, $z'_0 = 0.05$, for the cases of (a)-(b) purely elastic impact, $e = 1$, (c)-(d) inelastic impact $e = 0.8$.

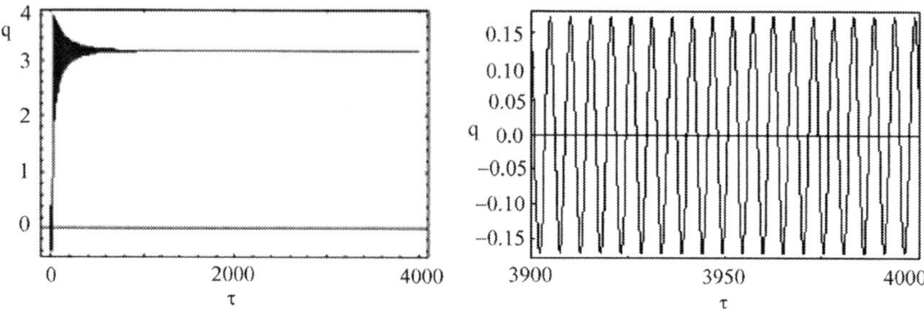

Fig. 3.27. Time history records of ship response under wave excitation frequency $\nu = 1.2$, amplitude a= 0.08, and initial conditions $z_0 = 0.29$, $z_0^{'} = 0.23$, for the cases of (a) purely elastic impact, $e = 1$ showing rotational motion, (b) inelastic impact $e = 0.8$.

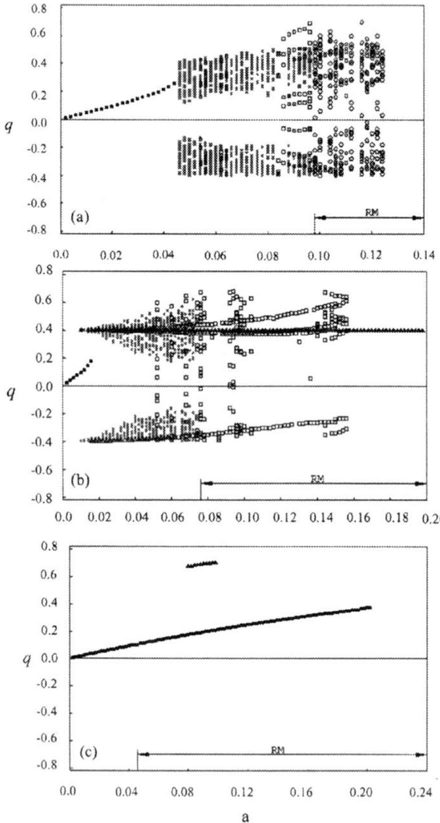

Fig. 3.28. Bifurcation diagrams for inelastic impact $e = 0.8$ and three different excitation frequencies (a) $\nu = 0.88$, (b) $\nu = 0.94$, and (c) $\nu = 1.2$. ■ period-1 non-impacting, ▲ period-1 impacting motion, grey squares modulated response, □ multi-period response, quintet symbol is for chaotic motion, RM rotational motion.

in Figs. 3.26(a)-(b) for the case of $e = 1$, while it reveals periodic non-impact oscillations with a smaller steady state amplitude for the case of $e = 0.8$ as shown in Figs. 3.26(c)-(d). Under a different set of initial conditions $z_0 = 0.29$, and $z_0^{'} = 0.23$, the ship experiences rotational motion for the case of elastic impact, $e = 1$, and periodic motion for the inelastic case, $e = 0.08$, as shown in Fig. 3.27(a) and (b), respectively.

For three different excitation frequencies, $\nu = 0.88$, 0.94, and 1.2, the dependence of the response amplitude on the excitation amplitude is shown in Figs. 3.28(a) – 3.28(c), respectively. These figures reveal the response bifurcation from one motion regime to another. These include period–1 motion, modulation motion, multi–period motion, chaotic response and rotational motion. There is also a coexistence of several response regimes that take place over a finite region of excitation amplitude depending on initial conditions. This is particularly manifested at excitation frequencies less than unity, i.e., $\nu < 1$. On the other hand, Fig. 3.28(c) shows a narrow region over which a non-impacting period–1 and impacting period–1 motions coexist in addition to the coexistence of rotational motion over a wide range of excitation amplitude. Generally, as the excitation frequency increases the excitation amplitude at which rotational motion occurs is decreased. As expected, the additional damping associated with inelastic impact is significant than the linear and nonlinear damping terms in equation (3.33).

3.6 Closing Remarks

This chapter presented a wide spectrum of problems modeled by one freedom motion against single– or double–sided barriers. The bouncing ball on an oscillating table has been recognized a key problem to several applications such as pneumatic hammers, riveting machines, peening guns, inertia shakers and pile drivers. The protection of operators and vibration isolation means of such systems will be discussed in Chapter 5. The conditions for the existence of periodic solutions of mass–spring–dashpot systems with restraints together with the stability boundaries were obtained in terms of the coefficient of restitution and system parameters. The dynamic characteristics of normal and inverted simple pendulums with one– or two–sided walls were found to share complex phenomena with the restraint mass–spring oscillator such as chaos and chattering. One of the main features of these classical systems is the occurrence of chattering, which is an inherent property of impacting systems. An important application of vibro–impact dynamics is the interaction of ship roll dynamics with icebergs in cold regions. The safety criterion of ship navigation in these regions has been established by estimating the stability index.

Chapter 4
Two– and Multi–Degree–of–Freedom Systems

4.1 Introduction

This chapter considers the problem of vibro–impact dynamics of lumped systems represented by two and more degrees of freedom in the presence of single and two barriers. Of particular interest of these systems is the prediction of bifurcation and chaos manifested by the occurrence of period–doubling and saddle–node bifurcations. These systems are mainly used as vibro–impact vibration absorbers, spherical pendulums, and coupled pendulums. In the presence of weak nonlinearities these systems can experience internal resonance conditions, and possibly parametric resonance when subjected to parametric excitation. Vibro–impact systems have also been used as passive vibration control such as vibro–impact dampers, which will be addressed in Chapter 8.

4.2 Two–Degree–of–Freedom Systems

4.2.1 Overview

Analytical and numerical simulations of two-degree-of-freedom systems may exhibit a stable impact vibration occurring over a wide range of excitation frequencies when the clearance between the two masses is close to the excitation amplitude. A sub–impact vibration was found to occur when the frequency is high, while a two–impact vibration occurs as the frequency becomes low. These systems were studied in the literature by many researchers ([655], [450], [482], [348], [187], [50], [499], [3], [832], [616], [618], [562], [621], [457], [1050], [1052], [1068], [1069], [1089], [198], [491], [200], [201], [1046], [792]). The harmonic excitation of a two–degree–of–freedom system with a clearance was examined by Masri [655] and Irie and Fukaya [450]. Awrejcewicz and Tomczak [43] considered the control of one– and two–degree–of–freedom vibro–impact systems with a delay loop with the purpose of stabilizing the vibro–impact periodic motion after the occurrence of disturbances.

R.A. Ibrahim: Vibro-Impact Dynamics: Model., Map. & Appl., LNACM 43, pp. 97–123.
springerlink.com

Blazejczyk–Okolewska et al [118] considered vibro–impact oscillators driven by harmonic excitation. They obtained the bifurcation diagram of the relative displacement of two oscillators versus the excitation frequency. It revealed complicated structure caused by jumps of the system trajectory from one type to another type. Three or four attractors of different types such as periodic, chaotic and two different quasi–periodic attractors were found to coexist. These jumps may occur in random and unpredictable fashion. Dynamical uncertainty introduced by these jumps was observed similar to the uncertainty introduced by riddled basins in coupled systems.

The problem of periodic impact of a bouncing mass on a coupled two–degree–of–freedom system in the presence of 1 : 4 strong resonance was studied by Luo et al ([628], [630]) using the center manifold theorem. Here the strong resonance follows the definition given by Ioos [447], Arnold [31], and Kuznetsov [560]. If the Jocobian of the system equations of motion possesses a pair of complex conjugate eigenvalues, which satisfy the condition $|\lambda_{1,2}| = 1$, as the control parameter passes the critical value, and $\lambda_{1,2}^m = 1$, $m = 1, 2, 3$, and 4 at the bifurcation point, it is possible that Hopf or subharmonic bifurcations in strong resonance cases (1 : 1, 1 : 2, 1 : 3 or 1 : 4) occur. On the other hand, weak resonance occurs when $\lambda_{1,2}^m = 1$, $m \geq 5$, and $\lambda_{1,2}^m \neq 1$, $m \geq 1$. Luo et al [630] found that two–parameter bifurcations of fixed points in their system near 1 : 4 strong resonance are characterized by quasi–periodic impact motion, stable and unstable 4 : 4 motions, i.e., four excitation periods and four impacts during one impact motion period. This concept of strong or weak resonance is different from that was originally introduced by Kunitsyn and Matveyev [556] and Kunitsyn and Muratov [557] who formulated the normal form of a dynamical system that contains the first nonlinear terms for an arbitrary number of non–interacting as well as interacting resonances of an odd order. They classified the internal resonance conditions into weak and strong. *Weak resonance* preserves the stability of the system, while *strong resonance* results in system's instability.

Aidanpää and Gupta [3] identified the system parameter ranges (such as damping, coefficient of restitution, distribution of masses and clearance) in a two–degree–of–freedom impact oscillator with proportional damping that result in stable periodic multiple impacts. The maximum displacement of one of the masses was limited to a threshold value by a rigid wall. Mikhlin [676] and Mikhlin et al [679] studied direct and inverse problems that arise in vibro–impact oscillators with two coordinates undergoing single– or double–sided impacts. It was found that these systems possess nonlinear vibro–impact localized and non–localized time–periodic motions, complicated bifurcation structures giving rise to new types of single– and double–sided impacting motions, mode instabilities, and chaotic responses.

The presence of friction in vibro–impact oscillators adds another degree of complexity to the modeling of these systems. Peterka ([820], [821]), Marghitu and Hurmuzlu [641], and Peterka and Szollos [831] considered different vibro–impact oscillators with frictional impact. Půst et al [873] presented an overview of the vibration of two degrees of freedom systems with impacts and dry friction.

Qian and Torres [875] examined the existence and multiplicity of non–trivial periodic bouncing solutions for linear and asymptotically linear impact oscillators by applying a generalized version of the Poincaré–Birkhoff theorem to an adequate Poincaré section called the successor map. Luo [611] considered two vibro–impact systems and analytically derived period$-n$ single–impact motions and Poincaré maps. It was found that near the point of codimension–two bifurcation, Hopf bifurcation of period–one single–impact motion can coexist with Hopf bifurcation of period–two double–impact motion. Period doubling bifurcation of period–one single–impact motion was found to exist near the point of codimension–two bifurcation. Luo et al ([631], [629], [627]) considered the periodic excitation of an oscillator with symmetrically placed rigid stops. They used the center manifold theory to analyze local codimension–two bifurcations associated with double Hopf bifurcation and interaction of Hopf and pitchfork bifurcation. It was found that near the value of double Hopf bifurcation there exist period–one double–impact symmetrical motion and quasi–periodic impact motions.

Pinnington [856] considered the problem of a single collision between two single–degree–of–freedom systems separated by a gap. The main parameters governing the motion of this system are the ratio of strain energy to kinetic energy at initial contact and the damping of the contact. A strong impact was found to give a half–sine displacement pulse, while a weak impact gives a truncated half–sine pulse. The coefficient of restitution describing the energy loss during collision was found to increase with the contact stiffness, damping, and the relative velocity. A time–stepping power balance algorithm was developed to calculate the work done due to oscillator internal hysteresis and the work done of collision between a series of colliding oscillators [857]. The energy exchange and dissipation from a collision of a pair of oscillators was studied by creating an equivalent oscillator pair, one has the energy of the in–phase motion and the other has the out–of–phase energy. It was found that the energy exchange between colliding oscillators is proportional to the initial kinetic energy difference of the oscillators. Furthermore, the work done during collision was found to be proportional to the out–of–phase energy.

The dynamic behavior of two linearly coupled masses in which one mass can have inelastic impacts with a fixed rigid stop was studied by Valente et al [1022]. The system was found to be governed by three types of motion: coupled harmonic oscillation, simple harmonic motion, and discrete rebounds. It was proven the existence of a non–zero measure set of orbits that lead to infinite impacts with the stop in a finite time. Pascal [792] considered a two–degree–of–freedom oscillator with a colliding component and analyzed the dynamic behavior of the system when the barrier stiffness changes from a finite value to an infinite one (see also [824]). For the case of rigid impact and in the absence of external excitation, a family of periodic solutions was obtained analytically. In the case of soft impact, with finite time duration of the impact, periodic solutions with an arbitrary value of the period were reported. Periodic motions were also obtained when the system is subjected to harmonic excitation.

Two harmonically excited systems having symmetrical rigid constraints were considered by Luo and Zhang [626]. The impact in one system was characterized by collisions with the two constraints, while one of components of the other system collides with rigid obstacles. The dynamic characteristics of these systems were studied with special attention to Neimark–Sacker bifurcations associated with several periodic–impact motions. Period–one double–impact symmetrical motions and associated Poincaré maps of two systems were derived analytically. Neimark–Sacker bifurcations associated with several periodic–impact motions were found by numerical simulation. It was found that the vibratory systems having symmetrical rigid amplitude constraints may exhibit complex and rich quasi–periodic impact behavior under different system parameter conditions.

4.2.2 Vibro–Impact Absorbers

The dynamics of a harmonically excited primary mass and a secondary mass moving in an inclined slot within the primary mass was studied by Heiman et al ([400], [401]). In particular, the dynamics of the secondary mass for a number of impacts during an integer number of cycles of the base motion was formulated in terms of a return map. It was shown that harmonic, subharmonic, and chaotic motions can exist for various values of system parameters. Different types of stable motions as well as chaotic motions were found to co–exist. In another treatment, the primary mass was constrained to move vertically within a massive hollow shell [202]. The shell was subjected to harmonic excitation while gravity and shell contact govern the dynamic response of the primary mass. The equations of motion for separate domains controlling the mass were expressed as functional relationships. Natsiavas ([717], [718]) examined the response of vibration absorbers with elastic stops. The absorber with stops was shown to possess superior performance characteristics. For example, it was possible to design nonlinear absorbers and to suppress vibration levels over broader forcing frequency ranges than the conventional absorber. However, high amplitude beating and chaotic response were found to arise near the original resonance, because the coexisting periodic response may become unstable due to Hopf bifurcation.

Nucera et al [740] considered the concept of nonlinear vibration absorbers by introducing a set of nonlinear energy absorbers that are locally attached to the main structure. The purpose was to passively absorb a significant part of the applied seismic energy, locally confining it and then dissipating it in the smallest possible time. It was demonstrated that it is possible to passively divert the applied seismic energy from the main structure to a set of preferential nonlinear substructures where this energy is locally dissipated at a time scale fast enough to be of practical use for seismic mitigation. They adopted a two degree–of–freedom primary linear system and studied seismic–induced vibration control through the use of vibro–impact nonlinear energy absorbers. Karayannis et al [493] considered different configurations of linear primary systems with vibro–impact attachments, which act as shock absorber over a wide frequency range. It was found that better energy dissipation is achieved for weak values of the coupling stiffness and relatively large values of mass. Pilipchuk [851] considered coupled nonlinear two–degree–of–freedom oscillators, whose characteristic is close to linear for low amplitudes but becomes infinitely growing as the amplitude approaches certain limit. In particular, the analysis predicted the evolution of vibration modes as the energy is gradually pumped into or dissipates out of the system. It was shown that the in–phase and out–of–phase motions may follow qualitatively different scenarios as the system energy increases. For example, the in–phase mode was found to absorb the energy with equi–partition between the masses. On the other hand, the out–of–phase mode provides equal energy distribution only until certain critical energy level. Above that level and as a result of bifurcation of the 1 : 1 resonance path, one of the masses was found to become a dominant energy absorber.

Zhang and Luo [1129] described briefly the main features of complex dynamic characteristics of two–degree–of–freedom systems contacting a single barrier. These include Hopf bifurcations, period–doubling bifurcations, singularities and chaos for a two–degree–of–freedom vibro–impact system. Later, Mikhlin and Reshetnikova ([677], [678]) considered a nonlinear two–degree–of–freedom system consisting of a linear oscillator with a relatively large mass, and an essentially nonlinear oscillator with a relatively small mass, which acts as an absorber of the main linear system vibrations. It was shown that a stable localized vibration mode exists in a large region of the system parameters. Frequency response of the system under external periodic force was obtained.

4.2.3 Dynamic Analysis

The complex dynamic behavior of two–degree–of–freedom systems with one sided–barrier as shown in Fig. 4.1(a) was considered by Luo and Xie [616] and Wen and Xie [1069]. The equations of motion in non–dimensional form was given in the form

$$\mu_m X_1^{''} + 2\zeta\left(1+\mu_c\right) X_1^{'} - 2\zeta X_2^{'} + \left(1+\mu_k\right) X_1 - X_2 = f_1 \sin(\nu\tau+\theta) \quad (4.1)$$

$$X_2^{''} + 2\zeta X_2^{'} - 2\zeta X_1^{'} - X_1 + X_2 = (1-f_1)\sin(\nu\tau+\theta) \quad (4.2)$$

where a prime denotes differentiation with respect to the non–dimensional time parameter $\tau = \omega_{22} t$, $\omega_{22} = \sqrt{k_2/m_2}$, $\nu = \Omega/\omega_{22}$, $X_i = x_i/x_0$, $i =$

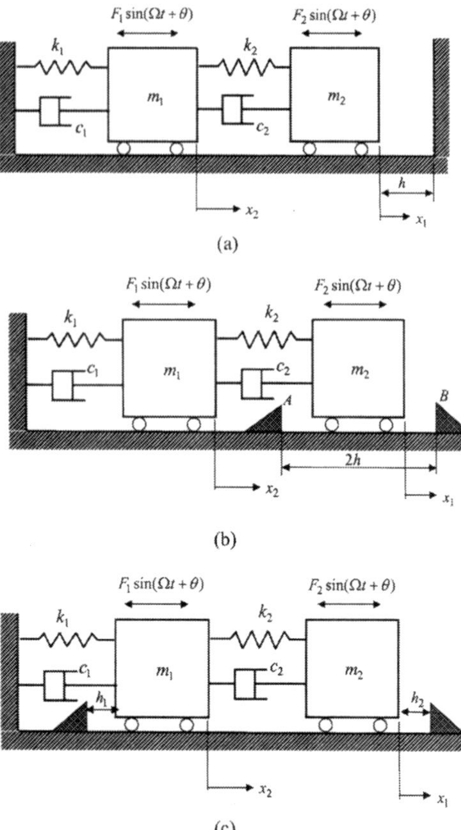

Fig. 4.1. Schematic diagrams of two-degree-of-freedom systems with (a) one-sided barrier for one mass, (b) two-sided barrier on one mass, (c) one-sided barrier on each mass.

$1, 2$, $x_0 = (F_1 + F_2)/k_2$, $\mu_m = m_1/m_2$, $\mu_k = k_1/k_2$, $\mu_c = c_1/c_2$, $f_1 = F_1/(F_1 + F_2)$, $\zeta = c_2/2\sqrt{k_2 m_2}$.

Equations (4.1) and (4.2) describe the system dynamics in the absence of any barrier. In the presence of the barrier to the second mass, the following impact law must be augmented:

$$X'_{2+} = -eX'_{2-} \tag{4.3}$$

where e is the coefficient of restitution.

While the second mass experiences non–smooth dynamics due to impact with the barrier, the first mass experiences periodic smooth velocity. The periodic motion implies

$$X_1(0) = X_1(2\pi/\nu), \quad X'_1(0) = X'_1(2\pi/\nu), \quad X_2(0) = H = h/x_0$$

$$X'_2(2\pi/\nu) = H, \quad X'_{2+}(0) = -eX'_{2-}(2\pi/\nu) \tag{4.4}$$

Luo and Xie [616] obtained the linearized Poincaré map at the fixed point. The stability of periodic impacts was determined by estimating the eigenvalues of the Jacobian of the Poincaré map, $Df(\nu, 0) = D\{f_1, f_2, f_3, f_4\}^T$, where T is the transpose, $f_1 = \Delta\widetilde{X}_{20}$ is the net displacement of the perturbed system trajectory from the unperturbed trajectory, $f_2 = \Delta\widetilde{X'}_{10}$, $f_3 = \Delta\widetilde{X'}_{2+}$, $f_4 = \Delta\widetilde{\theta}$. If all eigenvalues of $Df(\nu, 0)$ are inside the unit circle, then the periodic solution is stable, otherwise it is unstable. When the eigenvalues of $Df(\nu, 0)$, with the largest modulus, are on the unit circle, bifurcations occur in various ways according to their numbers and their positions on the unit circle. This results in qualitative changes in the system dynamics. Luo and Xie [616] considered the case of a single complex conjugate pair of simple non–real eigenvalues, crossing the unit circle with non–zero velocity as ν passes the critical value ν_c. For the system parameters $\mu_m = 2$, $\mu_k = 5$, $f_1 = 0$, $\zeta = 0$, $H = 1.5$, and $e = 0.8$, the eigenvalues of $Df(\nu, 0)$ were estimated over the range of $\nu = 0.7219$ to 0.7368 and were found to be strictly inside the unit circle. For $\nu = \nu_c = 0.7368$, it was found that there is a complex conjugate pair of eigenvalues on the unit circle and the remainder is inside the unit circle. At the critical excitation frequency ratio ν_c a supercritical Hopf bifurcation occurred. At $\nu = 0.7369$ a quasi–periodic response was reported and the corresponding Poincaré sections were estimated and plotted in Fig. 4.2(a). It was found that a single torus doubling begins to occur at $\nu = 0.744$, see Fig. 4.2(b), and at $\nu = 0.7519$ and 0.7555 the system settled into chaotic motion as shown in Figs. 4.2(c,d).

Wen and Xie [1069] considered the stability problem of the same system but for different system parameters, $\mu_m = 2$, $\mu_k = 6.1$, $f_1 = 0$, $\zeta = 0$, $H = 1.6$, and $e = 0.82$, for which $\nu_c = 0.745716$. For $\nu = 0.75$ the system was found to experience stable period–2 fixed points after 1500 impacts. As ν is allowed to increase the system exhibits stable period–4 points, above this excitation frequency ratio, period–doubling cascade stops and four Hopf circles emerge simultaneously by Hopf bifurcations of the corresponding period–4 points. As ν is increased further the system develops chaotic response through a finite number of times of torus–doubling process. Further increase of excitation frequency results in stable quasi–periodic impacts represented by a complex torus. The sequence settles after that to chaotic motion through torus–doubling bifurcation.

The system shown in Fig. 4.1(b) is a two–degree–of–freedom system with symmetrically placed rigid stops at the right mass was studied by Luo and Xie [620]. The period–one double–impact symmetrical motion and its Poincaré map were derived analytically. The routes from period–one double–impact symmetrical motion to chaos, via pitchfork bifurcations and period–doubling bifurcation, were numerically studied. Some non–typical routes to chaos, caused by grazing and Hopf bifurcation of period–2 four–impact motion, were studied. Hopf bifurcations of period–one double–impact symmetrical and

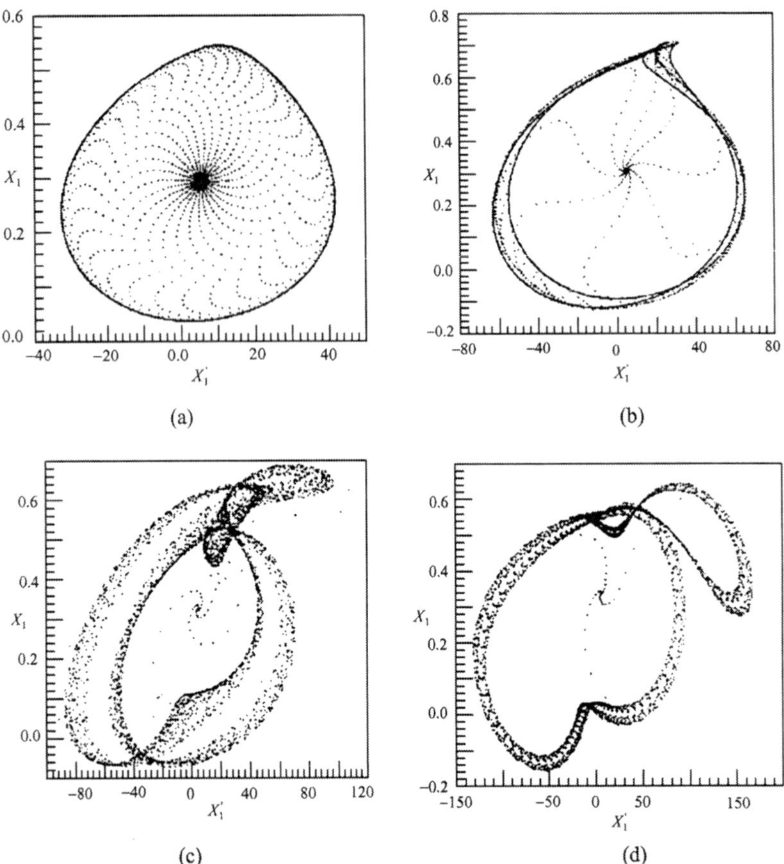

Fig. 4.2. Projection of Poincar é sections showing the quasi-periodic response for system parameters $\mu_m = 2$, $\mu_k = 5$, $f_1 = 0$, $H = 1.5$, $e = 0.8$, (a) $\nu = 0.7369$, (b) $\nu = 0.744$, (c) $\nu = 0.7519$, (d) $\nu = 0.7555$, [616].

antisymmetrical motions were shown to exist in the two–degree–of–freedom vibratory system with two–sided stops. Interesting feature like the period–one four–impact symmetrical motion was also found.

The existence and stability of period–one double–impact symmetrical motions were examined by Luo and Xie [620]. The vibro–impact system was considered for the system parameters $m_1/(m_1 + m_2) = 0.6667$, $k_1/(k_1 + k_2) = 0.8333$, $\zeta = 0.05$, $e = 0.8$, $f_1 = 0$, and $H = 0.1$. Fig. 4.3 shows phase portraits taken for different values of excitation frequency. The periodic motions of the system is characterized by the symbol $n - p - q$, where n is the forcing cycles, p and q are the number of impacts occurring at constraints B and A, respectively. As the excitation frequency increases from $\nu = 3.716962$, the $1 - 1 - 1$ symmetrical motion changes its stability, and pitchfork bifurcation of $1 - 1 - 1$ symmetrical motion occurs so that a pair of antisymmetrical

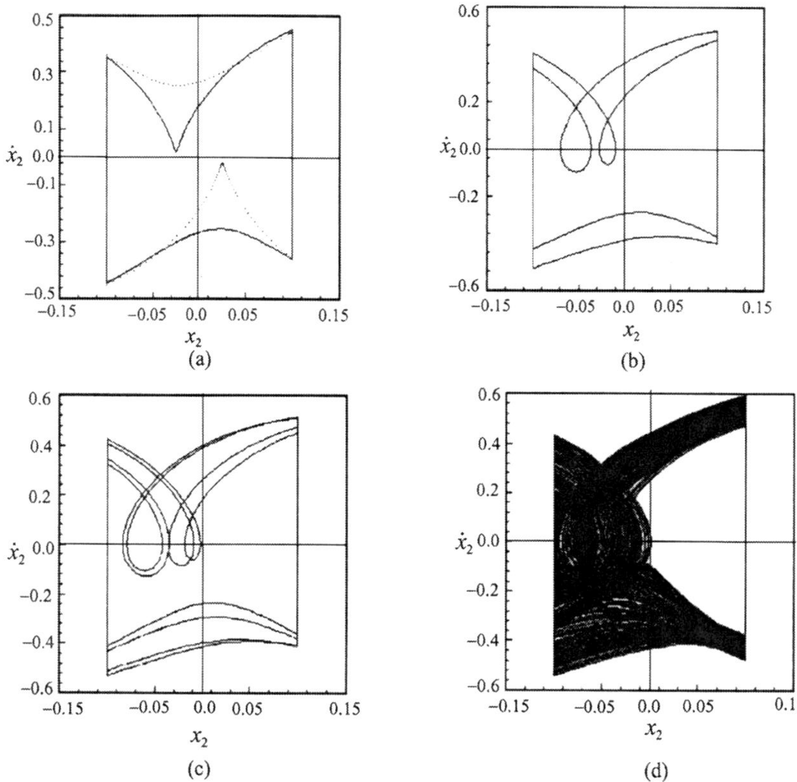

Fig. 4.3. Phase portraits of the impacting mass m_2 (a) Pair of antisymmetric period-1 double impact motions, $\nu = 3.5$ pair of antisymmetrical period-1 double-impact motions, (b) $2-2-2$ asymmetrical motion, $\nu = 3.1$, (c) $4-4-4$ asymmetrical motion, $\nu = 3.029$, (d) chaos, $\nu = 2.9$, [620].

double–impact orbits are born, as demonstrated in Fig. 4.3(a). The stability and local bifurcation of period–1 double–impact symmetrical motion were numerically confirmed. With further decrease in the excitation frequency ratio ν, the $1-1-1$ antisymmetrical motion became unstable. These motions then undergo a succession of period doubling bifurcations, which eventually result in apparently non–periodic, or chaotic motions. The $2-2-2$, and $4-4-4$ impact motions, and chaotic regime of the mass are shown in Figs. 4.3(b) through 4.3(d), respectively, in the form of phase plane portraits. Later, Yue and Xie [1120] considered a two–degree–of–freedom vibro–impact system having two–sided impact constraints. They predicted a symmetric period $n-2$ motion. The symmetric period $n-2$ motion corresponds to the symmetric fixed point of the Poincaré map. It was shown that the symmetry of the Poincaré map suppresses the period–doubling bifurcation, Hopf–flip bifurcation and pitchfork–flip bifurcation of the symmetric period $n-2$

motion. They proved that both the two antisymmetric period $n-2$ motions have the same stability.

In the design process of impact forming machines, it is important to achieve the desired periodic impact velocities. The global bifurcation diagrams for the relative impact velocities of the impact–forming system versus the excitation frequency enable the design engineer to select excitation frequency ranges in which stable period–1 single–impact response to occur, and to predict the peak–impact velocity and shorter impact period of such response. The dynamics of an impact–forming machine was analytically modeled by a lumped two–degree–of–freedom system by Luo [612]. Special attention was given to the stability of period–n single–impact motion, Hopf bifurcations in non–resonance and weak resonance cases, subharmonic and Hopf bifurcations in $1:4$ strong resonance, and codimension–two bifurcations and chaotic motions. Stability and local bifurcations of period–1 single–impact motion were analyzed by using the Poincaré map. Local bifurcation analyses and numerical simulation revealed that period–1 single–impact motion undergoes period doubling bifurcation or Hopf bifurcation with change of control parameters. Period–one single–impact motion undergoes either subharmonic or Hopf bifurcation in $1:4$ strong resonance case. The grazing instability was found to occur in the strong resonance case. On the grazing boundary of periodic–impact motion a new impact in the motion period was found to appear. Luo et al ([625], [622]) introduced a three–dimensional map with dynamical variables defined at the impact instants for a two–degree–of–freedom plastic impact oscillator. The piecewise nature of the system was caused by the transitions of free flight and sticking motions of two masses immediately after the impact, and the singularity of map was generated via the grazing contact of two masses and corresponding instability of periodic motions. These properties of the map were shown to exhibit particular types of sliding and grazing bifurcations of periodic–impact motions under parameter variation.

Wagg and Bishop [1050] considered the dynamics of a two–degree–of–freedom impact oscillator with a motion limiting constraint. Bifurcations occurring between differing regimes of impacting motion and in particular those occurring due to a grazing bifurcation was observed for a particular set of parameters. Both periodic and chaotic chatter motions and the regions of sticking were found to exist. The so–called "*rising phenomena*", which occur in sticking solutions of a two–degree of freedom impact oscillator with double barriers was considered by Wagg [1045]. The system is shown in Fig. 4.1(c), which is described by the equations of motion

$$\ddot{x}_1 + \frac{c}{m}\left(2\dot{x}_1 - \dot{x}_2\right) + \frac{k}{m}\left(2x_1 - x_2\right) = \frac{A_1}{m}\sin\Omega t \tag{4.5}$$

$$\ddot{x}_2 + \frac{c}{m}\left(\dot{x}_2 - \dot{x}_1\right) + \frac{k}{m}\left(x_2 - x_1\right) = \frac{A_2}{m}\sin\Omega t \tag{4.6}$$

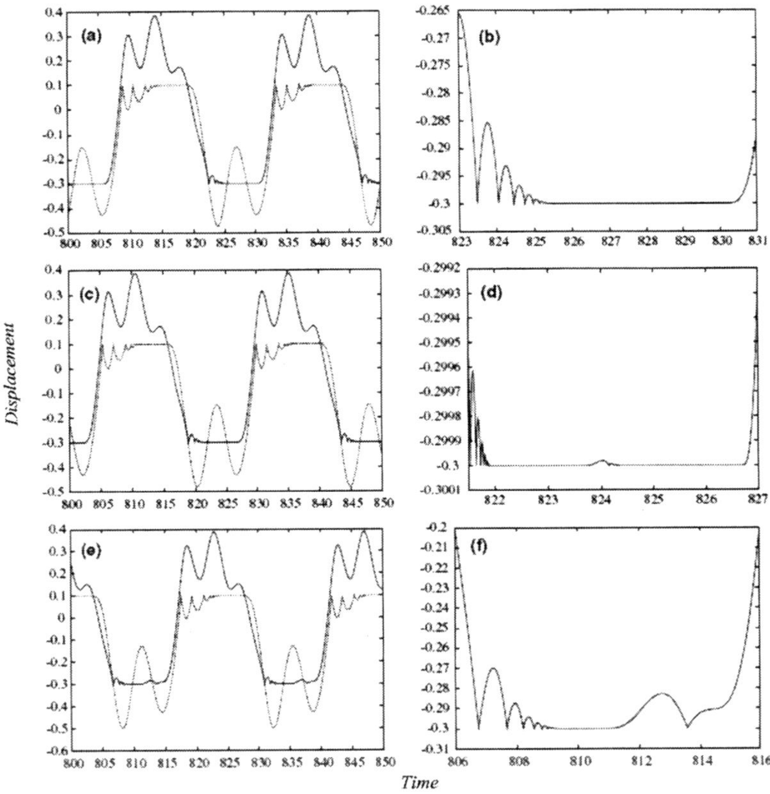

Fig. 4.4. Rising bifurcation in a two degree-of-freedom impact oscillator for system parameters $m_1 = m_2 = 1$, $k_1 = k_2 = 1$, $c_1 = c_2 = 0.1$, $s_1 = -0.3$, $s_2 = 0.1$, $e = 0.7$, $A_1 = 0.5$, $A_2 = 0.0$, (a,b) $\Omega = 0.255$, (c,d) $\Omega = 0.2561$, (e,f) $\Omega = 0.26$. ..... x_1, —— x_2, [1045].

where x_1 and x_2 are the displacements of masses $m_1 = m$ and $m_2 = m$, respectively. The spring stiffnesses are $k_1 = k_2 = k$ and the damping coefficients are $c_1 = c_2 = c$. A_1 and A_2 are the excitation amplitudes of the two masses and Ω is the excitation frequency. The barriers are positioned at distances h_1 and h_2 from the static equilibrium position. Equations (4.5) and (4.6) are subject to the following conditions for the free flight

$$(x_i - h_i) \lesseqgtr 0 \quad \textit{for all } h_i \gtreqless 0 \tag{4.7}$$

For system parameters $m_1 = m_2 = 1$, $k_1 = k_2 = 1$, $c_1 = c_2 = 0.1$, $h_1 = -0.3$, $h_2 = 0.1$, $e = 0.7$, and excitation amplitudes $A_1 = 0.5$ and $A_2 = 0.0$, Wagg [1045] carried out a series of numerical simulations and obtained the response time history records shown in Fig. 4.4 for excitation frequencies $\Omega = 0.255$ (shown in Figs. 4.4(a,b)), $\Omega = 0.2561$ (shown in

Figs. 4.4(c,d)), and $\Omega = 0.26$ (shown in Figs. 4.4(e,f)). It is seen that the mass lifts off, or rises, part way through the sticking phase of the motion. The mass then experiences a chatter sequence and then sticks to the stop again. As the excitation frequency increases, say, $\Omega = 0.26$, the amplitude rise grows and the second part of the sticking phase reduces until only a single impact remains. Toulemonde and Gontier [1011] indicated that the sudden drop in sticking in the sticking regime is one way of identifying that a rising even had occurred.

4.3 Spherical Pendulum under Liquid Flow

Flow–induced oscillations of rod elements in pressurized water reactors may interact with constraints. The planar impact oscillation of these elements was analytically and experimentally studied by Hennig and Grunwald [403]. The treatment dealt with the dependence of the pendulum elements oscillation on the water flow rate. Low–speed oblique elastic impacts frequently occur in heat exchangers and other equipment having loosely supported tubes or pipes. It was found that in addition to the planar or "polarized" impact self–oscillation, both gyratory spatial impact motions and almost non–impact pendulum sliding around the channel wall can coexist. Païdoussis et al [769] experimentally demonstrated that a cantilevered pipe conveying fluid, interacting with motion–limiting nonlinear constraints, exhibits regions of chaotic motions. In their analysis it was shown that the planar dynamics of the flexible pipe system revealed chaotic oscillations in the parameter space. The case of restraint pipes conveying fluid will be considered in Chapter 6.

Osakue and Rogers [754] developed a pendulum–type impact apparatus to study friction during impacts. A hardened steel sphere at the end of the pendulum collides with a flat steel surface for a range of approach velocities and angles. The normal and tangential contact force waveforms were measured using a tri–axial piezoelectric force transducer. The results showed that the tangential force is less than the limiting Amontons–Coulomb friction predictions at low impact angles. Two regimes of stick–slip and gross–slip friction were clearly distinguished by a new friction parameter called the "specific traction ratio". Tangential force reversal was observed at low impact angles indicating local tangential oscillations. The stick–slip results were found consistent with a partial slip model where the contact zone has a central sticking region surrounded by a ring area undergoing slip. This section deals with the flow–induced impact oscillations of a spherical pendulum whose motion is described in terms of two angular motions, θ and φ as shown in Fig. 4.5. The equations of motion of this system were written in the form [814]

$$\ddot{\varphi} - \dot{\theta}^2 \sin\varphi\cos\varphi + \omega_n^2 \sin\varphi = 0 \tag{4.8}$$

$$\ddot{\theta}\sin^2\varphi - 2\dot{\varphi}\dot{\theta}\sin\varphi\cos\varphi = 0 \tag{4.9}$$

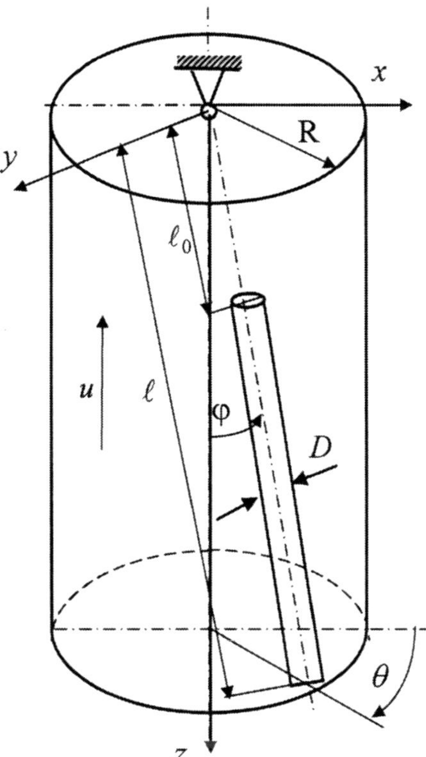

Fig. 4.5. Schematic diagram of the spherical pendulum suspended in a cylindrical tube showing the coordinate frame as considered by Peterka. [814].

where $\omega_n = \sqrt{\frac{3g(\rho-\rho_f)\left(\ell^2-\ell_0^2\right)}{2(\rho+c_M\rho_f)\left(\ell^3-\ell_0^3\right)}}$, ℓ and ℓ_0 are the lengths of the suspension massless arm and the total pendulum arm, respectively. ρ and ρ_f are densities of the pendulum and fluid, respectively, and c_M is the coefficient of the virtual mass of the fluid. In view of the restricted motion of the pendulum, the angle φ is very small and equations (4.8) and (4.9) can be linearized with respect to φ to take the form

$$\varphi'' + \left(1 - \theta'^2\right)\varphi = 0 \tag{4.10}$$

$$\theta'' + 2\theta'\varphi'/\varphi = 0 \tag{4.11}$$

where a prime denotes differentiation with respect to the non–dimensional time parameter $\tau = \omega_n t$. Peterka [814] introduced the effect of fluid hydrodynamic forces and damping forces due to fluid and structure. The resulting equations of motion are

$$\varphi'' + \varphi' F\left(\varphi, \varphi', \theta'\right) + \left(1 - \theta'^2 + \frac{J c_M U^2}{\rho + c_M \rho_f}\right)\varphi = 0 \qquad (4.12)$$

$$\theta'' + \theta' \left[2\left(\varphi'/\varphi\right) + F\left(\varphi, \varphi', \theta'\right)\right] = 0 \qquad (4.13)$$

where $F\left(\varphi, \varphi', \theta'\right) = \frac{\left[G\sqrt{\varphi'^2+\varphi^2\theta'^2}+H-(Kc_N+Lc_M)U\right]}{(\rho+c_M\rho_f)}$, $U = u/\left(\ell\omega_n\right)$, u is the fluid flow velocity. $G = \frac{3\alpha\left(\ell^4-\ell_0^4\right)(c_D)_0}{4A\left(\ell^3-\ell_0^3\right)}$, $H = \frac{3\beta\left(\ell^5-\ell_0^5\right)(c_k)_0}{5A\omega_n\left(\ell^3-\ell_0^3\right)}$, $J = \frac{3\alpha\rho_f D\ell^2\left(\ell^2-\ell_0^2\right)(c_D)_0}{4A\left(\ell^3-\ell_0^3\right)}$, $K = \frac{\ell D\rho_f}{2A}$, $L = \frac{3\rho_f\ell^2\left(\ell^2-\ell_0^2\right)}{4A\left(\ell^3-\ell_0^3\right)}$, $\alpha = c_D/\left(c_D\right)_0$, $\beta = c_k/\left(c_k\right)_0$, c_N, c_D, c_k are coefficients of hydrodynamic force, liquid damping, and structural damping, respectively. $\left(c_D\right)_0$ and $\left(c_k\right)_0$ are coefficients corresponding to zero flow velocity. D is the pendulum diameter.

Note that equation (4.13) includes singularity when $\varphi = 0$. In view of this singularity together with the sudden change of pendulum velocity at the instant of impact, Peterka [814] used a special algorithm for accurate numerical simulation. The dependence of the critical liquid flow velocity at which the pendulum experiences self oscillations on the viscous hydrodynamic force coefficient, c_N, is shown in Fig. 4.6 for different values of liquid virtual mass coefficient, c_M.

In the presence of a circular barrier of radius R, equations (4.12) and (4.13) must be augmented with the impact law and braking coefficient, respectively

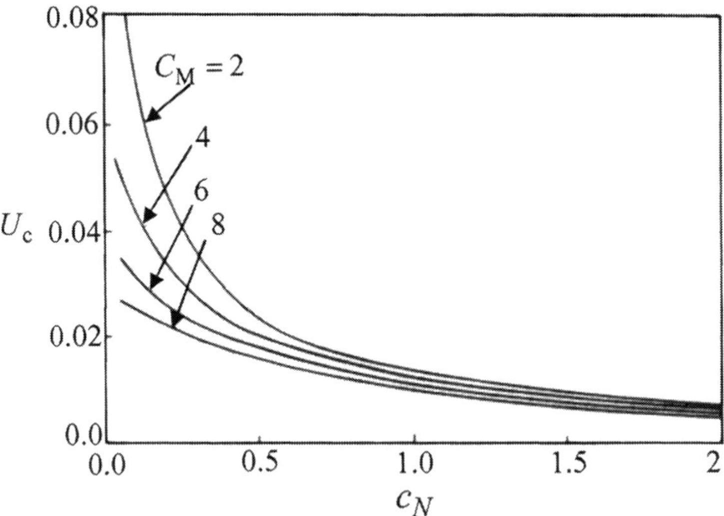

Fig. 4.6. Dependence of the critical flow velocity on the hydrodynamic force coefficient for different values of the coefficient of the virtual liquid mass, $\ell = 1.54m$, $\ell_0 = 0.4m$, $R = 4.5mm$, $\rho_f = 1{,}000kg/m^3$, $\rho = 11{,}800kg/m^3$, $(c_D)_0 = 0.127kg/m^2$, $(c_k)_0 = 0.2kg/sm^3$, [814].

$$\varphi'_{+} = -e\varphi'_{-}, \qquad and\ \theta'_{+} = (1 - B)\,\theta'_{-} \tag{4.14}$$

where e is the coefficient of restitution of the normal part of the impact, and B is known as the "braking" coefficient of the circumferential part of the impact velocity (see Section 5.3 for more details). The numerical simulation of equations (4.12) and (4.13) subject to conditions (4.14) was found to exhibit self–excited oscillations of the pendulum with periodically stabilized impact regardless of the initial conditions of the gyratory motion. Fig. 4.7(a) shows the locus of the stabilized motion by solid curves. The dependence of the pendulum angular circumferential displacement, $\Delta\theta$, and pendulum velocity, φ'_{-}, on the braking coefficient, B, is shown in Fig. 4.7(b). It is seen that for low values of B, the values of $\Delta\theta$ and φ'_{-} reach their low values and the gyratory motion approaches that of the pendulum sliding around the channel wall. As B increases both $\Delta\theta$ and φ'_{-} increase until B reaches its critical value, B_c, at which the gyratory self–excited oscillation turns into a

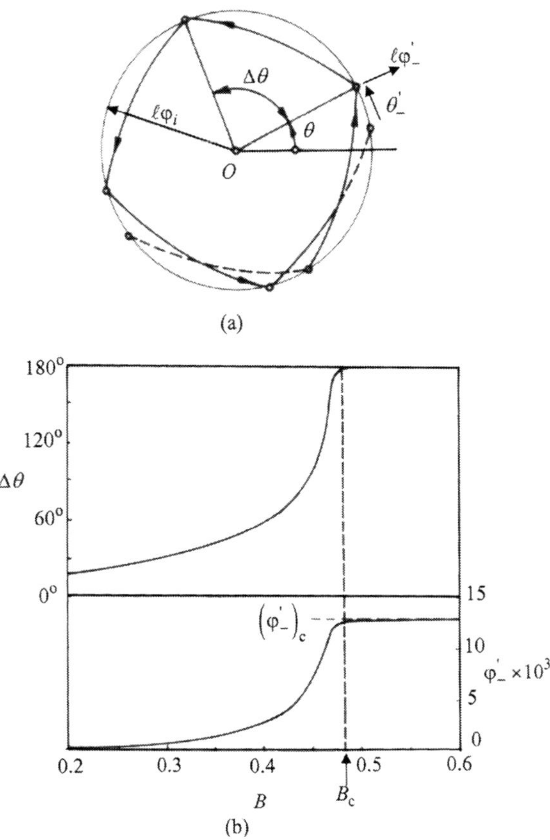

Fig. 4.7. Periodic impact of self-excited oscillation (a) stabilized motion shown by solid paths, (b) Dependence of the angular difference $\Delta\theta$ and the velocity before impact φ'_{-} on the braking coefficient B, [814].

dangerous polarized oscillation along the channel diameter $\Delta\theta = 180^o$ with intensive impacts of the pendulum against the opposite sides of the channel wall. The polarized oscillations become stabilized regardless of the value of the braking coefficient, B, when the initial conditions do not provide the pendulum rotational motion, i.e., $\theta^{'}(0) = 0$.

The bi–planar vibration of the beam experiencing oblique impacts was described in two perpendicular planes which intersect in the beam longitudinal axis ([815], [816]). Three types of oblique impacts were assumed. These are: 1) impact on the plane stop, 2) impact in a circular hole eccentrically placed with respect to the beam axis, and 3) impacts between neighboring beams in the bundle. The velocity component in the direction normal to the contact surfaces of impacting masses is changed by means of the restitution coefficient. The tangential component of the velocity is changed using the braking coefficient which depends on the dry friction coefficient, the restitution coefficient and on the incidence angle.

4.4 Pendulum Simulating Liquid Sloshing Impact

An impulsive acceleration to a liquid container can result in impact hydrodynamic pressure of the free liquid surface on the tank walls. It can also occur during maneuvering or docking of spacecraft in an essentially low gravity field. When hydraulic jumps or traveling waves are present extremely high impact pressures can occur on the tank walls in gasoline tankers and ship cargo tanks [192]. Typical pressure traces recorded under this sloshing condition were reported by Cox et al [192]. Rumyantsev [899] analyzed the collision of a body containing a viscous liquid with a rigid barrier. Liquid sloshing impact can be more severe longitudinally than laterally if no transverse baffles are introduced. The longitudinal acceleration peaks are larger than the lateral ones. The liquid impact is probably more severe to the structure for longitudinal than for lateral sloshing. Ye and Birk [1102] measured the fluid pressure in horizontal partially filled cylindrical tanks with different length to diameter ratios when suddenly accelerated by impact along the longitudinal axis. The peak pressure on the end of the tank was strongly affected by the fill level and the tank length–diameter ratio.

The hydrodynamic pressure distribution of impact loads is an important factor in studying the integrity of storage tanks and related safety problems. Milgram [680] experimentally studied the sloshing impact pressure in roofed liquid tanks. The sloshing roof impact problem was studied by Kurihara, et al [559]. Minowa et al [682] conducted a series of shaking table tests of a rectangular tank to measure roof impact pressures, natural frequencies and modes of bulging vibrations. Their measured results showed that the roof impact pressures possess great potential damage to tank as the pressure reached as high as 30 *psi* under 400 *gal*[1] El–Centro seismic excitation. An improved

[1] The *gal*, or *galileo*, is a unit of acceleration, centimeter per second squared used extensively in the science of gravitational field.

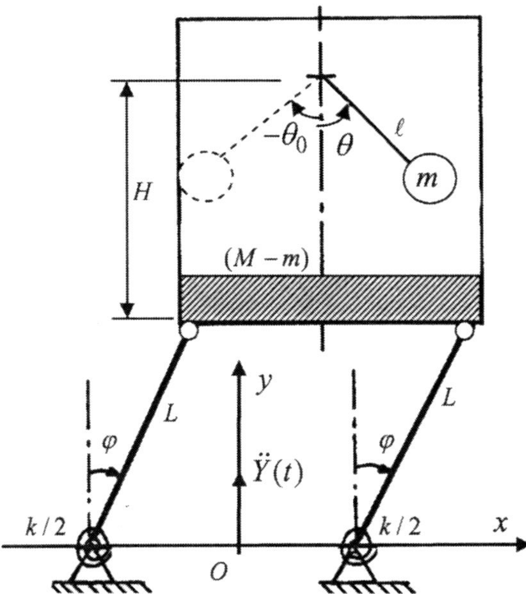

Fig. 4.8. Schematic diagram of a liquid sloshing impact represented by a simple pendulum inter-acting with its support structure represented by inverted pendulum.

numerical algorithm, which permits repeated liquid impacts, was developed by Su and Kang ([965], [966], [967]). Later, Su and Wang ([968], [969]) developed a three–dimensional finite difference scheme for large liquid amplitude motion in rectangular and cylindrical containers subjected to vertical and arbitrary excitations. Minowa [681] studied the large sloshing amplitude under impulsive earthquake inputs. He observed that the high sloshing waves were produced in the vicinity of surface centers.

The equivalent mechanical model of liquid impact loading is a pendulum describing impacts with the tank walls. Pilipchuk and Ibrahim [852] introduced this modeling into the equations of motion of a nonlinear system simulating liquid sloshing impact in tanks supported by an elastic structure. The system comprised of a liquid container supported by four massless rods of length L, which are restrained by four torsional springs of stiffness k at the base as shown in Fig. 4.8. The base is subjected to vertical base acceleration $\ddot{Y}(t)$. The total mass of the container including liquid is M and the equivalent sloshing mass of the first asymmetric mode of the liquid is m. The fluid free surface is modeled as a pendulum of length ℓ. The equivalent pendulum parameters for different types of container geometry are documented in Ibrahim [436]. The pendulum can reach the walls of the tank if its angle with the vertical axis is $\theta = \pm\theta_0$. Following the idea of non–smooth coordinate transformation due to Zhuravlev [1137], Pilipchuk and Ibrahim [854] introduced the new coordinate $x = x(t)$ through the coordinate transformation

$$\theta = \theta_0 S(x), \tag{4.15}$$

where $S(x)$ is a periodic saw–tooth piecewise linear function defined as

$$S(x) = \begin{cases} x & if \; -1 \le x \le 1 \\ 2-x & if \; 1 \le x \le 3 \end{cases} \qquad S(x+4n) = S(x), \quad n = 1, 2, ... \tag{4.16}$$

The system kinetic energy may be written in the form

$$KE = \frac{1}{2}\left\{ (M-m)\, L^2 \dot{\varphi}^2 + m \left[L^2 \dot{\varphi}^2 + 2\ell L \dot{\varphi}\dot{\theta} \cos(\varphi + \theta) + \ell^2 \dot{\theta}^2 \right] \right\} \tag{4.17}$$

The transformation (4.15) produces singular terms ($\delta-$ functions) in the equations of motion due to the product $\dot{\varphi}\dot{\theta}$. Alternatively, the following transformation may be introduced

$$\begin{Bmatrix} \theta \\ \varphi \end{Bmatrix} = \theta_0 \begin{bmatrix} 1 & 0 \\ a & b \end{bmatrix} \begin{Bmatrix} S(x) \\ y \end{Bmatrix}. \tag{4.18}$$

Introducing transformation (4.18) into equation (4.17) gives

$$KE = \left[\left(\frac{ML^2}{2} a^2 + m\ell La + \frac{m\ell^2}{2} \right) \dot{x}^2 + (MLa + m\ell)\, bLS'(x) \dot{x}\dot{y} + \frac{ML^2}{2} b \dot{y}^2 \right] \theta_0^2, \tag{4.19}$$

where the equality $\left[S'(x) \right]^2 = 1$ has been used and a prime denotes differentiation with respect to x. The constants a and b are selected such that the product $S'(x)\dot{x}\dot{y}$ is eliminated, i.e.,

$$a = -\frac{m\ell}{ML}, \quad \text{and } b^2 = \left(1 - \frac{m}{M}\right) \frac{m\ell^2}{ML^2}. \tag{4.20}$$

The following symmetric form of the kinetic energy is obtained, i.e.,

$$KE = \frac{1}{2}\left(1 - \frac{m}{M}\right) m\ell^2 \theta_0^2 \left(\dot{x}^2 + \dot{y}^2 \right). \tag{4.21}$$

The potential energy is

$$\Pi = \frac{1}{2}\left[(k - MgL)\, \varphi^2 + mg\ell\theta^2 \right]. \tag{4.22}$$

In terms of the non–smooth coordinates, equation (4.22) takes the form

$$\Pi = \frac{1}{2}\left\{ (k - MgL) \left[aS(x) + by \right]^2 + mg\ell S^2(x) \right\} \theta_0^2. \tag{4.23}$$

The equations of motion in terms of the non–smooth coordinates are

$$\begin{aligned}&\ell^2 m\left(1-\frac{m}{M}\right)\ddot{x}+\left(a^2 c_1+c_2\right)\dot{x}+\left\{\left(g-\ddot{Y}\right)\ell m\right.\\&\left.+a^2\left[k-\left(g-\ddot{Y}\right)LM\right]\right\}S(x)S^{'}(x)\\&=-ab\left\{\left[k-\left(g-\ddot{Y}\right)LM\right]y+c_1\dot{y}\right\}S^{'}(x),\end{aligned} \tag{4.24}$$

$$\begin{aligned}&\ell^2 m\left(1-\frac{m}{M}\right)\ddot{y}+b^2 c_1\dot{y}+b^2\left[k-\left(g-\ddot{Y}\right)LM\right]y\\&=-abc_1 S^{'}(x)-ab\left[k-\left(g-\ddot{Y}\right)LM\right]S(x),\end{aligned} \tag{4.25}$$

where c_1 and c_2 are the damping coefficients associated with the φ and θ oscillations, respectively. These equations may be written in the non–dimensional form

$$\begin{aligned}&\frac{d^2x}{d\tau^2}+\zeta_2\frac{dx}{d\tau}+\frac{1}{(1-\mu)}\left[\left(1+\mu\nu^2\right)-(1-\lambda\mu)A_Y(\tau)\right]S(x)S^{'}(x)\\&=\sqrt{\frac{\mu}{1-\mu}}\left[\left[\nu^2+\lambda A_Y(\tau)\right]y+\zeta_1\frac{dy}{d\tau}\right]S^{'}(x),\end{aligned} \tag{4.26}$$

$$\begin{aligned}&\frac{d^2y}{d\tau^2}+\zeta_1\frac{dy}{d\tau}+\left(\nu^2+\lambda A_Y(\tau)\right)y\\&=\sqrt{\frac{\mu}{1-\mu}}\left\{\frac{A_Y(\tau)}{\mu}+\zeta_1\frac{dx}{d\tau}S^{'}(x)+\left[\nu^2+\lambda A_Y(\tau)\right]S(x)\right\},\end{aligned} \tag{4.27}$$

where $\mu=m/M$, $\lambda=\ell/L$, $\tau=t\sqrt{g/\ell}$, $A_Y(\tau)=\ddot{Y}(t)/g\theta_0$, $\nu=\omega_L/\omega_\ell$, $\omega_L=\sqrt{\frac{k-MgL}{ML^2}}$, $\omega_\ell=\sqrt{g/\ell}$, $\zeta_1=\frac{c_1}{ML^2\omega_\ell}$, and $\zeta_2=\frac{\mu\zeta_1}{1-\mu}\left(1+\frac{c_2}{c_1\lambda^2\mu^2}\right)$.

In the absence of damping equations (4.26) and (4.27) take the form

$$\frac{d^2x}{d\tau^2}+\varpi^2 S(x)S^{'}(x)=\left\{a_1 y+\left[a_2 S(x)+a_3 y\right]A_Y(\tau)\right\}S^{'}(x), \tag{4.28}$$

$$\frac{d^2y}{d\tau^2}+\nu^2 y=a_1 S(x)+\left[a_3 S(x)-\lambda y\right]A_Y(\tau), \tag{4.29}$$

where $\varpi^2=\left(1+\mu\nu^2\right)/(1-\mu)$, $a_1=\nu^2\sqrt{\mu/(1-\mu)}$, $a_2=(1-\lambda\mu)/(1-\mu)$, $a_3=\lambda\sqrt{\mu/(1-\lambda\mu)}$.

For small amplitude oscillations, i.e., when $|x|\leq 1$ one has $S(x)=x$ and $S^{'}(x)=1$ and equations (4.28) and (4.29) are linearly coupled oscillators under parametric excitation. Above the critical value $|x| = 1$, the identities $S(x) = x$ and $S^{'}(x) = 1$ do not hold and the system becomes essentially nonlinear.

4.4.1 Unperturbed System Dynamics

In the absence of parametric excitation the undamped system oscillations depend on the initial conditions $x(0)$, $y(0)$ and its dynamics is governed by the autonomous equations

$$\frac{d^2x}{d\tau^2} + \frac{\partial \Pi(x,y)}{\partial x} = 0, \tag{4.30}$$

$$\frac{d^2y}{d\tau^2} + \frac{\partial \Pi(x,y)}{\partial y} = 0, \tag{4.31}$$

where the potential energy $\Pi(x,y)$ is

$$\Pi(x,y) = \frac{\mu\nu^2}{2(1-\mu)}\left[S(x) - \sqrt{\frac{1-\mu}{\mu}}y\right]^2 + \frac{S^2(x)}{2(1-\mu)}. \tag{4.32}$$

Equations (4.30) and (4.31) admit the energy integral

$$E = \Pi(x,y) + \frac{1}{2}\left[\left(\frac{dx}{d\tau}\right)^2 + \left(\frac{dy}{d\tau}\right)^2\right]. \tag{4.33}$$

Equation (4.33) reveals that the system cannot leave the region $D = [(x,y):\ \Pi(x,y) \leq E]$ bounded by the curves

$$\Pi(x,y) = E = const. \tag{4.34}$$

For different levels of initial energy (initial conditions) the system may experience non–impact periodic motion, grazing impact, and impact motion. Fig. 4.9 shows a family of curves for different levels of initial energy, E. For sufficiently small energy, the family consists of periodic set of separated ellipses in the configuration plane (x,y). If the energy is localized initially in only the in–phase mode the system will remain inside the cell where the motion started at $t = 0$. The onset of windows is provided by the system's critical energy level, E_c, which is sufficient to reach the barrier with zero velocity, at which grazing impact occurs. The critical energy corresponding to the grazing impact can be determined from the energy integral (4.33). When the system reaches the barriers, one has $S(x) = \pm 1$, and the minimum total energy is given by equation (4.33) when the kinetic energy is zero and $y = \pm\sqrt{\mu/(1-\mu)}$. Substituting this value of y, one obtains the critical energy

$$E_c = \frac{1}{2(1-\mu)}. \tag{4.35}$$

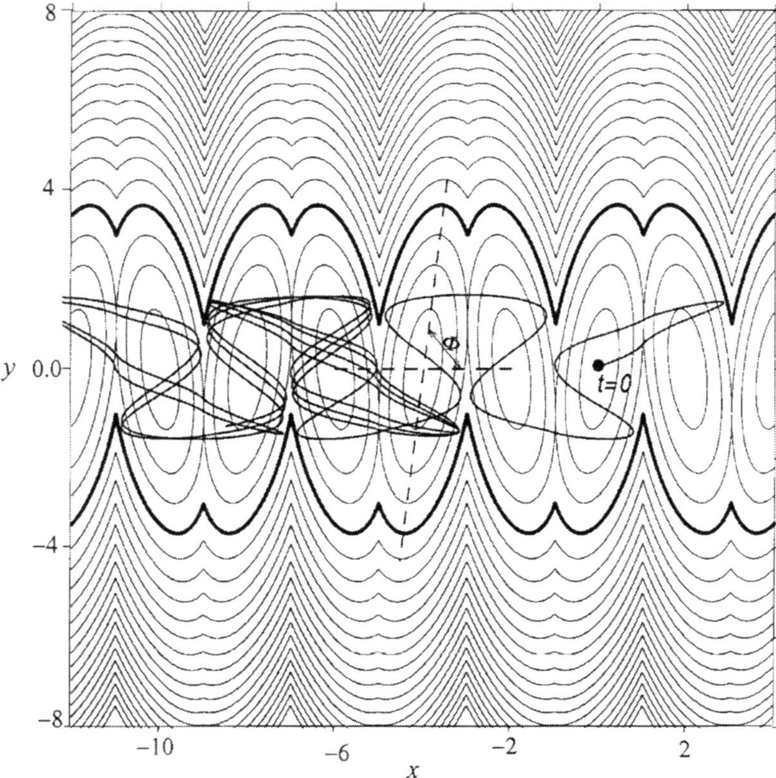

Fig. 4.9. Contours of constant energy levels $E = const$ and trajectory of the model on the configuration plane. Closed orbits belong to non-impact periodic motion, open curves belong to impact motion, grazing motion is identified by contacted ellipses, [854].

Above this critical energy level, the barriers become reachable at non–zero velocity. In this case, impact occurs and penetration through a potential window geometrically becomes possible. However, the condition $E > E_c$ is not sufficient and does not guarantee impact. The sufficient condition depends on both the total energy and its distribution among the modes as well. For example, Fig. 4.9 reveals the linear in–phase mode associated with the major axis (shown by the inclined dashed straight line) of the truncated ellipses does not cross the potential window. It means that the system will never reach the windows if the initial energy is localized in the in–phase mode.

Fig. 4.9 demonstrates essential features that depend on a combination of the system parameters. For example, the potential curves of the linear system with no barriers are "*structurally stable*", i.e., they possess the same geometrical properties for any energy level. In the presence of barriers the family of potential curves experience symmetry breaking as indicated by the family of incomplete ellipses. These curves intersect each other at cusps which are aligned along a vertical line at a particular value of x. Thus, it is important to select some essential parameters to characterize the geometrical properties of the non–homogeneous potential field. An analysis of Fig. 4.9 shows that such a natural parameter can be associated with inclination of the ellipses to the x–axis. This inclination is defined by the angle between the major axis of the ellipse and the x–axis

$$\tan \Phi = \frac{2\nu^2 \sqrt{\mu - \mu^2}}{\sqrt{(1-\nu^2)^2 + 4\mu\nu^2} - 1 + \nu^2(1-2\mu)}. \tag{4.36}$$

The angle Φ indicates the directions of the normal modes to the barriers and thus it represents one of the two possible parameters by which different systems of the class considered can be distinguished.

4.4.2 Perturbed System Dynamics

In the linear domain of the system dynamics, one has to set $S(x) = x$, and $S^{'}(x) = 1$ in equations (4.28) and (4.29), which may be written in the compact matrix form

$$\frac{d^2\mathbf{u}}{d\tau^2} + \mathbf{K}\mathbf{u} = \mathbf{P}\mathbf{u}A_Y(\tau), \tag{4.37}$$

where $\mathbf{u} = \begin{Bmatrix} x \\ y \end{Bmatrix}$, $\mathbf{K} = \begin{bmatrix} \varpi^2 & -a_1 \\ -a_1 & -\lambda \end{bmatrix}$, and $\mathbf{P} = \begin{bmatrix} a_2 & a_3 \\ a_3 & -\lambda \end{bmatrix}$.

By setting the right–hand side of equation (4.37) to zero, one can study the linear normal mode natural frequencies. The corresponding normalized natural frequencies and modal fractions are, respectively,

$$\omega_{1,2}^2 = \frac{1}{2}\left[\left(\nu^2 + \varpi^2\right) \mp \sqrt{\left(\nu^2 - \varpi^2\right)^2 + 4a_1^2}\right], \tag{4.38}$$

and

$$\{\phi_{1,2}\} = \frac{1}{\sqrt{\left(\nu^2 - \omega_{1,2}^2\right)^2 + a_1^2}} \begin{Bmatrix} \nu^2 - \omega_{1,2}^2 \\ a_1 \end{Bmatrix},$$

$$with\ \phi_k^T \phi_n = \begin{cases} 1\ if\ k = n \\ 0\ if\ k \neq n \end{cases}. \tag{4.39}$$

The first modal vector corresponds to the lowest natural frequency ω_1 and is directed along the major axes of the ellipses shown in Fig. 4.9. The direction can be defined by the angle Φ, i.e., $\tan \Phi = a_1 / \left(\nu^2 - \omega_1^2\right)$ which is equivalent to equation (4.36). Introducing the transformation to the principal coordinates

$$\begin{Bmatrix} x \\ y \end{Bmatrix} = \begin{bmatrix} \phi_1 \ \phi_2 \end{bmatrix} \begin{Bmatrix} q_1 \\ q_2 \end{Bmatrix}. \tag{4.40}$$

the equation of motion may be written in the form

$$\frac{d^2 q_1}{d\tau^2} + \omega_1^2 q_1 = (p_{11} q_1 + p_{12} q_2)\, A_Y(\tau), \tag{4.41}$$

$$\frac{d^2 q_2}{d\tau^2} + \omega_2^2 q_2 = (p_{21} q_1 + p_{22} q_2)\, A_Y(\tau), \tag{4.42}$$

where $p_{kn} = \{\phi_k\}^T \mathbf{P} \{\phi_n\}$.

Under harmonic parametric excitation, $A_Y(\tau) = F \cos \Omega\tau$. One can determine the response characteristics under three different parametric resonance conditions. These are 1) first mode excitation: $\Omega = 2\omega_1$, 2) second mode excitation: $\Omega = 2\omega_2$, and 3) mixed mode excitation $\Omega = \omega_1 + \omega_2$. The response characteristics of these three cases were determined using the method of averaging and the temporal evolution of the system energy was estimated for each case.

For the case of first mode parametric excitation in the neighborhood of the resonance condition $\Omega = 2\omega_1$, Figs. 4.10(a), and 4.10(b) show qualitatively different energy time history records of the in–phase mode for undamped and damped cases, respectively. It is seen that for the undamped case, the energy response tends to grow with time as shown in Fig. 4.10(a) whereas for the damped case the energy becomes bounded. The suppression of the energy growing is due to a cooperative role of the damping and impacts. Over the time period $0 \leq \tau \leq 2500$ the system behavior is governed by parametric resonance as shown in Fig. 4.10(b). Above this period impact occurs and the system energy behaves randomly with bounded levels.

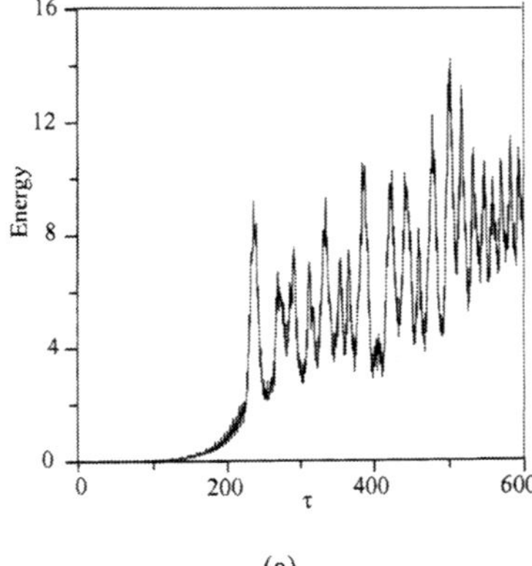

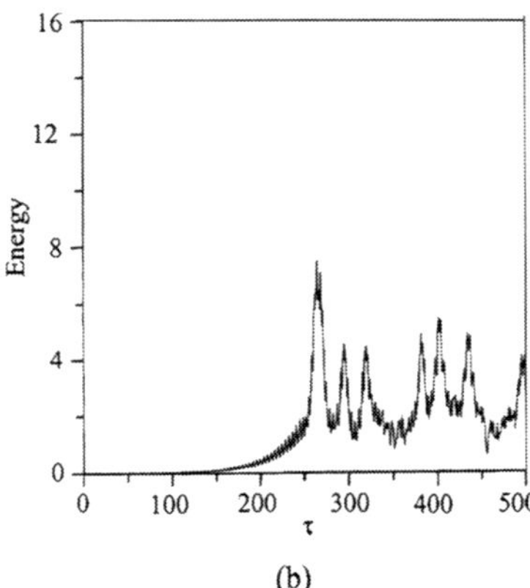

Fig. 4.10. System response energy time history records under first mode parametric excitation for $\lambda = 0.5$, $\mu = 0.5$, $\nu = 1.0$, $F = 0.7$, $\Phi = 1.01722$: (a) $\zeta_1 = \zeta_2 = 0.0$, (b) $\zeta_1 = \zeta_2 = 0.01$, [854].

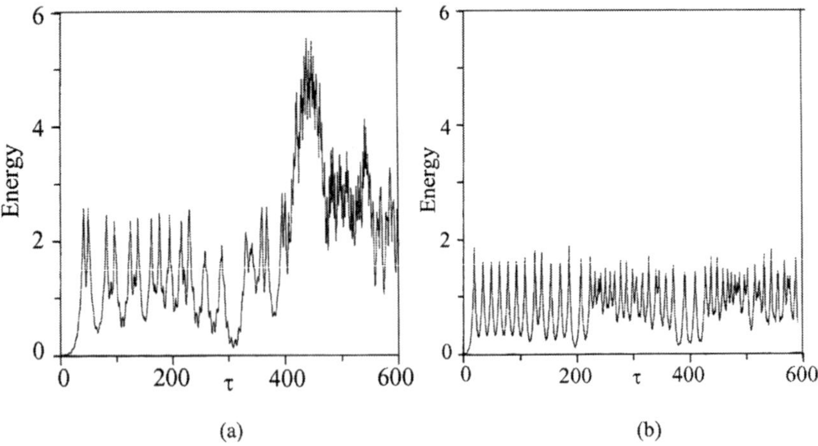

Fig. 4.11. System response energy time history records under second mode parametric excitation for $\lambda = 0.5$, $F = 0.7$, $\zeta_1 = \zeta_2 = 0.0$: (a) $\mu = 0.5$, $\nu = 0.5$, $\Phi = 1.01722$, (b) $\mu = 0.5$, $\nu = 1.0$, $\Phi = 1.33897$, [854].

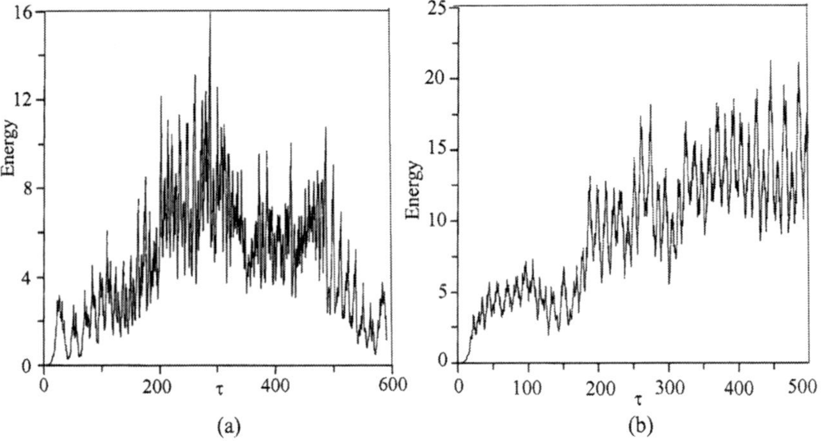

Fig. 4.12. System response energy time history records under mixed mode parametric excitation for $\lambda = 0.5$, $F = 0.7$, $\zeta_1 = \zeta_2 = 0.0$: (a) $\mu = 0.5$, $\nu = 0.5$, (b) $\mu = 0.5$, $\nu = 1.0$, [854].

For the case of second mode parametric excitation in the neighborhood of the resonance condition $\Omega = 2\omega_2$, the scenario is quite opposite. As the angle Φ increases the impacts become more beneficial. Figs. 4.11(a) and 4.11(b) show two time history records of the system response energy for two values of Φ. It is seen that the energy level decreases as the angle Φ increases. The damping is set to zero to demonstrate the role of impacts in suppressing the influence of parametric resonance.

For the case of mixed mode excitation, in the neighborhood of combination parametric resonance, $\Omega = \omega_1 + \omega_2$, Figs. 4.12(a) and 4.12(b) show two different response energy time history records for two different values of frequency parameters, $\nu = 0.5$ and 1.0, respectively. Comparing Figs. 4.10(b) and 4.11(a), obtained for the same parameters, reveals that the system energy response is greater for the case of combination resonance than for the case of the principal resonance. The same comparison is also observed in Figs. 4.11(a) and 4.12(b). Fig. 4.12(b) reveals that the energy is growing with time with random fluctuations. To this end, these results show that the system response under combination parametric resonance involves combined features of both in–phase and out–of–phase modes.

4.4.3 Influence of Internal Resonance

One can phenomenologically describe the interaction between the pendulum and the tank walls with a power function as $F_{impact} = b(\theta/\theta_0)^{2n-1}$, where $n \gg 1$ is an integer and is a positive constant parameter usually measured experimentally. The localized dissipative force may be approximated by the expression $F_d = d(\theta/\theta_0)^{2p}\dot{\theta}$, where d is a constant, p is a positive integer (generally $p \neq n$), and a dot denotes differentiation with respect to time t. With reference to Fig. 4.8, the equations of motion of this system are

$$\ddot{\theta} + \frac{g}{\ell}\theta - \frac{L}{\ell}\ddot{\varphi} = \frac{L}{2\ell}\ddot{\varphi}(\theta+\varphi)^2 + \frac{L}{\ell}\dot{\varphi}^2(\theta+\varphi) + \frac{\ddot{Y}(t)}{\ell}\theta + \frac{g}{6\ell}\theta^3 - \frac{d}{m\ell^2}\left(\frac{\theta}{\theta_0}\right)^{2p}\dot{\theta} - \frac{b}{m\ell^2}\left(\frac{\theta}{\theta_0}\right)^{2n-1}, \tag{4.43}$$

$$\ddot{\varphi} + \left(\frac{k}{ML^2} - \frac{g}{\ell}\right)\varphi + \frac{m\ell}{ML}\ddot{\theta} = \frac{m\ell}{2ML}\ddot{\theta}(\theta+\varphi)^2 - \frac{g}{6\ell}\varphi^3 + \frac{m\ell}{ML}\dot{\theta}^2(\theta+\varphi) - \frac{\ddot{Y}(t)}{L}\varphi, \tag{4.44}$$

where the approximations $\sin\theta \approx \theta - \theta^3/3!$ and $\cos\theta \approx 1 - \theta^2/2$ have been introduced. In addition to the parametric resonance conditions there arises also an internal resonance condition of fourth–order, i.e., $\omega_3 = 3\omega_1$.

Pilipchuk and Ibrahim [852] employed the saw–tooth time transformation to describe the in–phase and out–of phase nonlinear periodic regimes. Based on explicit forms of analytical solutions, all basic characteristics of nonlinear free and forced response regimes were estimated. It was found that a high frequency out–of–phase nonlinear mode takes place with relatively small tank amplitude and is more stable than the in–phase oscillation mode under small perturbations. The in–phase mode was found to possess relatively large tank amplitudes and does not preserve its symmetry under periodic parametric excitation.

The same system was later examined using the multiple scale method by El–Sayad et al [289] in the neighborhood of the three parametric resonance conditions. When the first normal mode was parametrically excited the system exhibits hard nonlinear behavior and the impact loading reduced the response amplitude. On the other hand, when the second mode was parametrically excited, the impact loading results in complex response behavior characterized by multiple steady–state solutions where the response switches from soft to hard nonlinear characteristics. Under combined parametric resonance the system behaves like a soft system in the absence of impact and as a hard system in the presence of impact.

Under simultaneous parametric and internal resonance conditions the system response was studied using the multiple scales method by Ibrahim and El–Sayad [439] and by applying the Lie group transformations by Pilipchuk and Ibrahim [853]. Both studies predicted the same system response characteristics. For example, under first– and mixed–mode parametric excitation, the normal modes interact through internal resonance. Depending on initial conditions and internal detuning parameter, the response can be quasi–periodic or chaotic with irregular jumps between two unstable equilibria. In the presence of impact forces, the system preserves fixed response amplitude response within a small range of internal detuning parameter. Beyond that range, the response exhibits quasi–periodic motion mainly governed by initial conditions, internal detuning parameter, damping ratios and excitation level. Under second mode parametric excitation the second mode reaches fixed response amplitude, depending on initial conditions, with no energy sharing with the first mode. However, the phase angles were found to vary with time. Under combination parametric resonance, and in the absence of impact forces, the response was found to be sensitive to initial conditions.

4.5 Multi–Degree–of–Freedom Systems

The analysis of three and multi-degree-of-freedom systems in the presence of constraints is more involved than the single and two-degree-of-freedom systems. However a number of attempts have been reported in the literature (see, e.g., [195], [196], [370], [1010], [1011], [870], [1049], [431], [1047]). Gontier and Toulemonde [371] employed a continuation method and showed that the dynamics of multiple impact periodic responses was manifested in a typical cascade of subharmonic bifurcation pattern. Zones of chaotic motion and coexisting attractors were identified. Natsiavas [718] determined the exact steady state response for a class of strongly nonlinear multi–degree–of–freedom oscillators involving colliding components. These oscillators were represented in the form of an arbitrary number of degrees of freedom and configuration, incorporating a component with a geometric nonlinearity. The analysis defined the location of harmonic and subharmonic responses characterized by one abrupt change of the nonlinear element parameters per response cycle. Performance of vibration absorbers with elastic stops was also considered. Furthermore, examples of oscillators exhibiting 2 : 1 and 3 : 1 internal resonance conditions were treated. The response characteristics were compared with those of systems with continuous nonlinearities.

Fredriksson and Nordmark [342] considered a class of impact oscillators with several degrees of freedom described by non–smooth nonlinear equations of motion. The impact is due to the motion of one body, constrained by a motion limiter. The velocities of the system were assumed to change instantaneously at impact. By defining a discontinuity mapping, it was shown how Poincaré mapping can be obtained as an expansion in a local coordinate. This gives the mapping the desired form, thus making it possible to employ standard techniques.

Luo et al [613] considered a multi–degree–of–freedom system in which the maximum displacement of one of the masses is limited to a threshold value by the symmetrical rigid stops. Double Neimark–Sacker bifurcation of the system was analyzed using the center manifold and normal form method of maps. The period–one double–impact symmetrical motion and disturbed map of the system were derived analytically. A center manifold theorem technique was applied to reduce the Poincaré map to a four–dimensional one together with the normal form map associated with double Neimark–Sacker bifurcation. For the case of a three–degree–of–freedom system with symmetrical stops, the existence and stability of period–one double–impact symmetrical motion were obtained. Near the value of double Neimark–Sacker bifurcation it was found that the system possesses period–one double–impact symmetrical motion and quasi–periodic impact motions. With change of system parameters, the quasi–periodic impact motions usually lead to chaos via "tire–like" torus doubling.

The rising phenomenon was considered for multi–modal systems by Wagg [1048] using energy balance techniques for an arbitrary contact interval the effects of modal vibration can be included. The energy balance was used to obtain a relationship between the coefficient of restitution and the modal energy during the contact period. This allows one to study the effects of impact-induced vibration.

4.6 Closing Remarks

The dynamic behavior of vibro–impact two– and mutli–degree–of–freedom lumped systems under free and forced excitation revealed new nonlinear phenomena not observed in single–degree–of–freedom systems. These phenomena include the occurrence of strong resonance of different orders, the so–called rising phenomena that occur in sticking solutions. An important advantage of these systems is their utilization as nonlinear vibration absorbers. Under parametric excitation, the absorbing phenomena are only restricted for the out–of–phase modes. The presence of internal resonance may add more complexity in the system behavior in which a steady state solution may not be achieved.

Chapter 5
Non–Classical Lumped Systems

5.1 Introduction

Traditional design tools and regulations are based on zero–clearance analysis and are inadequate to account for the influence of contact loss. The recent developments of the theory of vibro–impact dynamics form an excellent foundation to modify the design tools of mechanical and electromechanical systems with clearances, backlash and free–play nonlinearities. Examples of such systems include mechanical joints in space truss structures, free–play in aerospace control surfaces, rub–impact in rotating machinery, and impact micro–actuators. The recent developments of the dynamic characteristics of these systems will be assessed in this chapter. Another important application is the human vocal fold collision and its effect on the tissue damage. Some nonlinear two–dimensional models of vocal folds will be discussed. This chapter is closed by a brief review of other applications including cutting tools and machines, gear rattling, and multi–body systems.

5.2 Mechanical Joints

5.2.1 Overview

Mechanical joints exist in different forms and are essential in design and construction of mechanisms, multi–body systems, space truss structures, and aircraft wing–store coupling. Impact in joint gaps may cause a transfer of vibration energy from low–frequency global modes to low–frequency local modes. Contact loss in mechanisms' joints may give rise to system degrading impacts when contact is re–established. Pinned joints normally have a small amount of clearance between pins and clevis/tang fittings. The small gap can cause very significant changes in the dynamics and damping of truss structures, particularly when the joint can traverse the dead–band region.

Very small gaps in joints usually create nonlinearities that may lead to chaotic dynamics. Chaotic dynamics in space structures imposes some

R.A. Ibrahim: Vibro-Impact Dynamics: Model., Map. & Appl., LNACM 43, pp. 125–149.
springerlink.com

difficulties in the design of active control systems to damp out transient oscillations ([518], [689]). Moon and Li [689] studied the dynamics of a pin–jointed space truss structure. Their experimental results exhibited broad–band chaotic–like vibrations under sinusoidal excitation. When a tension cable was added to place the structure under compressive loads, the level of chaos was reduced. Dynamic contacts in an elastic joint were simulated by a nonlinear joint model comprised of a nonlinear spring and damper ([689], [413], [414]).

Dubowsky and Freudenstein ([267], [268]) formulated an analytical model of an elastic mechanical joint with clearances, encountered in mechanical and electromechanical systems. They estimated the dynamic force amplification, frequency response, time–displacement characteristics, and other dynamic characteristics. It was shown that the elastic impact–pair model exhibited a variety of dynamic characteristics. Heiman et al [400] considered the dynamics of an inclined impact pair consisting of an oscillatory base and a secondary mass constrained to move in an inclined slot with the base mass. The dynamics of the secondary mass for alternating impacts was formulated in terms of a map over one period of the base motion. Steady state $2:1$ motions, their stability and subsequent period doubling bifurcations were estimated.

Available analytical tools such as the massless link model are complex and computationally expensive. Earles and Seneviratne [282] studied the phenomenon of contact loss using the massless link model. The massless link basically models the joint clearance as an extra degree of freedom. They derived the equations of motion of a four–bar mechanism with a clearance joint. The equations were then decoupled to yield an expression for the clearance joint force magnitude in terms of the system kinematic variables and the clearance link response. It was shown that there exists a single zero–clearance dimensionless parameter which governs the contact loss condition at the clearance joint. Pereira and Nikravesh [804] presented a computational scheme for the analysis of mechanical systems that undergo intermittent motion. A canonical form of the equations of motion was derived with a minimal set of coordinates. These equations were used in a procedure for balancing the momenta of the system over the period of impact, calculating the jump in the body momentum, velocity discontinuities and rebounds. The work was extended to open and closed–loop mechanical systems where jumps in the constraints' momenta were identified.

Folkman et al [338] measured damping characteristics of a three–bay truss with pinned joints. Test results demonstrated that structures using pinned joints can have damping that is dependent on gravity. Damping rates were found to change by a factor of up to five due to variation of gravity induced loads. This variation could be very significant when characterizing space structures in a $1-g$ environment. As the joint gap opens and closes the resulting impact was found to transfer vibration energy from low–frequency global modes to high–frequency local modes. This was associated with an increase in the rate of energy dissipation through material damping. Berzi et al

[108] developed an identification approach for the nonlinear dissipative spring mass model of a coupling sleeve joint. Stiffness, friction and mass parameters were determined by minimizing the difference between simulated and measured responses of the joint. Flores et al [331] studied the influence of the joint clearance of spatial joints on the kinematics and dynamics of multi–body systems. An analytical approach was developed for the revolute joint in which the basic elements are the journal and bearing. Under certain working conditions these two mechanical elements collide with each other. A continuous contact force model was adopted to evaluate the contact–impact force.

5.2.2 Free–Play in Aerospace Structures

In aerospace structures, free–play nonlinearity effects have been the subject several studies ([566], [1084], [505], [1098], [9]). For example, Laurenson and Trn [566] investigated the flutter of a missile with control surfaces having free–play nonlinearity. At a particular flight speed, the amplitude of oscillation, caused by external excitation, was found to build up. Due to the presence of free–play nonlinearity in combination with increasing amplitude of oscillation, the effective stiffness of the system was found to increase and the motion becomes stable at some limited amplitude. Kim and Lee [505] found that responses involving limit cycle oscillation (LCO) and chaotic motion are highly influenced by the pitch–to–plunge frequency ratio in an airfoil with free–play nonlinearity. Experimental studies of a wing model with free–play nonlinearity in pitch showed the appearance of double LCO [1098]. Alighanbari [9] studied three–degree–of–freedom airfoil–aileron dynamics with free–play nonlinearity in the aileron hinge moment. Bifurcation analysis indicated various LCO solutions for velocities well below the linear flutter boundary. Depending on the initial conditions and air speed, quasi–periodic and chaotic oscillations were reported for the aileron motion.

In a series of papers, Bae and Lee [66] and Bae et al ([69], [68], [67]) considered the influence of structural nonlinearities, represented by free–play and bilinear, on various types of LCO and periodic motions. Zentner and Poirion [1122] developed a reduced order model based on mode synthesis where the degrees of freedom that might have contact are separated from the regular substructures. This formulation allows for a direct application of Moreau's theory for non–smooth dynamics of contact problems. Their approach was applied to a fluid–structure coupled system as given by an aircraft wing with free–play in the control surface connection.

A nonlinear gust response analysis of a typical airfoil section with control surface free–play was presented by Tang et al [984]. A two–degree–of–freedom wing section with structural free–play type nonlinearity in the pitching freedom was experimentally examined in subsonic wind tunnel by Marsden and Price [643]. The effect of the free–play on the aeroelastic response was studied and the flutter speed was found to decrease with increasing the free–play length. An experimental model with a wing–store model with and without free–play was designed by Tang and Dowell [982] for the study of flutter and LCOs. The wing was modeled as a simple plate of constant thickness. The store was modeled as a slender rigid body that is jointed with the wing through two support points. The fore support point was articulated to wing and the aft support point connected with the wing through a spring with a free–gap. This design allows the store to have motion only in pitch. In this arrangement the store nonlinearity is mainly due to the free–play. Tang and Dowell [983] experimentally and analytically examined the effect of free–play in a wing–store model on the wing response amplitude. It was found that the wing gust response amplitude increases as the free–play gap or the gust angle increases and are almost independent for the store pitch initial conditions.

5.3 Rub–Impact Dynamics of Rotors

5.3.1 Overview

In operating rotating machinery, excessive rubbing between rotating and stationary parts can lead to destructive instability of rotors. During rubbing, dynamic phenomena such as impact associated with friction, stiffening and coupling effects may occur. Occasional partial rubbing between rotor/stator systems forms a strong nonlinear system. Nonlinear rotor systems involving bearing clearances were extensively studied in the literature (see, e.g., [480], [117], [107], [723], [168], [169], [496], [497], [498], [100], [908], [1094], [175], [176], [177], [178], [179], [761], [713], [284], [285], [510]). These studies focused on estimating super and subharmonic responses using perturbation techniques, harmonic balance method together with a fast Fourier transform procedure. An assessment of early work on rub–related phenomena in rotating machinery was presented by Muszynska [701]. This serious problem was claimed to cause about 10% of all aircraft engine failures. The main cause of rubbing and impacting owes its origin to the existence of a small clearance between the rotor and the casing and the existence of mass imbalance of the rotor and its eccentricity.

The dry friction between the rotor and casing upon contact forms another factor. Muszynska et al ([702], [703]) conducted analytical and experimental investigations to examine the problem of rotor–to–stationary element rubbing in rotating machines simulating the space shuttle main engine high pressure fuel turbo–pump. They developed an analytical model of rotor/bearing/seal system under rub condition and numerically estimated the dynamic rotor responses. The estimated results were found to be in good agreement with experimental measurements.

Goldman and Muszynska ([366], [367]) presented some experimental results from a simulator of a rotating machine with clearances and impacting. Bifurcation diagrams were generated numerically and revealed stable and chaotic response boundaries. They established the existence of main and higher order resonances in structures with clearances and impacting under external periodic excitations. Transition from non–contact to contact states was found to result in variable stiffness and damping, impacting, and intermittent friction. For the case of rotator and stator contact, Goldman and Muszynska [367] developed a model to correlate the local radial and tangential effects with global behavior of the system. The results of numerical simulations of a simple rotor/stator system based on that model were presented in the form of bifurcation diagrams, rotor lateral vibration time–base waves, and orbits. The system was observed to exhibit an additional subharmonic regime of vibration due to the stiffness asymmetry. Later, Goldman and Muszynska [368] considered the lateral vibration response of a rotor, experiencing periodic contact with a non–rotating machine component. For the case of inelastic impact, the system was described by a piecewise step–changing stiffness. A special coordinate transformation was introduced to convert discontinuities to smooth continuous forces. This transformation enabled them to apply traditional methods such as the average technique to examine a variety of unbalance–related resonant regimes of rotor lateral motion.

Wei et al [1067] highlighted the main features of the rub interaction between rotor and stator of rotating machinery. The contact forces induced by rotor/stator rub were found to increase the system stiffness. The occurrence of impact to the system created severe transient and nonlinear characteristics. Furthermore, the friction force due to the relative motion between the rotor and stator in contact may cause the reverse whirl motion of the rotor and severe friction affecting the normal stress conditions and generates local thermal distortion of the rotor. Furthermore, the contact introduces coupling effect which feeds impact force back to the system and further complicates the operating conditions of rotating machinery. Wei et al [1067] proposed an analytical model of the rotor–stator rubbing by taking into account the feedback of the impact forces. The response characteristics of rubbing at different speed values were examined based on a simple rotor/stator system. Some experimental results were compared with those predicted numerically.

Chaotic motion was experimentally observed in a rubbing rotor system by Piccoli and Weber [840]. Chu and Zhang ([181], [182]) identified three routes to chaos following the occurrence of rub–impact as the rotating speed increases. These were period doubling bifurcation, grazing bifurcation and a sudden transition from periodic motion to chaos. Quasi–periodic motions were also reported. The stability of the rub solutions of a general model of a Jeffcott rotor system with rub–impact for micro–rotating machinery in power microelectromechanical systems (MEMS) was studied by Zhang and Meng [1125] and Zhang et al ([1126], [1127]). The effects of rotating speed, imbalance, damping coefficient, and friction coefficient on the micro–rotor system responses were examined. It was shown that the rub–impact in micro–rotor system with the scale effects in friction alternates among periodic, quasi–periodic and chaotic motions as the system parameters vary. Popprath and Ecker [864] examined the intermittent contact of a Jeffcott rotor model with a stator interacting with the rotor model through nonlinear contact forces. Banakh and Nikiforov [75] studied the vibro–impact interaction between a rotor and floating sealing ring in the presence of hydrodynamic forces in the clearance between the rotor and the ring as well as dry friction between the ring and casing. Han et al [389] showed that the rub–impact of a rotor system with a fixed limiter resulted in periodic, quasi–periodic, and complex motions.

Azeez and Vakakis [49] studied the vibro–impact dynamics of a rotor–dynamic system using the method of proper orthogonal decomposition to extract dominant coherent structures from the time–series data. Conditions for the existence of single rub–impacting period–n motions for a rotor system with rigid constraints were analytically derived by Li and Lu [589] and Lu et al [602]. The scalogram and the wavelet phase spectrum were introduced to characterize the dynamical behavior of rotor–stator systems with rub–impact by Peng et al [803]. Acoustic emission due to rub–impact between the rotor and the stator is attributed to the elastic strain in the rubbing location. Such acoustic emission was used to identify and diagnose the rub–impact fault by Yongyong et al [1113]. They analyzed the acoustic emission characteristics of the rub–impact in the rotor–bearing system experimentally. The acoustic emission and the vibration signals were compared and analyzed to demonstrate the superiority of the acoustic emission based method for the rub–impact identification. The wavelet scalogram was used to analyze the time–frequency features, propagation characteristic and frequency dispersion characteristic of the rub–impact acoustic emission. The results revealed that the rub–impact acoustic emission is due to the multi–modal elastic wave, which mainly includes the flexural wave and extensional wave.

5.3.2 Case Study

Li and Païdoussis [586] studied the rub–impact problem in a rotor–casing system modeled by two degrees of freedom. The abrupt change in the translational velocity of the rotor was modeled by introducing a coefficient of restitution, e, and dry friction coefficient, μ. With reference to Fig. 5.1 One can decompose each velocity vector in terms of the normal (radial, $\mathbf{e}_r$) and tangential (transversal, $\mathbf{e}_\theta$) components, i.e.,

$$\mathbf{V} = \mathbf{V}^n \mathbf{e}_r + \mathbf{V}^t \mathbf{e}_\theta \tag{5.1}$$

The normal velocities immediately before, $\mathbf{V}_-^n$, and after, $\mathbf{V}_+^n$, impact are related by the impact law,

$$\mathbf{V}_+^n = -e\mathbf{V}_-^n \tag{5.2}$$

The tangential velocity after impact was given in terms of the braking coefficient, B, by applying the principle of linear impulse to the rotor in both normal and tangential directions, i.e.,

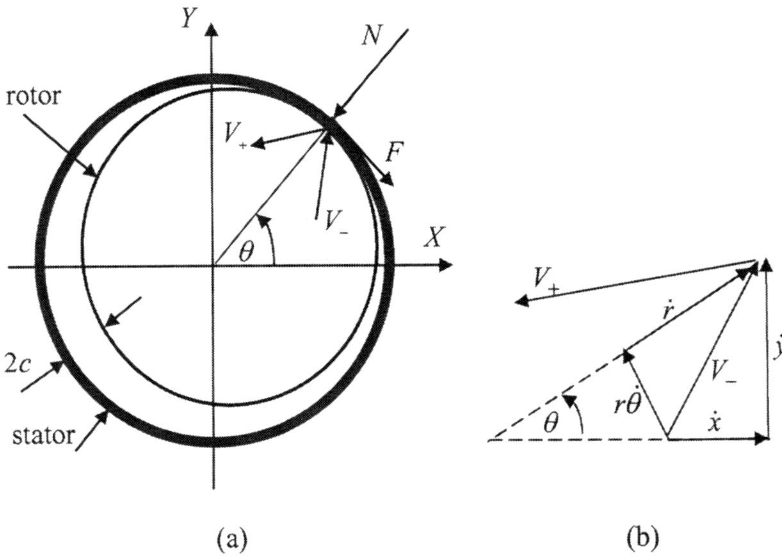

Fig. 5.1. The model and velocity components of the rotor, (a) velocity vectors before and after impacts showing the normal and tangential forces, (b) velocity components in polar and rectangular coordinates.

$$\int_{-}^{+} N dt = m\left(\mathbf{V}_{-}^{n} - \mathbf{V}_{+}^{n}\right), \tag{5.3a}$$

$$\int_{-}^{+} F dt = m\left(\mathbf{V}_{-}^{t} - \mathbf{V}_{+}^{t}\right). \tag{5.3b}$$

Introducing the definition of the Coulomb friction law, i.e., $F = \mu N$, where F and N are friction and normal forces of the rotor at the instant of impact, one can establish the tangential velocity after impact, i.e.,

$$\mathbf{V}_{+}^{t} = (1 - B)\mathbf{V}_{-}^{t}, \tag{5.4}$$

where $B = \mu(1+e)\mathbf{V}_{-}^{n}/\mathbf{V}_{-}^{t}$ is the brake coefficient. If $B \geq 1$, there is an inversion of $\mathbf{V}_{-}^{t}$ upon impact. The expression of B reveals that it can possess both positive and negative values. For impacting to occur, $\mathbf{V}_{-}^{n}$ must be positive, and thus $\mathbf{V}_{-}^{t}$ can be either positive (forward whirl) or negative (backward whirl). If $\mathbf{V}_{-}^{t} > 0$ the direction of the precession velocity of the whirl motion is the same as that of the rotor. On the other hand, if $\mathbf{V}_{-}^{t} < 0$, the rotor center motion is accelerated in the clockwise direction and the energy comes from the external driving torque.

Under the influence of the rotor mass imbalance and zero initial conditions, Li and Païdoussis [586] obtained the regions of different steady state motions of the rotor in terms of eccentricity parameter and the dry friction coefficient

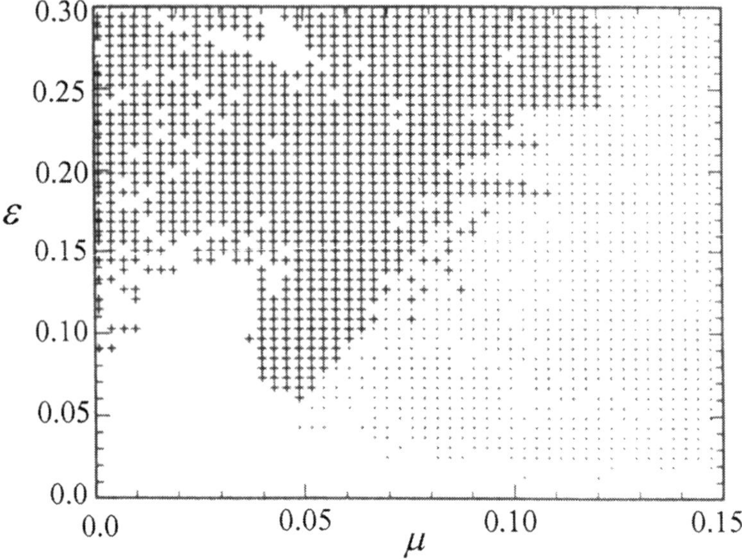

Fig. 5.2. Regions of different steady state rotor motions: Rubbing (+, counter-clockwise, •, clockwise), and impacting (empty space) for $e = 0.9$, [586].

for a fixed value of coefficient of restitution $e = 0.9$. Fig. 5.2 shows three regions. The first region occupied by dots represents a clockwise rubbing of the rotor against the stator. This motion is called "dry whip" or backward rub. The second region is occupied by crosses and each cross represents a parameter pair resulting in rubbing in the same direction of the rotor motion (counter–clockwise) or forward rubs. The empty regions represent impacting motion. Fig. 5.3 shows a bifurcation diagram constructed by recording the pre–impact steady state radial velocity $\dot{r}$ for each value of the friction coefficient, μ, for $e = 0.075$. For $\mu \leq 0.019$ only finitely many values of the impact velocity exist, indicating that the solutions are periodic or quasi–periodic. In the neighborhood of $\mu = 0.02$ the rotor motion is chaotic. A further increase in the friction coefficient brings the rotor back to periodic over the friction coefficient range $0.0212 \leq \mu \leq 0.0242$. Above that range the radial velocity begins to assume multi–values. For $\mu \geq 0.02875$ another set of bifurcations occur, leading to chaotic motion.

Qin et al [877] studied the contact between the rotor and its stator with bearings. It was found that rubbing creates grazing bifurcation. With variation of system parameters, such as rotating speed, imbalance and external damping, complex response characteristics were observed such as period–doubling bifurcation and torus bifurcation. A general analytical model of a rub–impact rotor–bearing system with mass imbalance was developed by Shen et al [935]. The model consists of radial elastic impact and tangential

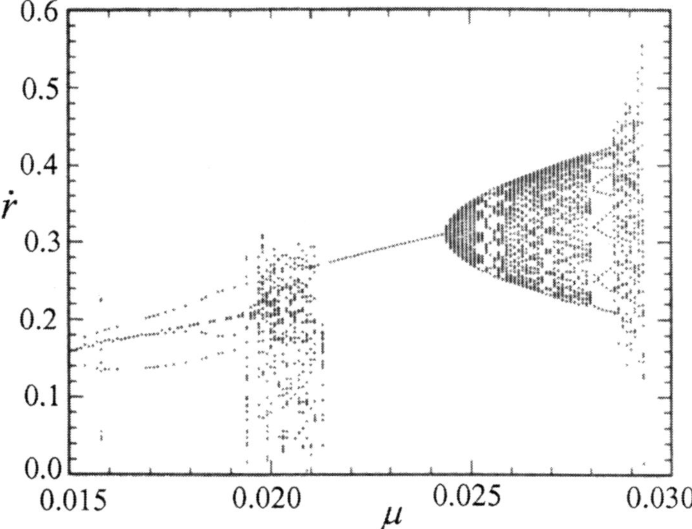

Fig. 5.3. Bifurcation diagram showing the dependence of radial velocity on the dry friction coefficient for eccentricity $\varepsilon = 0.075$, [586].

Coulomb friction. Bifurcation diagrams in the presence of mass eccentricity were generated by taking the rotating speed as the control parameter. It was found that the mass eccentricity of the rotor has a substantial effect on the motion characteristics of the rub–impact rotor–bearing system.

Yue–Gang et al [1119] considered the influence of the nonlinear oil–film force and nonlinear friction force in studying the rotor–bearing rub–impact dynamics. It was shown that within the subcritical speed range both quasi–periodic and chaotic motion regions increase with increasing the coefficient of relative speed. Within the supercritical speed range, the chaotic motion region was found to decrease. However, the rub–impact forces increase and the quasi–period motions become gradually of period–3. These features were used as a theoretical basis for the failure diagnostics of rotor–bearing systems at rub–impact fault. The rubbing between the stator structure and rotor blades of turbo–machinery was examined by Sinha [947] and Chu and Lu [180]. The nonlinear dynamic effect of repeated tip impact of a rotating Timoshenko beam spinning with constant angular velocity was studied by Sinha [947]. Due to the discontinuity of parameters in the differential equations of vibro–impact systems, the transient response exhibited complicated dynamic behavior. An axial rub–impact force model corresponding to the six degrees of freedom of the Jeffcott rotor was developed by Yuan et al [1117]. The mass imbalance and axial rub–impact effects were considered. It was concluded that the dynamic behavior of axial rub–impact is quite distinct from those of radial ones, and that the axial vibrations are essential to the diagnostics of axial rub–impact.

5.4 Micro–Actuators

Some micro–actuators are designed to perform repeated impacts to generate large–scale displacements ([695], [683]). For example, a smooth impact linear motor is one of the superior actuators. Despite its simple construction, this linear actuator demonstrates a long stroke and a high positioning resolution because of its two modes of operation. One mode is for the long stroke drive, and the other is for the positioning drive. Yamagata and Higuchi [1092] proposed a positioning system comprised of a motion mechanism driven by impact force generated by piezoelectric actuator. The system utilizes dry friction and impulsive inertia force.

In micro– or nano–robotics, high precision movement in two or more degrees of freedom is one of the main requirements. Zesch et al [1123] described some basic driving mechanisms for the Switzerland ETHZ Nanorobot Project, in which two new piezoelectric devices were developed. "Abalone" is a three–degree–of–freedom system that relies on the impact drive principle. Within the actuator's local range of 6 μm fine positioning is possible with a resolution better than 10 nm. "NanoCrab" is a bearingless rotational micromotor relying on the stick–slip effect. A new precise actuator with two degrees of freedom for translational and rotational motions actuated by the impact force of an end–loaded piezoelectric cantilever was proposed by Zhang et al [1124]. The actuator performance on translational and rotational motions were measured and revealed large travel range, strong driving ability, and high positioning resolution.

Mita et al [683] developed a micro–machined actuator to produce a precise and unlimited displacement. Their impact actuator is composed of a movable mass, which is supported by suspensions, driving electrodes, stoppers and a frame as shown in Fig. 5.4. The impact of the movable mass against the stopper is the source of the action. The movable mass is accelerated by the electrostatic force between the mass and electrodes. It collides with the stopper that is fixed to the base of the actuator. When the impact force exceeds the maximum static friction, the actuator begins to move. The actuator stops eventually due to the resistance of frictional forces. The suspended silicon micro–mass is encapsulated between two glass plates and driven by electrostatic force. Dankowicz and Zhao [207], Zhao et al ([1132], [1133]) and Zhao and Dankowicz ([1130], [1131]) analyzed the near–grazing dynamics of the impact micro–actuator developed originally by Mita et al [683]. The bifurcation behavior of the non–impacting and impacting dynamics of the micro–actuator was discussed in terms of period–1 impacting orbits with one impacting and one non–impacting loop per period. Kang et al [489] developed a computational toolbox for the bifurcation analysis of micro–actuators

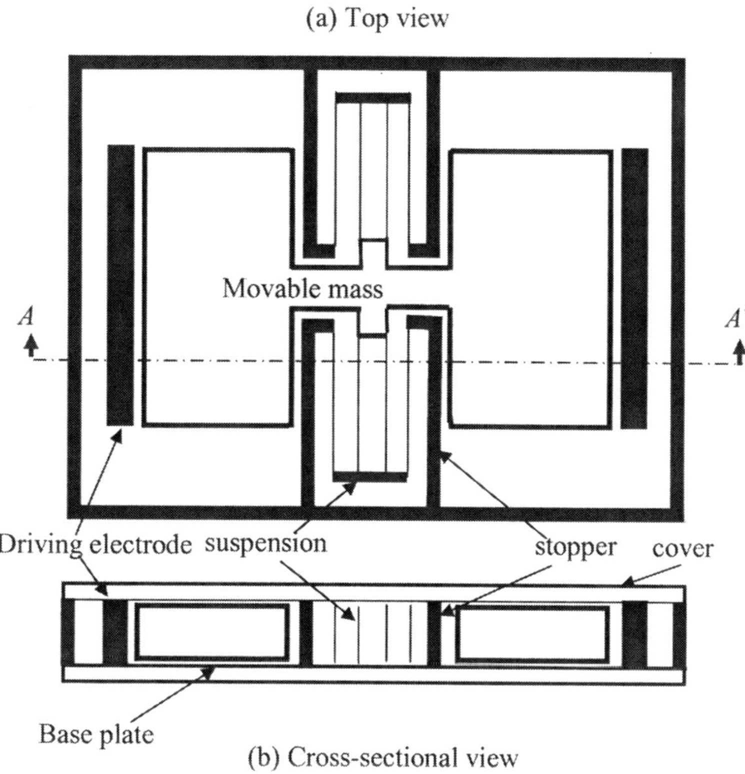

Fig. 5.4. Schematic diagram of impact actuator, [683].

interrupted by discrete–in–time events. Branches of multi–segment periodic solution trajectories were continued under variations in a single parameter. It also provided bifurcation curves corresponding to the locus of period–doubling, saddle node, and grazing bifurcation points were continued under variations in two parameters.

Zhang et al [1128] proposed a piecewise nonlinear model to understand the impact dynamics of micro–oscillators such as MEMS switches and tapping–mode atomic–force–microscopy. The deformation was considered large and the tapping occurred in the nonlinear frequency response region. Both softening and hardening effects were considered. It was found that the nonlinearity not only shifts the bifurcation area, but also changes the bandwidth of the tapping event. The comparison between experimental and numerical results demonstrated the validity of the considered piecewise nonlinear model.

The feedback control of a class of complementary–slackness hybrid mechanical systems composed of an uncontrollable part and a controlled one, linked by a unilateral constraint and an impact rule was studied by Brogliato and Zavala–Rio [142]. The approach was based on a nontrivial extension of

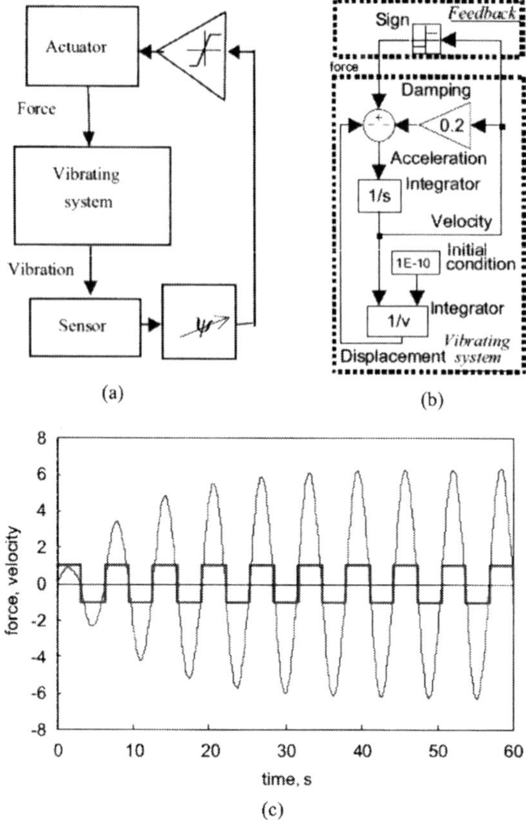

Fig. 5.5. Autoresonant system showing (a) feedback design, (b) Matlab-Simulink model, and (c) force (rectangular pulse) and velocity (sinusoidal) time history record, [950].

the one degree–of–freedom juggler control design proposed by Zavala–Rio and Brogliato [1121]. The role of various physical and control properties characteristic of the system on its stabilization properties was considered. Later, Menini and Tornambe [670] considered a particular class of jugglers, under actuated linear mechanical systems, subject to non–smooth impacts. It was shown that the overall system is controllable and that just the impacts can be used to solve a dead–beat regulation problem.

Sokolov et al [950] introduced a new approach known as "*autoresonant excitation*" to design resonant vibratory equipment as a *self–sustained* oscillating system with electronic and electro–mechanical feedback using an actuator of the synchronous type. Synchronous actuators produce an excitation force whose frequency does not depend on load but only on the frequency of the power supply ([54], [36]). It was demonstrated that autoresonant approach can overcome the intrinsic drawback in the electromagnetic actuators for vibrating and vibro–impact machines. The basic idea of the autoresonant system is shown in Fig. 5.5. Fig. 5.5(a) shows the feedback design scheme, which shifts the phase of the vibration signal from the sensor and amplifies its power with limitation. The actuator transforms the signal into an excitation force. With a given phase shift in the feedback circuit the resulting vibration would have the same frequency as the natural frequency of the mechanical subsystem without feedback. Fig. 5.5(b) is the Matlab–Simulink model in which the rectangular pulse force is used for excitation while the velocity signal is used in the feedback. The time history records of both the rectangular pulse excitation and velocity are shown in Fig. 5.5(c). Physical phase–shift was not required to provide resonant mode of vibration. Later, Babitsky and Sokolov [64] and Voronina et al [1041] considered the autoresonance control in a simple vibro–impact system with an elemental electromagnetic actuator and ultrasonic transducers for machining applications. It was indicated that autoresonant approach enhances the application of electromagnetic actuators in vibrating machines.

5.5 Vocal Folds

When air passes the space between vibrating vocal folds (called glottis), the vibration of the vocal folds is excited as a result of a fluid–structure interaction mechanism. The glottal oscillations serve as the main generator of the acoustic excitation of the whole human vocal tract, which finally results in phonation. The collision between the vocal folds involves the interaction of aerodynamic pressure with inertia and elastic forces of the folds. The contact force produced by vocal fold collision increases the risk of tissue damage ([1004], [470], [471]). The study of vocal fold impact is important for understanding the mechanism of phonation and determining potential vocal fold pathological development [987]. Furthermore, stresses and strains within the vocal fold tissue may play a critical role in voice fatigue, tissue damage and resulting voice disorders, and tissue healing.

Reviews of some models are given by Sorokin [952] and Kob [517]. These models are valuable for understanding the mechanics of voice production and in some applications such as speech synthesis and recognition [329] and voice pathology [408]. The generation of voiced sound is a highly nonlinear process. The nonlinearities owe their origin to the nonlinear stress–strain characteristics of vocal fold tissue [1003], collision of the folds [451], nonlinear interaction of the airflow, and glotted area [1026]. In view of these nonlinearities, complex bifurcation patterns and low–dimensional chaos were reported in the literature ([1003], [1004], [669], [410], [564], [472]). Herzel and Knudsen [409] proposed an autonomous fourth–order model of vocal fold vibrations. Each fold was represented by a lower and upper mass, and the aerodynamic forces were derived from a modified Bernoulli equation. The model exhibited many features of normal phonation in a wide parameter region. At the borderlines

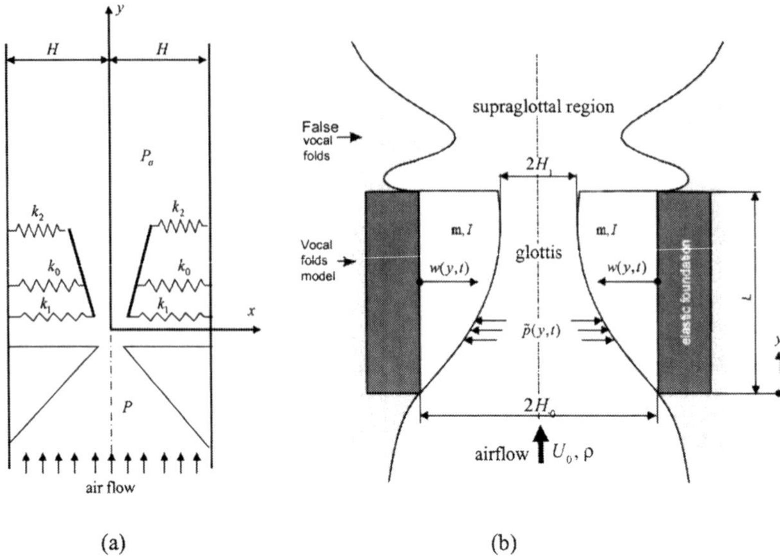

Fig. 5.6. Schematic diagrams of two-degree-of-freedom models of the vocal folds as proposed by (a) Landa [564], (b) Horáček et al, [426].

of this region coexistence of limit cycles, period–doubling and chaos were observed.

The mechanics underlying collision between vocal folds is associated with aerodynamic variables that are linked to voice quality changes [422]. Gunter [381] modeled the tissue mechanics governing vocal–fold closure and collision during phonation. The role of elastic forces in glottal closure and in the development of stresses that may be a risk factor for pathology development was identified. The model involved nonlinear dynamic contact that incorporated three–dimensional, linear elastic, finite–element representation of a single vocal fold, a rigid midline surface, and quasi–static air pressure boundary conditions. Contact force between vocal folds was found to be directly proportional to compressive stress, vertical shear stress, and Von Mises stress in the tissue.

The vocal folds are living soft tissues of a complicated material structure composed of several tissue layers, and their motion is generally a three–dimensional motion of a continuous viscoelastic system. However, they have been frequently modeled with only a few degrees of freedom [1004]. For example, Flanagan and Landgraf [330] represented each vocal fold by a simple harmonic oscillator single–mass model. Landa [564] proposed a model that involves two absolutely rigid plates suspended by springs to the walls of a tube with rigid angled cross–section as shown in Fig. 5.6(a). Under air flow the system is akin to that of the bending–torsion flutter of an aircraft wing. On

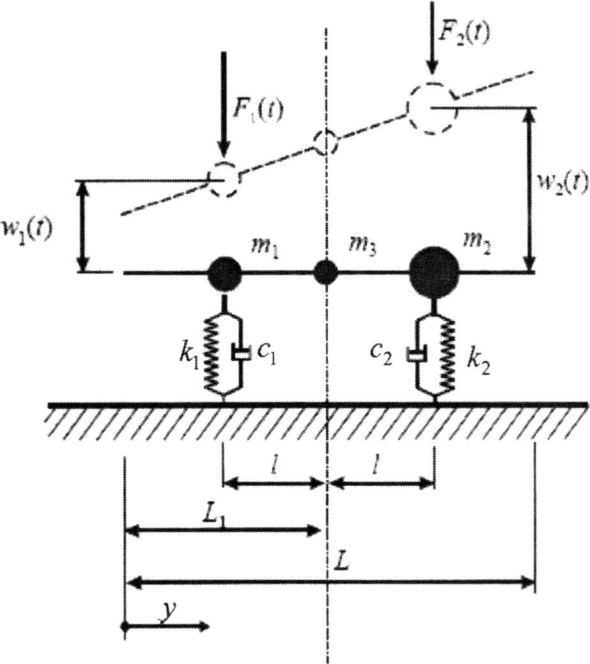

Fig. 5.7. Schematic diagram of two-degree-of-freedom vocal folds model, [426].

the other hand, Horáček et al [426] considered another similar two–degree–of–freedom model shown in Fig. 5.6(b).

With reference to Fig. 5.6(b) the model of the glottis with the vocal folds is represented by a channel with planar symmetry conveying air. The length of the channel, L, is measured parallel to both the plane of symmetry and the direction of air–flow. The channel walls are shown by two vocal–fold–shaped rigid bodies of mass m and moment of inertia I, which are vibrating symmetrically in the opposite phase identical amplitudes on an elastic foundation. The rigid bodies oscillate in the fluid of density ρ_a flowing in the channel with a mean flow velocity U_0 at the inlet ($y = 0$). The width of the cross–section of the channel is $2H_0$ at the inlet, and the minimum cross–section of the channel, known as the glottal width is $2H_1$. The schematic diagram of the lumped two–degree–of–freedom model proposed by Horáček et al [426] is shown in Fig. 5.7 and the equations of motion of this model were developed by Horáček and Švec [425] in the matrix form

$$\mathbf{M}\ddot{\mathbf{V}} + \mathbf{B}\dot{\mathbf{V}} + \mathbf{K}\mathbf{V} = \mathbf{F}(t), \tag{5.5}$$

where
$\mathbf{V} = \begin{Bmatrix} V_1 \\ V_2 \end{Bmatrix} = \begin{Bmatrix} \frac{w_2 - w_1}{2l} \\ \frac{w_2 + w_1}{2l} \end{Bmatrix}$, $\mathbf{M} = \begin{bmatrix} -lm_1 & m_1 + \frac{m_3}{2} \\ lm_2 & m_2 + \frac{m_3}{2} \end{bmatrix}$, $\mathbf{K} = \begin{bmatrix} -lk_1 & k_1 \\ lk_2 & k_2 \end{bmatrix}$, $\mathbf{F}(t) = \begin{Bmatrix} F_1(t) \\ F_2(t) \end{Bmatrix}$, $\mathbf{B}$ is a proportional structural damping matrix given in the form $\mathbf{B} = \epsilon_1 \mathbf{M} + \epsilon_2 \mathbf{K}$, where ϵ_1 and ϵ_1 are constants related to the damping coefficients c_1 and c_2. When the glottis is open, the equivalent excitation aerodynamic forces $F_1(t)$ and $F_2(t)$ acting on the vocal folds are

$$F_{1,2}(t) = \frac{h}{2} \int \left(1 \mp \frac{x}{l} \pm \frac{L_1}{l}\right) \widetilde{p}(y,t) dy, \tag{5.6}$$

where the upper sign is belonging to the first force and the lower sign to the second force. h is the width of the channel (identical to the width of the rigid body) and is measured perpendicular to the direction of the air flow and parallel to the plane of symmetry. $\widetilde{p}(y,t)$ is the air pressure along the vibrating body surface. The masses are given by the expressions

$$\begin{aligned} m_{1,2} &= \frac{1}{2l^2} \left(I + m\ell^2 \pm m\ell l\right), \\ m_3 &= m \left[1 - \left(\frac{\ell}{l}\right)^2\right] - \frac{I}{l^2} \end{aligned} \tag{5.7}$$

where ℓ is the location of the center of mass of the rigid body from the axis passing through m_3.

The unsteady aerodynamic forces for open glottis were developed based on the unsteady continuity and one–dimensional Euler equations for incompressible fluid. The complete expressions of aerodynamic forces $F_{1,2}(t)$ are lengthy and the reader can consult Horáček and Švec [425] for complete derivation. The Hertz model of impact was introduced to account for vocal–fold collisions and was written in the form

$$F_H = k_H \delta^{3/2} \left(1 + b_H \dot{\delta}\right) \tag{5.8}$$

where $k_H \approxeq \frac{4}{3} \frac{E\sqrt{r}}{(1-\varepsilon^2)}$, δ is the penetration of the vocal–fold element through the contact plane, E is Young's modulus, and ε is the Poisson ratio. r is the radius of curvature of the impacting body surfaces of the vocal–fold model in the contact point according to the well–known relationship of the radius of curvature $\frac{1}{r} = \frac{\left|d^2 f(y)/dy^2\right|}{[1+(df(y)/dy)^2]^{3/2}}$. This surface was approximated by the function $f(y) = a_1 y + \frac{a_2}{2} y^2$, where $a_1 = 1.858$ and $a_2 = 319.722\ m^{-1}$.

The moving surface of the vibrating vocal–fold element was expressed by the function

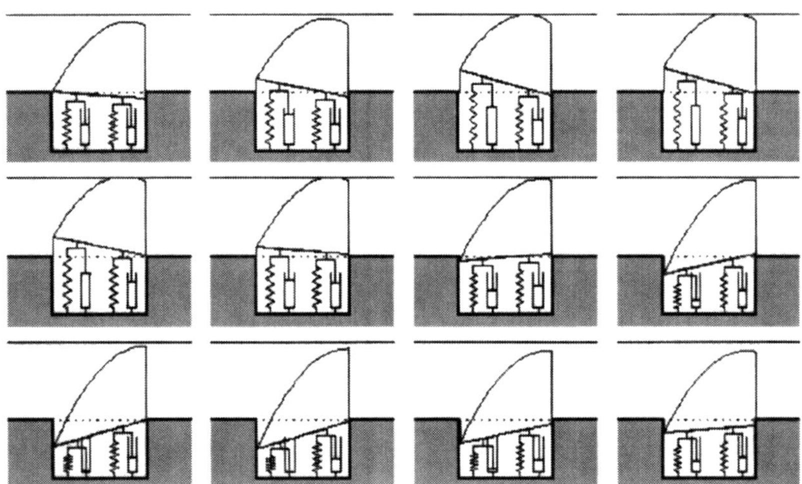

Fig. 5.8. Sequence of the vocal-fold motion during one period of oscillation for air flow velocity: $U_0 = 1.6m/s$, glottal half-gap, $H_1 = 0.2mm$, natural frequencies $f_1 = 100Hz$, $f_2 = 105Hz$, lung pressure $p_{lung} = 380$, mean glottal flow volume rate $Q = 0.181$, [426].

$$\begin{aligned} X(y,t) &= f(y) + w(y,t) \\ &= f(y) + (y - L_1)\, V_1 + V_2, \qquad y \in \langle 0, L \rangle \end{aligned} \tag{5.9}$$

The coordinates of the contact point were obtained in the form

$$x_{\max} = f(y_{\max}) + (y_{\max} - L_1)\, V_1 + V_2, \tag{5.10a}$$

$$y_{\max} = \min\{L,\ \max[0,\ -(V_1 + a_1)/a_2]\} \tag{5.10b}$$

During impact, when the glottis is closed, the aerodynamic forces, $F_{1,2}(t)$, are switched off. In this case, three types of forces acting on the vocal–fold shaped element were considered. These are 1) the Hertz contact force, 2) the sub–glottal pressure, which acts on the sub–glottal part of the element surface, and 3) the supra–glottal pressure, which acts on the supra–glottal part of the element surface. That pressure was set to zero. The numerical simulations performed by Horáček et al [426] provide useful information in understanding the mechanism of vocal–fold collision. Fig. 5.8 shows a sequence of the vocal–fold motion during one period of oscillation. The impact forces of the vocal folds resulting from the Hertz model were related to the impact stress, which can be measured in the real vocal folds. The impact stress σ was evaluated by dividing the impact force by the contact area. The maximum impact stress $\sigma_{\max}$ was calculated as the maximum value in one oscillation period according to the formula

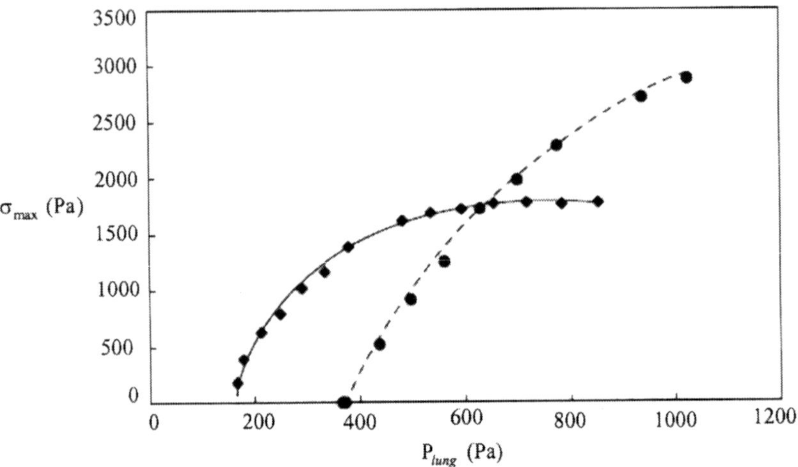

Fig. 5.9. Dependence of the maximum impact stress during self-oscillations on the pressure in lings for $f_1 = 100Hz$, $f_2 = 105Hz$, and glottal half-gaps: ♦ $H_1 = 0.2mm$, • $H_1 = 0.5mm$, [426].

$$\sigma_{\max} = \frac{3}{2}\frac{F_{H,\max}}{\pi R^2} \tag{5.11}$$

where $F_{H,\max}$ is the maximum contact force during one period of oscillation, and R is the maximum contact radius. Fig. 5.9 shows the dependence of the maximum impact stress $\sigma_{\max}$, for two glottal half–gaps, on the pressure in the lungs. Each curve ends at the phonation instability pressure. The threshold pressure increases with the pressure in lungs, P_{lung}. The calculated values for the maximum impact stress shown in Fig. 5.9 were found to be in good agreement with impact stress values measured on real vocal folds by Jiang and Titz [471], Hess et al [415] and Verdolini et al [1037].

Van Hirtum et al [1027] summarized preliminary experimental results on physical modeling of vocal folds models and the collision of the two folds. The distribution of the transverse strain component was used to examine the influence of vocal fold collision on potential tissue damage [639]. In the position of maximum opening the vocal folds are deformed by a combination of a bulging–type deformation and the opening movement. At this time instance, the transverse strains at the medial surface were negative, an indication of Poisson's deformation. During the closing stage, vocal folds collide and simultaneously a mode 3 vibration pattern emerges. Closure of the glottal opening is not complete and two incomplete closure areas are formed during the closure stage. These open areas are located at the anterior and posterior ends of the model larynx.

De Oliveira Rosa et al [216] developed a three–dimensional model to simulate the larynx during vocalization. A contact–impact algorithm was incorporated to deal with the physics of the collision between true vocal folds. The results revealed that the simulated larynx can reproduce the vertical and horizontal phase difference in the tissue movements and that the false vocal folds affect the pressure distribution over the larynx surfaces. A self–oscillating finite–element model that combines aerodynamic properties, tissue mechanics, airflow–tissue interactions, and vocal fold collisions, was developed by Tao et al [987] and Tao and Jiang [986]. The model simulated the vocal fold vibration during phonation. The spatial and temporal characteristics of mechanical stress in the vocal folds were predicted. It was found that mechanical stress periodically undulates with vibration of the vocal folds and that vocal fold impact causes a jump in the normal stress value. Spatially, the normal stress was found to be significantly higher on the vocal fold surface than inside of the vocal folds.

Spencer et al [954] conducted a series of experiments to determine the mechanical fields on the superior surface of a self–oscillating physical model of the human vocal folds. They used a three–dimensional digital image correlation obtained by using a high–speed camera together with a mirror system to measure displacement fields. The strains, strain rates, and stresses on the superior surface of the model vocal folds were computed. The dependence of these variables on flow rate was established. A Hertzian impact model was used to estimate the contact pressure on the medial surface from superior surface strains. A tensile stress dominated state was observed on the superior surface during collision between the model folds. Collision between the model vocal folds was found to limit the medial–lateral stress levels on the superior surface.

5.6 Vibration Protection under Vibro–Impact

Systems subjected to impact and shock loads exhibit severe vibration and they need special isolation means. Impact loading is encountered in many mechanical applications such as pneumatic hammers, slamming loads on water waves acting on ships and ocean structures, and vibro–impact systems with rigid or elastic stops. In order to protect a given object against these undesirable disturbances, a vibrating protection system may be placed between the vibration source and the object. In their research monograph, Alabuzhev et al [6] introduced a number of vibration protection systems with quasi–zero stiffness. Fig. 5.10 shows schematic diagrams of selected systems whose load bearing elastic elements possess constant positive stiffness as well as devices with negative–stiffness. This type of isolators has been used for vibration isolation of operators' seats in vehicles [896], impact action hand–held machines (see, e.g., [1147], [7], [962]), and railway car suspensions ([361], [1099]). Systems with negative–stiffness have been treated by Gerner et al [361] and Yashin et al [1099], and the performance of such systems in the chaotic motion regime was examined by Goverdovskiy et al [372], Lee and Goverdovskiy [570], and Lee et al [571].

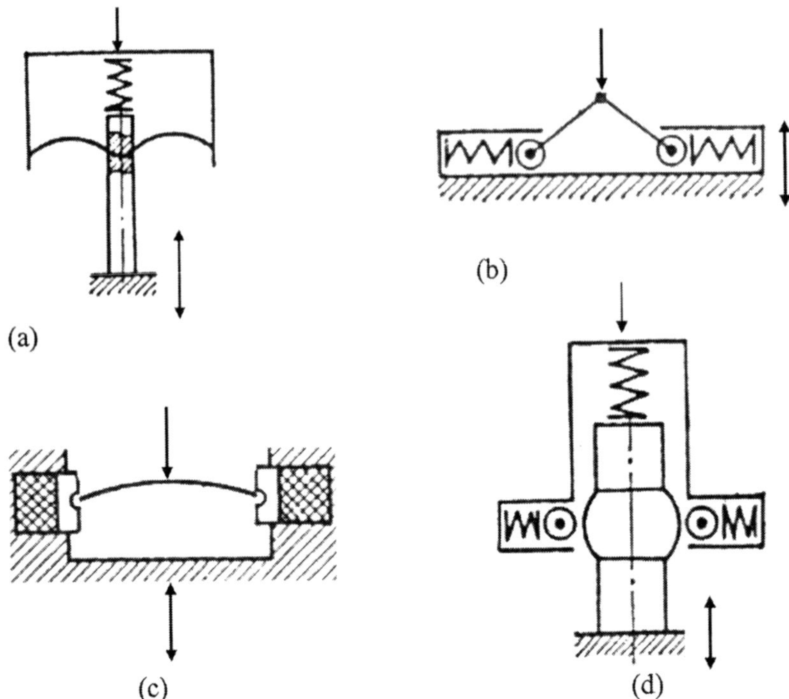

Fig. 5.10. Schematic diagrams of typical vibration protecting mechanisms: (a) Vertical coil spring with two buckled beams, (b) Links restraint against horizontal springs, (c) Buckled beam isolator, (d) Vertical and horizontal spring system, [6].

A comprehensive assessment of recent developments of nonlinear isolators in the absence of active control means was presented by Ibrahim [437]. Some modifications of the nonlinear characteristics of isolating mounts carrying rigid structures subjected to impact loads were proposed by Dufour et al [276]. The modifications were adapted with respect to impact vibrations to achieve a well design behavior. The protection of workers against vibrations generated from hand–held tools requires special vibration isolation means. Dobry and Brzezinski [262] developed a strong elastic nonlinear isolator to minimize the interaction force between the tool and the handle. In an effort to prevent injuries from impact impulse loads, Balandin et al [73] presented a review of research activities dealing with the limiting performance analysis of impact isolation systems. Shu and Shen [941] and Zhiqing and Pilkey [1134] conducted a limiting performance analysis to study the optimal shock and impact isolation of mechanical systems via wavelet transform.

A bumpered vibration protection arrangement of a gimbaled electro–optical device was developed by Veprik et al [1035]. This device was designed based on a split Sterling cry cooler for the cooling of an infrared focal array. The installation of bumpers with enlarged travel reduces the probability of accidental impacts and effectively reduces the excessive deflections. However, the presence of bumpers turns the vibration isolation arrangement into a potential strongly nonlinear vibro–impact system with unfavorable characteristics [53]. In an effort to eliminate these characteristics, Babitsky and Veprik [62] introduced a new concept based on the cooperation use of an undamped, low frequency vibration isolator in combination with optimally damped bumpers installed with minimal free travel distance.

Some isolators experience a sudden change in the values of their parameters, which can be represented as piecewise linear or nonlinear functions. It is known that soft isolators are best for isolation. However, nonlinear hardening is required to minimize relative displacement at high amplitude ([466], [467], [712]). Patrick and Jazar ([793], [794]) introduced a secondary suspension to limit high relative displacements. Mahinfala et al [634] adopted a hyperbolic–tangent saturation function to study the frequency response of vibration isolators with saturating spring elements. Deshpande et al ([217], [218]) employed an adapted averaging approach to study highly nonlinear systems described by piecewise linear representation. They obtained an implicit function for frequency response of a bilinear system under steady state. This function was examined for jump–avoidance and a condition was derived to ensure that the undesirable phenomenon of "jump" does not occur and the system response is functional and unique.

Orzechowski et al [753] studied the behavior of nonlinear isolation mounts and measured their response to impulsive inputs. Their results showed the limitation of mounting linearity in reducing the transient portion of the response due to impact without adding damping to the system. The presence of barriers is known to prevent a vibratory system from exceeding the relative amplitude particularly in the neighborhood of resonance. The effect of end stops in an isolator was studied by Narimani and Golnaraghi [711] and Narimani et al [712]. They adopted a piecewise linear system and applied an averaging method to identify the range of the parameters, which minimize the relative displacement of the system. It was found that the damping ratio plays a dominant role than stiffness in piecewise linear vibration isolators.

5.7 Other Applications

There are many other applications of vibro–impact systems and the purpose of this section is to summarize them. These applications include the ultrasonic cutting as a vibro–impact process ([34], [35], [865]), hand–held percussion machines, drills and forming machines ([55], [96], [97], [612], [629]), bumpered vibration isolators ([62], [1031], [1032]), impact loading in disc brakes and clutches ([1118], [120], [65]), impact–noise generation due to wheel and rail discontinuities ([1036], [886], [511]), and systems described by a Duffing oscillator under impact loading ([85], [112]). It was shown that excitation of the vibro–impact mode of tool–workpiece interaction is the most effective way of using ultrasonic influence on dynamical characteristics of machining.

5.7.1 Drill–Strings

The dynamics of drill–strings used for oil well drilling can present complex vibrational states ([1105], [1106], [1012]). For example, Yigit and Christoforou [1105] developed a model to study the transverse vibrations of drill–strings caused by axial loading and impact with the well–bore wall. In a later work, Yigit and Christoforou [1106] extended their model to account also for torsional vibrations. Trindade et al [1012] examined the oscillations of a vertical slender beam, clamped in its upper extreme, pinned in its lower one and constrained inside an outer cylinder in its lower portion. It was shown that one should account for the axial displacement dynamics, using nonlinear strain–displacement relations, since the coupling of axial–bending dynamics may be very important in the dynamical behavior of general slender beams such as a drill–string. In particular, the micro–impacts, accompanying the bottom portion due to the beam compressive softening were represented only when using a nonlinear axial–bending coupling.

5.7.2 Machine Tools

The flexibility of tools and work pieces, the high spindle frequencies, and the inherent impact nonlinearities in the milling process can lead to complicated dynamic tool–workpiece interactions. Davies and Balachandran [210] conducted an experimental investigation to study the vibrations of a thin–walled part during milling. From the time series, power spectra, autocorrelations, auto–bispectra, and phase portraits, it was inferred that stiffness and damping nonlinearities due to the intermittent cutting action have a pronounced effect on the dynamics of the workpiece. A mechanics–based model with impact nonlinearities was developed to explain the observed results. The predicted results were found to be in good agreement with the experimental observations.

Neumann and Sattel [724] proposed a model for a piezoelectric device for drilling brittle materials. The motion of the piezoelectric actuator tip follows a prescribed harmonic vibration. Experimental tests with a prototype device as well as simulations with simple models revealed irregular motion of the impacting mass. The temporal behavior of this system was examined using the set–oriented numerical methods. These methods are based on an adaptive subdivision technique for cell–mapping to approximate attractors and invariant measures. The contact conditions between subsystems were described by complementary kinematics and force relations and Newton's impact law, respectively. Periodic and chaotic orbits were detected and parameter ranges for the occurrence of different types of solutions were determined. Information on the probability of attaining a particular attractor was obtained by quantifying its connected basin of attraction.

Peterka [823] modeled a forming machine by a double impact oscillator consisting of two symmetrically arranged single impact oscillators. The anti–phase impact motion of this system was found identical to the dynamics of single–degree–of–freedom systems with constraints. The in–phase motion and the influence of asymmetries of the system parameters were numerically studied and verified experimentally. Dynamic models of percussive–rotary drilling have been developed by Batako et al ([95], [96]) and Batako and Pioroinen [98]. This nonlinear system exhibits friction–induced vibration resulting in impact excitation, which influences the parameters of stick–slip motion. The model incorporates the friction force as a function of sliding velocity. The sliding velocity allows for the self–excitation of the coupled vibration of the rotating bit and striker, which tends to a steady state periodic cycle. It was found that the contact stiffness does not strongly affect the system dynamic behavior, rather it defines the magnitude of the impact force.

5.7.3 Printer Actuators

Hendriks [402] presented a lumped–parameter description of an impact printer actuator of the stored–energy type. The equations of motion were integrated both for single– and multiple–current pulse excitation. It was shown that for low repetition rates, each impact is distinct and independent, but at higher rates of impacts interaction. The interaction manifested itself initially as flight–time and print–force variations where strict periodicity of the actuator motion is lost. At extremely high repetition rates it was found that the actuator 'hangs up' and the backstop no longer participates in the actuator dynamics. Tung and Shaw ([1019], [1020]) considered the chaotic dynamics of a simple print hammer model in the form of a piecewise linear oscillator. The chaotic motion was characterized by Lyapunov exponents and by the existence of a set of strange attractors. Later, Tung [1018] examined the dynamic response of a single–degree–of–freedom system subjected to non–harmonic excitation. The amplitude and stability of the periodic responses together with Lyapunov exponents were estimated over a range of forcing periods. Jerrelind and Dankowicz [469] proposed a control methodology that introduces discrete changes in the position of a system discontinuity during the printer hammer motion while the hammer is away from the discontinuity. A forced, piecewise smooth, single–degree–of–freedom model of a Braille impact hammer is used to illustrate the methodology and to yield representative numerical results.

5.7.4 Gear Rattling

Gear rattling occurs due to repetitive impacts of teeth and is manifested by a vibration signature, which corresponds to the bands of frequencies due to torsional engine oscillations, meshing frequencies, and impact characteristics of lubricant conjunctions. Brindeu [134] considered the case with one–degree–of–freedom model realized by the rod–crank mechanism. Rattling in gear boxes and machines due to backlash has been studied extensively in the literature (see, e.g., [48], [1095], [946], [188], [554], [555], [170], [492], [486], [894], [894], [760], [759], [135], [137], [220], [506], [990], [508], [647], [977]). Halse et al [386] studied the nonlinear dynamics of lightly damped pair of meshing spur gears modeled as a single–degree–of–freedom oscillator with backlash that exhibits undesired noise and vibration. The analysis included both large finite and infinite stiffness values. It was shown that the permanent contact solution can coexist with many other stable rattling solutions, which were computed analytically. The regions of existence and stability of different families of rattling solutions on two–parameter bifurcation diagrams revealed that the large finite and infinite stiffness models give the same results. Tangasawi et al [985] presented numerical models for a gear pair contact in an attempt to study idle rattle conditions. It was found that gear rattle manifests itself as a band of frequencies, which shift towards lower spectral regions as the lubricant temperature rises.

5.7.5 Multi–Body Dynamics

In multi-body dynamics, the effects of zero velocity impacts were the subject of many studies ([364], [136], [138], [140], [340]). By introducing a mapping, it was possible to isolate the contribution to the local dynamics that comes from the grazing impact. When a multi–body system collides with a single body or with another multi–body system, impact dynamics with friction should be considered. Han and Gilmore [388] presented a general computer oriented analysis of impact dynamics incorporating friction. The presence of friction between sliding contacts during the impact processes makes the problem difficult since events such as reverse sliding or sticking must be determined. This may occur at different times throughout the impact, must be determined. The boundary representations of the bodies were used to solve for velocities at the points of contact. Barauskas [93] developed a numerical algorithm for solving the structural equations of motion with unilateral constraints involving normal, oblique impact and friction interaction points. The algorithm was an extension of the generalized Newmark scheme where Lagrange multipliers and a minimum work approach were employed at each time step. The reduction of the number of dynamic degrees of freedom of the unilaterally constrained structures was carried out by representing the equations of motion in modal coordinates of the unconstrained structure and truncating the dynamic contribution of higher modes. The algorithm was applied to the problem of free longitudinal vibro–impact motion of an elastic vibro–converter and free longitudinal and bending vibration of a vibro–converter interacting with a moving rigid body by oblique impact and friction forces.

The dynamics of oblique impact in vibrating systems was considered in several studies ([585], [802], [390], [391]). Numerical simulations of these systems were carried out using the incremental impulse method. The results of numerical simulations revealed dynamic phenomena such as periodic and chaotic vibro–impacts. Dynamic systems for the vibrational alignment of components during automated assembly may experience lose of contact. In this case, vibrational impact motion regime occurs and the mobile component repeatedly and systematically impact with other components. Baksys and Puodziuniene ([70], [71]) presented a two–degree–of–freedom vibro–impact system in which the moving body experiences oblique impact. Vibro–impact motion of the body starts from the static equilibrium to the sate of dynamic equilibrium under transient regimes. The distance between these equilibrium positions determines the maximum error of the mutual interpositions of assembly components. It was concluded that impact body motion can occur only for certain combinations of system and excitation parameters.

Complementarity conditions describe those conditions at the contact location and govern the velocities before and after impact ([837], [158]).Brogliato et al [140] presented an overview of some problems related to numerical simulation of finite dimensional non–smooth multibody mechanical systems. This class of systems involves complementarity conditions and impact phenomena, which make its study and numerical analysis a difficult task that cannot be solved by using known ordinary differential equation or differential–algebraic equation integrators only.

5.8 Closing Remarks

Vibro–impact of non–classical systems constitutes a wide range of engineering problems. Engineering applications include the free–play in joints of mechanisms, trusses, aerospace applications, rotor–stator rubbing of rotating machinery, gear rattling, machine tools, and multi–degree–of–freedom systems. Isolation of severe vibration due to impact loading is an important problem that seeks effective solutions. In bioengineering, the vocal folds vibro–impact plays a major role in sound generation. There is a strong need to develop accurate analytical modeling and numerical algorithms under random excitation. This issue will be addressed in Chapter 7.

Chapter 6
Continuous Systems

6.1 Introduction

Continuous structural elements differ from lumped systems in their analytical modeling. Continuous systems are usually described by partial differential equations together with appropriate boundary conditions in addition to the impact law. The early work of vibro–impact of continuous systems such as beams and plates is believed to be treated by Erington [298] for different boundary conditions. The Hertz law of impact at the point of contact leads to nonlinear integral equation for contact force in all cases of transverse impact. A thin cantilever type plate impacting against a barrier is found in many practical applications such as automatic reed type valves in small refrigeration compressors or engines, switches in electrical relays, etc. The failure problem of these systems owes its origin to impact stresses. This chapter addresses the vibro–impact dynamics in strings, beams, constrained pipes conveying fluid, nuclear reactors and heat exchangers, plates, and slamming waves on elastic structures.

6.2 Strings

Strings are simple elements and rich in their dynamics. Their dynamics is complex when their motion is limited by point stops. The impact of strings with point stops was studied by Krupenin ([538], [539], [540]), Moon et al [688], Astashev et al [39], Astashev and Krupenin ([37], [38]), and Murphy and Morrison [700]. Murphy and Morrison [700] considered sinusoidally excited strings subject to an amplitude restraint. They included initiation of both periodic trapezoidal standing waves and standing waves of other types characterized by wave profiles composed of segments of straight lines. The vibro–impact dynamics of a string carrying an array of masses was studied experimentally by Moon et al [688]. A spatial pattern return map was then used to observe the change in spatial patterns with time.

R.A. Ibrahim: Vibro-Impact Dynamics: Model., Map. & Appl., LNACM 43, pp. 151–191.
springerlink.com

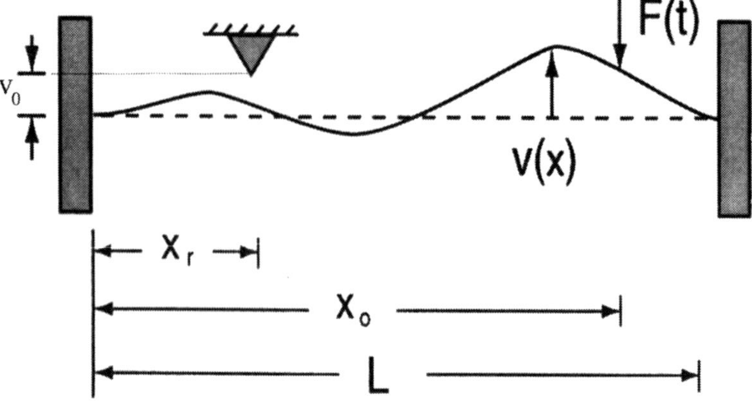

Fig. 6.1. Schematic diagram of a vibrating string against a knife-edge amplitude constraint, [700].

In a series of papers ([39], [37], [38], [536], [537], [541], [542], [543], [544], [545], [546], [547], [549], [550], [551], [1033], [1034]) some interesting phenomena associated with discrete and distributed vibro–impact systems were reported. These studies include multiple nonlinear effects related to the formation of specific nonlinear waves of trapezoidal profiles, advent of localization of intensive impacts, generation of higher harmonic components, and generation of non–synchronic and chaotic oscillations. The hypothesis of instantaneous impact that takes place in impact elements made up of strings and absolutely rigid elastic stops was described by Krupenin [536]. Two types of periodic standing waves of string dynamics were predicted. The waves of the first type are observed upon passing linear resonance, while the second type occurs when increasing the excitation amplitude, excitation frequency (frequency pulling), or the gap (amplitude pulling). For regimes with trapezoidal profiles of claps, the usual dynamic effects for "impact vibrators" were detected together with aperiodic waves of a more complicated nature.

Fig. 6.1 shows a schematic diagram of a vibrating string against a knife-edge amplitude constraint modeled as localized stiff spring positioned as v_0 from the string equilibrium position. The governing partial differential equation of a string in non–dimensional form may be written in the form, [700],

$$\frac{\partial^2 V}{\partial \tau^2} + 2\zeta\omega_1 \frac{\partial V}{\partial \tau} - \left[\int \frac{1}{2}\frac{EA}{\pi^2 P}\left(\frac{\partial V}{\partial \xi}\right)^2 d\xi\right]\frac{\partial^2 V}{\partial \xi^2}$$

$$+\frac{KL^2}{\pi^2 P}\delta\left(\xi-\xi_r\right)\left[V(\xi)-V_0\right] = \frac{\delta\left(\xi-\xi_r\right)}{\pi^2 P}F_0 \sin\nu\tau \qquad (6.1)$$

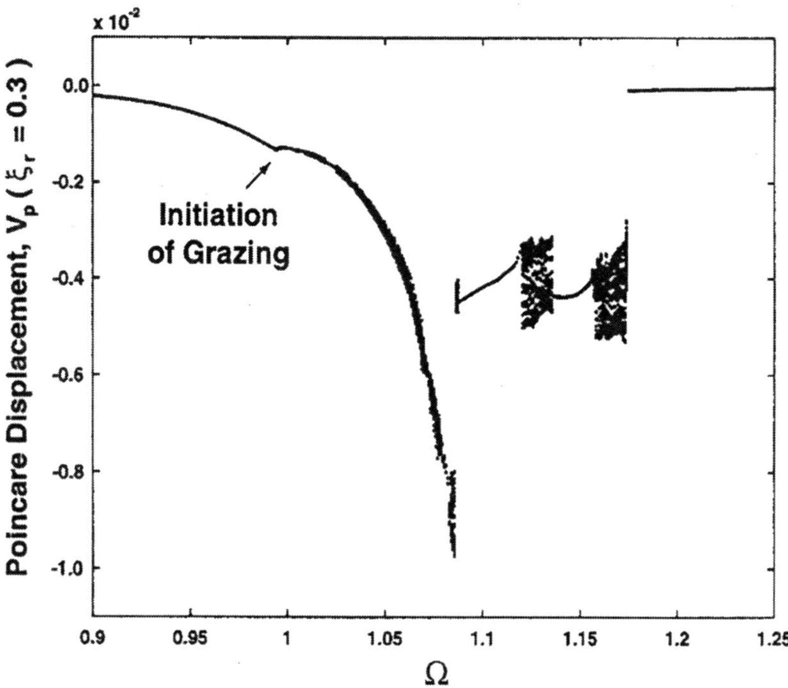

Fig. 6.2. A numerically generated amplitude versus excitation frequency response diagram, which includes impacts. Note the apparently chaotic portions as well as the periodic windows in the response, [700].

where $\xi = x/L$, L is the length of the string, P is the applied axial tension, $V(\xi) = v(x)/L$, $V_0(\xi = \xi_r) = v_0(x = x_r)/L$ is the gap size, $\tau = t(\pi/L)\sqrt{P/m}$, F_0 is the excitation amplitude, $\nu = \Omega/\omega_1$, $\delta(\cdot)$ is the Dirac delta function, K is the constraint stiffness given by

$$K = \begin{cases} 0 & -\infty \\ \leq \infty & v_0(x_r) < \infty \end{cases} \tag{6.2}$$

Equation (6.1) was discretized using Galerkin's method through the modal expansion

$$V(\xi, \tau) = \sum_{i=1}^{n} a_i(\tau) \sin(i\pi\xi) \tag{6.3}$$

The resulting ith normal mode equation was obtained in the form

$$\ddot{a}_i + 2\zeta\omega_i\dot{a}_i + \left[\frac{EA\pi^4}{4P} \sum_{j=1}^{n} j^4 a_j^2 + i^2 \right] a_i$$

$$+ \frac{KL^2}{\pi^2 P} \left[\sum_{j=1}^{n} a_j \sin(j\pi\xi_r) - V_0 \right] \sin(i\pi\xi_r) = \frac{2F_0}{\pi^2 P} \sin(i\pi\xi_r) \sin \nu\tau \tag{6.4}$$

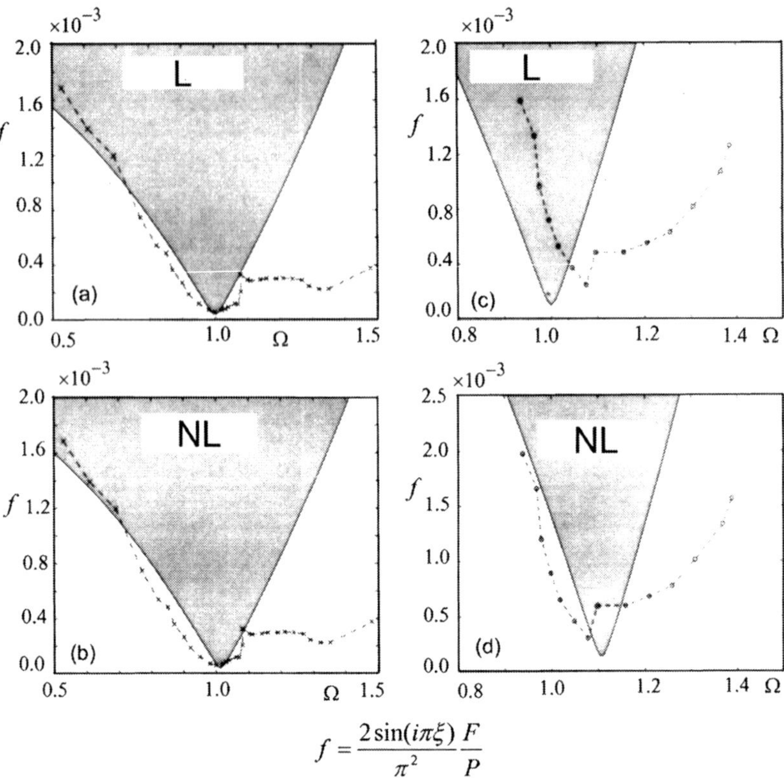

$$f = \frac{2\sin(i\pi\xi)}{\pi^2}\frac{F}{P}$$

Fig. 6.3. Boundaries of impact motion governed by excitation amplitude f and frequency Ω. (a) and (b) correspond to a small gap size $\overline{\sigma} = 0.00167$ (linear and nonlinear respectively), while (c) and (d) correspond to a larger gap size $\overline{\sigma} = 0.005$ (linear and nonlinear, respectively) L: Impact predicted by linear model, NL: Impact predicted by nonlinear model, [700].

The first summation on the left–hand side represents geometric nonlinearity associated with large deflection. The last expression on the left–hand side arises from strong stiffness nonlinearity of the barrier. $\omega_i = (i\pi/L)\sqrt{P/m}$ is the ith mode linear natural frequency. Murphy and Morrison [700] carried out numerical simulations for the first 20 terms of equation (6.4). In view of the large stiffness associated with impact, equation (6.4) experiences numerical stiffness since the eigenvalues of the local Jacobian matrix ranges over several orders of magnitude. A hybrid integration routine employing Geer's method during non–impact portion of the trajectory was adopted. This was then switched to a backward difference method. The problem of determining the exact time instance of impact was overcome by recasting the equations of motion according to Hénon's scheme [404]. Fig. 6.2 shows the dependence of the string steady state response on the excitation frequency ratio at the stop. The response amplitude was Poincaré sampled once per forcing frequency at zero–phase relative to the excitation. Over an excitation frequency range $0.9 < \nu < 0.9936$, the response is periodic with period–one and non–impacting. At $\nu = 0.9936$ grazing occurs with high–period motion then followed by chaotic motion. The boundaries separating between impact and non–impact oscillations were generated using the multiple–scales method where the response amplitude was set equal to the stop gap. On the plane of excitation amplitude–frequency these boundaries are shown in Fig. 6.3 for two different values of stop gap according to linear and nonlinear analyses. The experimental results are shown by crosses ($\times$) and empty circles ($\bigcirc$). This figure reveals that the experimental results are in good agreement with the predicted ones only for excitation frequency ratio below resonance above which the results drastically differ. The observed deviation may be attributed to the inaccuracy of the method of multiple scales. This could be verified by conducting a numerical simulation.

6.3 Beams

6.3.1 Overview

Structural elements such as beams are usually subjected to dynamic excitations including impact loading. The dynamics of beam elements involving amplitude constraint barriers was extensively studied in the literature ([1062], [1063], [208], [834], [690], [926], [596], [197], [113], [114], [115], [1053], [1051], [1012], [645], [646]). Maezawa and Watanabe [632] found that the coefficient of restitution is not a constant parameter and the duration of collision is not zero. The forced vibration of a structure with an added spring constraint acting at a point was considered by Davies and Rogers [209]. A set of constrained vibration modes was obtained in terms of the assumed known modes of the unconstrained structure. It was shown that the unconstrained modes form a complete set for the constrained beam. The response was described in terms of constrained or unconstrained modes. The two descriptions were shown to be equivalent only if the damping is independent of the mode number.

Masri [658] and Masri et al [663] conducted analytical and experimental investigations of the response of a cantilever beam whose free–end oscillates against a single–sided elastic spring and its clamped end is subjected to harmonic excitation. The predicted and measured results were compared and the qualitative difference between the two results was found in the response of the system with increasing and decreasing the excitation frequency. A computer algorithm was developed by Fathi et al [309] to assess appropriate clearances and stiffness parameters of a given number of vibro–impact stops in a homogeneous and uniform beam colliding with a single stop. The dynamic response of a cantilever beam with its free end oscillates against double sided barriers was studied by Chattopadhyay and Saxena [162] under base harmonic excitation of the clamped end. The response was measured at the loose support for different values of excitation frequency for given initial gap and gap stiffness. The response was found to exhibit discontinuous resonance curves and multiple valued responses typical of nonlinear systems. Fathi and Popplewell [308] described computational strategies to accurately calculate the contact forces and their peaks when a beam impacts a resilient stop. The modified equation of motion of Timoshenko beam was developed by Chen et al [163] who obtained the impact response formulae under the impact of a lumped mass.

Lo [599] studied the contact chatter of a cantilever beam with the free end pressed against a barrier. The beam was modeled by the Bernoulli–Euler beam theory and the resulting integral equation was solved by the small time increment technique. Deflections, contact force, and chatter were calculated. The number of modes included in the solution, which depends on the fixed contact stiffness, was found to have a great effect on the details of chatter, but to have little effect on the overall pattern of chatter or deflection. Moon and Shaw [690] showed that forced vibrations of an elastic beam with nonlinear boundary conditions exhibit chaotic behavior of the strange attractor type under a sinusoidal forcing excitation. The beam was clamped at one end, and the other end was pinned for the tip displacement less than some fixed value and was free for displacements greater than this value. Subharmonic oscillations were dominating these types of motion. For certain values of forcing frequency and amplitude, the periodic motion was found to be unstable and non–periodic with bounded vibrations. These chaotic motions have a narrow band spectrum of frequency components near the subharmonic frequencies. Later, Shaw ([926], [927]) presented experimental results and compared them with those predicted analytically by Moon and Shaw [690]. Bishop [113] confirmed similar results for an impacting beam driven near its linear natural frequency.

The dynamic response of a thin cantilever beam impacting against an elastic stop of general three–dimensional geometry was studied by Wang and Kim ([1055], [1056]). The contact area between the beam and barrier was taken as the control parameter. Variation of the contact area was found to result in significant changes of the contact force. Furthermore, it was shown that at a point relatively far from the contact center the dynamic response is slightly influenced by the assumed area. It was also reported that fatigue failures of thin beams usually occur near the free edge, relatively far from the contact point, where the stress wave is reflected. Impact fatigue caused by a cyclic repetition of low energy and low velocity impacts can have detrimental effect on the performance and reliability of structural components ([463], [479]). For example it has the potential to initiate a crack and to cause its rapid propagation. Silberschmidt et al [943] analyzed the impact fatigue and associated mechanisms for specific features of cracking in adhesively bonded joints.

Fang and Wickert [307], Van de Vorst et al [1023], and Yagci et al [1091] conducted experimental investigations to measure the response of a cantilever beam carrying an end mass and that could impact a vibrating surface. The response exhibited a recurring pattern of resonances, period doubling bifurcations, and regions of aperiodic motions. Ervin and Wickert ([299], [300]) conducted another experimental study on a beam structure carrying a rigid body and impacts against compliant base structure subjected to sinusoidal excitation. It was demonstrated that the response characteristics associated with repetitive impact become increasingly complex as the eccentricity of the impact location and gap clearance grow. The contact force, impulse, and displacement were found to exhibit complex response characteristics as a function of the excitation frequency.

Numerical and experimental studies were conducted by Emaci et al [294] and Emad et al [295] to examine the nonlinear motion confinement phenomena in a nonlinear flexible assembly of two coupled cantilever beams whose motion is constrained by rigid barriers. The impact nonlinearities were simulated by clearance nonlinearities with steep stiffness characteristics. The predicted results confirmed the properties of the phenomenon of motion confinement. It was shown that under certain conditions the vibrational energy of the system is passively confined to only one of the two beams. The experimental results confirmed analytical predictions both in the transient and steady state regimes.

The impact problem of a helicopter blade droop stop was analyzed by Han et al [387]. It was shown that the coupling of elastic deflections with rigid motions is distinctive. When the rotation angle is large, couplings increase distinctly and the differences between the results of two different computational methods increase. Furthermore, the bending stiffness was found to have a direct influence on the coupling such that the coupling was found to increase with decreasing the bending stiffness.

Yin et al [1107] examined the transient behavior of a cantilever beam subjected to periodic excitation against a rod–like barrier. As impact and separation phase take place alternately, the transient waves induced either by impacts or by separations were found to travel in more complicated ways. In both impact and separation phases, the transient wave propagations were solved by the expansion of transient wave functions in a series of eigenfunctions (wave modes). Numerical results showed the convergence of the time–step size and truncation number of wave modes in the calculations of impact force. The results also revealed several transient phenomena involving the propagation of transient impact–induced waves, sub–impact phases, long–term impact motion, chatter, sticking motion, synchronous impact, non–synchronous impact (including asynchronous impact) and impact loss.

The case of a cantilever beam with its free end constraint between one (asymmetric collision) or two stops (symmetric collision) has received extensive analytical, numerical and experimental investigations ([690], [926], [21], [22]). Contrary to the case of continuous systems with symmetrical collision characteristics, the resonance curves of nonlinear response of approximate solution were shown as discontinuous. Bishop and Xu [116] performed numerical simulations to examine the behavior of a mathematical model approximating the response of a beam held vertically and clamped at its base. The beam was driven to impact against a motion limiting constraint. Control techniques have been used to eliminate chaotic motion followed impacts by tracking lower periodic solutions. The stabilized motions near the grazing incidence have lower impact velocities than the naturally existing stable solutions leading to a significant reduction of the impact force. Balachandran [72] studied experimentally and numerically the dynamics of a cantilever beam subjected to impact excitations in the form of harmonic and aharmonic functions of the form $Y_0|\cos\Omega t|$. During harmonic impacting motions, period–doubled motions, incomplete period–doubling sequences, aperiodic motions, and multiple responses were observed. Furthermore, during half–sine impact motions the response exhibited period–3 motions and modulated motions.

6.3.2 Analysis

With reference to Fig. 6.4(a), the equation of motion of the beam may be written in the form [22]

$$\frac{\partial^2 z}{\partial t^2}+\frac{EI}{\rho A}\frac{\partial^4 z}{\partial x^4}=Y_0\Omega^2\cos\Omega t, \tag{6.5}$$

where $z(x,t)$ is the beam lateral displacement, ρ is the beam mass density, A is the beam cross–section area, EI is the beam flexural rigidity, and $y = z + Y_0\cos\Omega t$. Note that the origin of the nonlinear boundary condition at the beam free end is mainly due to the collision with the barrier and is represented by the piecewise condition

$$EI\frac{\partial^3 z}{\partial x^3}|_{x=L}=\begin{cases}K\left(z_L-z_0\right) & if\ z_L\geq z_0\\ 0 & if\ z_L\leq z_0\end{cases}, \tag{6.6}$$

where K is the contact stiffness between the beam end and the barrier, and z_0 is the clearance. Based on the assumption of one impact per one excitation period, Aoki and Watanabe [22] represented the nonlinear impact force by a periodic function, $g(\theta)$ with $\theta=\Omega t$ of period 2π, as shown in Fig. 6.4(b), i.e.,

$$g(\theta)=\begin{cases}K\left(z_L-z_0\right) & -\frac{\theta_0}{2}\leq\theta\leq\frac{\theta_0}{2}\\ 0 & \frac{\theta_0}{2}\leq\theta\leq 2\pi-\frac{\theta_0}{2}\end{cases}, \tag{6.7}$$

where θ_0 denotes the value of the phase angle over which the contact of the beam end with the stop takes place. The representation of equation (6.7) states that one period of θ, i.e., 2π, of the resulting vibration is divided into two intervals. The first is of length θ_0 over which the beam end moves in contact as shown in Fig. 6.4(b). The second interval is of length $2\pi-\frac{\theta_0}{2}$ over which the beam end experiences free flight. The function $g(\theta)$ was then expanded into a Fourier series and the first two terms were retained to give

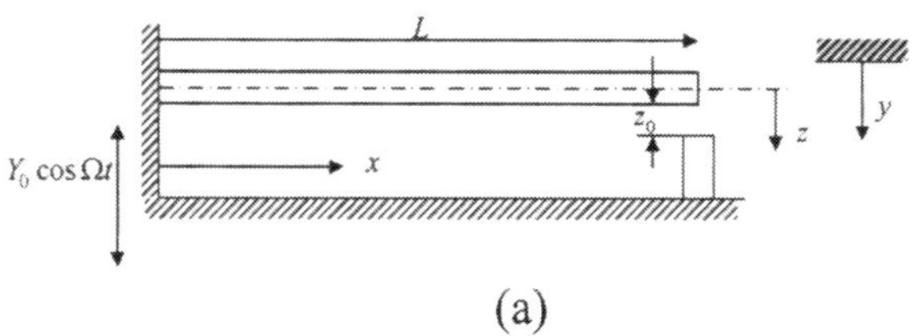

(a)

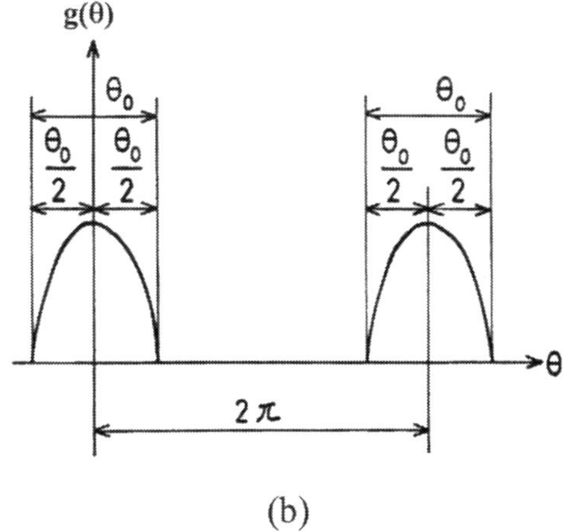

(b)

Fig. 6.4. Cantilver beam with one-sided restraint at the free end: (a) schematic diagram of a beam with unsymmetric boundary condition, and (b) periodicity of boundary force, [22].

$$g(\theta) = \frac{a_0}{2} + a_1 \cos \theta. \tag{6.8}$$

Watanabe et al [1065] and Watanabe and Shebata [1064] showed that when $g(\theta)$ is approximated by the fundamental term of the Fourier series, the approximate solution agrees with the exact solution for low values of the stiffness ratio $KL^3/(3EI) = K/k$, where $k = 3EI/L^3$. Applying the boundary conditions including equation (6.6), Aoki and Watanabe [22] obtained the following solution

$$\begin{aligned} z(x, \theta) &= Y_0(\cos \lambda x - 1) \cos \theta \\ &\quad + [A_1 (\cosh \lambda x - \cos \lambda x) + B_1 (\sinh \lambda x - \sin \lambda x)] \cos \theta, \end{aligned} \tag{6.9}$$

where

$$\lambda = \tfrac{1.8751}{L} \sqrt{\Omega/\omega_1}, \; \omega_1 = \tfrac{3.516}{L^2} \sqrt{EI/(\rho A)},$$

$$A_1 = \frac{1}{\Delta}[Y_0 (1 + \cosh \lambda x \cos \lambda x + \sinh \lambda x \sin \lambda x) - \frac{(a_0/2) + a_1 \cos\theta}{EI\lambda^3 \cos\theta} (\sinh \lambda x + \sin \lambda x)\Big\} ,$$

$$B_1 = \frac{1}{\Delta}[-Y_0 (\cosh \lambda x \sin \lambda x + \sinh \lambda x \cos \lambda x) - \frac{(a_0/2) + a_1 \cos\theta}{EI\lambda^3 \cos\theta} (\cosh \lambda x + \cos \lambda x)\Big\} , \quad and$$

$$\Delta = 2 (1 + \cosh \lambda x \cos \lambda x) .$$

The displacement of the beam end, $z(L, \theta)$, was obtained in the form

$$x(l, \theta) = \frac{Y_0 N_L \cos\theta}{1 - M_1 (a_1 L^3/(3EI\Gamma))} + \frac{M_1}{(3EI/L^3)} \cdot \frac{3\lambda_1 L g(\theta)}{(1.8751)^2(\Omega/\omega_1)^2}, \tag{6.10}$$

where

$$\Gamma = Y_0 N_L + \frac{M_1 a_1}{(3EI/L^3)} , \; N_1 = \frac{(1-\cos\lambda_1 L)(\cosh\lambda_1 L-1)}{1+\cos(\lambda_1 L)\cosh(\lambda_1 L)},$$
$$M_1 = \frac{\cos(\lambda_1 L)\sinh(\lambda_1 L)-\cosh(\lambda_1 L)\sin(\lambda_1 L)}{1+\cos(\lambda_1 L)\cos(\lambda_1 L)}.$$

For a given frequency ratio Ω/ω_1, Aoki and Watanabe (1998) obtained the following cubic equation in θ_0

$$\frac{M_1 K}{12\pi k}\left(1 + \frac{Y_0 N_L}{2Z_0}\right)\theta_0^3 - \frac{N_L Y_0}{8Z_0}\theta_0^2 + \frac{N_L Y_0}{Z_0} - 1 = 0 \tag{6.11}$$

Equation (6.11) was found to possess only one real root which establishes the interval over which impact occurs. θ_0 was obtained for the case of $\Gamma/Z_0 > 1$. Typical values of θ_0 were found to be less than $\pi/2$. The impact was assumed to take place once in one period of the beam vibration. In the presence of energy dissipation in the form of hysteresis loop, Aoki and Watanabe [22] obtained an algebraic equation of order 8 in θ_0. Bishop et al [114] used a simple mathematical model utilizing a coefficient of restitution rule to capture qualitative behavior of an experimental apparatus allowing the parameter space to be divided into zones according to their behavior type. Their results identified the zones, which separate regular period–1 impacting solutions from irregular, apparently chaotic, impacting and non–impacting motions.

Bishop et al [115] considered the dynamical response of a constrained thin beam held fixed at one end under external excitation. It was demonstrated that the parameter space around the natural frequency of the beam can be divided into two main regimes separated by a codimension–two bifurcation. It was shown that a smooth transition from non–impacting to one impact per period is possible by winding frequency up or down through a

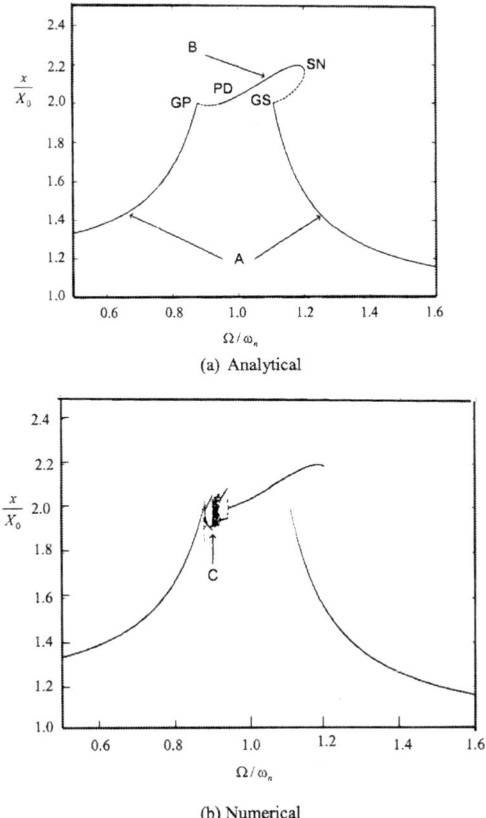

Fig. 6.5. Amplitude-frequency response of a cantlever beam restrained at the free end: (a) analytical, and (b) numerical predictions, solid curves belong to stable response of one impact per period while dashed curves belong to unstable for $F = 0.25$, [115].

critical value of the excitation amplitude. Above this critical value, jump phenomenon took place between impacting and non–impacting solutions. Figs. 6.5(a) and 6.5(b) show the amplitude–frequency response curves as predicted analytically and numerically, respectively for $F = 0.25$. Curves 'A' belong to non–impacting response, while curve 'B' belongs to one impact per period responses. Stable period–1 non–impacting motion was found to exist as the excitation frequency was increasing from zero. As Ω increases the response amplitude grows until it approaches the barrier at which grazing occurs at the point GP, where $\Omega/\omega_1 = 0.875$ producing a zero–velocity impact. Beyond the grazing point, the response loses its stability. Fig. 6.5(b) shows the subsequent motion may either be chaotic or experience complicated sequence of bifurcations. At point PD the response stabilizes to one impact per period via period–doubling bifurcation. This trend continues until the response reaches a saddle–node bifurcation at SN and the motion jumps to a

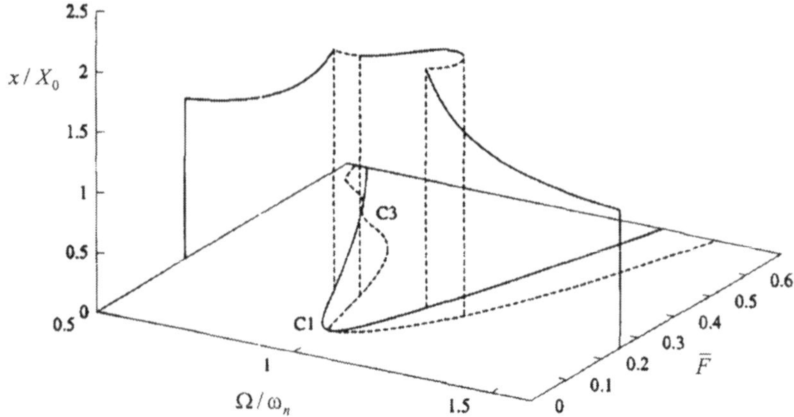

Fig. 6.6. Dependence of response amplitude ratio on excitation amplitude and frequency, [115].

non–impacting period–1 solution. As the excitation frequency decreases from $\Omega/\omega_1 = 1.6$, the non–impacting period–1 response grows until the grazing point GS at frequency ratio 1.11. This is followed by an impacting motion in the form of a hysteresis loop between GS and SN similar to the scenario described by Foale and Bishop ([334], [335]). Fig. 6.6 shows the locus of grazing bifurcations in the three–dimensional parameter space. The locus of the first grazing bifurcations is shown by the solid curve. The dashed curve between points '$C1$' and '$C3$' is the locus of period–doubling bifurcations. The other dashed curves represent the loci of saddle–node bifurcations. Points '$C1$' and '$C3$' belong to grazing, period–doubling and saddle–node loci meet, forming a codimension–two bifurcation ([454], [332]).

Wagg et al [1053] measured impact loads using an impact load cell. The impact load was observed to form spike train–type data. For vibro–impact motion of the beam, the duration of impacts was examined using a time of contact measure. The implications were discussed for vibro–impact systems mathematically modeled by using instantaneous impact assumptions (coefficient of restitution). The influence of noise and the data–acquisition process on the impact process was examined using numerical simulations of the experimental data. Later, Wagg and Bishop [1051] and Wagg [1044] applied a non–smooth dynamics approach to a cantilever beam discretized into several modes. Numerical simulations were compared with experimentally measured data for a flexible beam constrained to impact on one side.

6.4 Constrained Pipes Conveying Liquid

6.4.1 Two–Dimensional Dynamics

Pipes conveying fluid may experience fluidelastic instability in the form of static buckling (divergence) or unstable oscillations (flutter) similar to those encountered in aeroelastic structures ([764], [765], [766]). The case of divergence may occur in pipes supported at both ends, while flutter instability may occur for cantilevered pipes. Experimental investigations conducted by Gregory and Païdoussis [378] revealed that at sufficiently high flow speed the cantilevered pipe loses stability by flutter (Hopf bifurcation), leading to a stable limit–cycle planar motion. The orientation of this planar motion is usually dictated by the inherent imperfections in the axial symmetry of the pipe. The amplitude of oscillations was found to increase with flow velocity, so that if the motion constraints are appropriately positioned, then above a certain flow speed the pipe bangs on one or both constraints. In other cases, this planar impact oscillation deteriorates into a three–dimensional oscillation. Chen [164] considered a fluid conveying tube clamped at the upstream end and supported by a displacement spring at the other end. Stability and instability boundaries were constructed. The spring was found to have a destabilizing effect on the tube for certain ranges of the system parameters. Other early studies ([468], [970]) considered the case of pipes supported on a linear spring. It was found that the system exhibits instability in the form of either divergence or flutter. Edelstein and Chen [283] and Makrides and Edelstein [635] considered the same system but with a variable knife–edge support at some interior point of the free end. The support position was found to have a significant effect on the stability of the system.

Païdoussis and Moon [771] studied the dynamic behavior of a cantilevered pipe restricted by nonlinear motion restraints. For flow velocities sufficiently higher than the critical value for the Hopf bifurcation, the limit cycle motion was found to be large enough for the pipe to experience impact. As the flow speed was increased, a series of period doubling bifurcations led to chaos. It was argued that the mechanism leading to chaos was related to the interaction of limit–cycle oscillation occurring beyond Hopf bifurcation and potential wells associated with incipient divergence of the pipe with the motion barriers. Another study by Tang and Dowell [981] considered a pipe–beam system with two permanent magnets placed on both sides of the free end of the system. The force of the magnetic field was modeled by linear and cubic terms. It was found that chaotic oscillations may exist due to flutter instability alone when the flow velocity reaches a certain value and other parameters have appropriate values. Under forced external sinusoidal excitation, the characteristics of pipe–beam dynamics were found to be sensitive to both the damping and the number of degrees of freedom used in the model. By increasing the number of degrees of freedom, Païdoussis et al ([769], [770], [767]) and Païdoussis and Semler [772] showed that convergence is achieved with four or five degrees of freedom, which was confirmed experimentally. It was shown experimentally that a cantilevered pipe interacting with motion–limiting nonlinear constraints exhibits regions of chaotic motions. Motions of the system, sensed by an optical tracking system, were analyzed and the values of the fractal dimension of the system 1.03, 1.53, and 3.20 were obtained in the period–1, 'fuzzy' period–2 and chaotic regimes of oscillation, respectively. A four–dimensional analytical model was found to capture the essential dynamical features of observed behavior.

The chaotic dynamics of heat exchanger tubes impacting on loose baffle plates was studied by using an analytic model that involves delay differential equations [768]. The critical flow velocity for the local instability near the static equilibrium position of the flexible cylinder was obtained by assuming a harmonic solution in the discretized linearized model. It was found that by increasing the flow velocity beyond the critical value, the amplitude of oscillation grows until impacting with the loose support occurs. More complex motions then arise, leading to chaos for a sufficiently high flow velocity. A Lyapunov exponent technique was developed for delay differential equations and showed definitely that chaotic motions do occur. Fredriksson et al [343] reported some experimental results of a cantilevered pipe conveying fluid unilaterally constrained. The transition from stable periodic non–impacting motion to impacting motion, due to variations of parameters, was observed over a wide range of vibro–impact systems. The transition was found to emerge in a supercritical Hopf bifurcation. If the onset of impacting motion is close to the Hopf bifurcation, the impacting motion would exhibit chaotic behavior.

Païdoussis and Semler [773] considered the case of an intermediate spring support, and studied the stability of the original equilibrium. The regions in the parameter space where the system is stable or loses stability by divergence or flutter were determined. The stability of the other fixed points that emerge with increasing flow velocity was obtained for various system parameters. It revealed a very rich bifurcational behavior. The dynamics in the presence of harmonic perturbations in the flow was examined in the neighborhood of the double degeneracy, where heteroclinic orbits arise, and chaotic regions were shown to exist.

De Langre et al [213] and De Langre and Lebreton [214] developed a simplified model of loosely supported tubes impacting on elastic supports. Impact forces and sliding velocities were found to experience significant changes according to the vibratory regime. The system exhibited chaotic behavior due to impact nonlinearities at the supports. This phenomenon was studied both experimentally and numerically in order to test the ability of computer methods to predict chaos in such systems. The range of physical parameters where chaotic motion occurs was identified using spectral analysis at low frequency. A numerical time–stepping integration of the equation of motion was adopted and the amplitudes as well as chaotic regions of parameters were found in good agreement with experimental results.

The dynamics of a slightly modified pipe with the motion–limiting constraints and a linear spring support was studied by Jin [473] and Jin and Zou [474]. It was shown that for small flow velocities the static equilibrium was found stable for any value of the spring stiffness. However, when the flow velocity is relatively large, the pipe loses stability either by divergence if the spring stiffness is relatively large, or by flutter if it is relatively small. The numerical simulation revealed seven sub–regions in the flutter region. In each of these sub–regions a different behavior was found to emerge including the chaotic motions of the pipe. It was also found that there exist the quasi–periodic motions and route to chaos through breakup of the quasi–periodic torus surface in some parameter region of the system, which differs from that of periodic–doubling bifurcation route.

Steady nonlinear vibratory tube motions are possible due to the limiting effect of the support clearances. A convenient method was developed for the electro–mechanical simulation of unstable conditions, and the results were reported on vibro–impact tests using a planar system. Vento et al [1030] conducted experimental investigation allowing for two–dimensional tube motions. Experimental tests utilized a straight tube with an instrumented circular clearance–support at mid span and obtained a limited number of vibro–impact tube responses. The characteristics of steady vibrations were found to depend on the instability level, tube–support eccentricity and the initial conditions. In most of the experiments, tube responses became almost planar after some impacts. However, for a few tests, more complex orbital motions were observed.

Yau et al [1101] designed an active vibration control system to suppress the undesirable chaotic vibration in a constrained flexible pipe conveying fluid, which exhibits regions of flutter and chaotic motions at sufficiently high flow velocity. A four–dimensional analytical model obtained was utilized for designing the control law. Yau et al [1101] applied an optimal regulator theory to obtain feedback gains to stabilize the system with full state information. A state observer was added to estimate the required state signals. Furthermore, a robust controller based on quantitative feedback theory (QFT) scheme was developed. It was found that the QFT scheme provided stability robustness with respect to flow velocity variations.

6.4.2 Three–Dimensional Dynamics

The three–dimensional dynamics of a cantilevered pipe with an intermediate rotationally symmetric spring support was studies by Steindl and Troger ([955], [956], [957], [958], [959]). The three–dimensional transversal motion was constrained by an elastic support, which has the symmetry of the square. Kirchhoff's rod theory and the Kelvin–Voigt viscoelastic law were used to derive the pipe equations under the assumption of large displacement but small strain. The loss of stability of the trivial equilibrium position was studied.

In a three–part study, Païdoussis and his co–workers ([1043], [774], [684]) considered the three–dimensional nonlinear dynamics of unrestrained and restrained cantilevered pipes conveying fluid. The full derivation of the equations of motion of a cantilevered pipe using a modified version of Hamilton's principle, was adopted. Intermediate nonlinear spring constraints were incorporated into the equations of motion via the method of virtual work. A point mass fixed at the free end of the pipe was also added to the system. The effects of arrays of four or two springs or a single spring at a point along the pipe length with different spring configurations, and points of attachment on the pipe dynamics were examined. The main generic difference was found that in some cases, the system loses stability by planar flutter, and thereafter performs two–dimensional or three–dimensional periodic, quasi–periodic and chaotic oscillations. In other cases, the system was found to lose stability by divergence, followed at higher flows by oscillations in the plane of divergence or perpendicular to it, in the form of periodic, quasi–periodic or chaotic motion. These results were verified experimentally. When an additional "point" mass was attached at the free end, the dynamical behavior was analyzed in the form of a bifurcation diagram, along with time history records, phase–plane plots, power spectra and Poincaré maps. The results revealed planar periodic, quasi–periodic and chaotic oscillations, followed by three–dimensional quasi–periodic and chaotic motions.

The flow–induced vibration of a nonlinear restrained curved pipe conveying fluid was studied by Wang et al [1058] and Wang and Ni [1057]. The nonlinear partial–differential equation governing the pipe in–plane vibration was discretized using the differential quadrature. The numerical simulation revealed several complex dynamic characteristics, such as limit cycle oscillations, chaotic motion, and static buckling, depending on the fluid flow velocity parameter. The same system was also considered by Qiao et al [876] and Lin and Qiao [597]. Under harmonic excitation, the numerical results indicated that the pipe without motion–limiting constraints behaves as an ordinary linear system. On the other hand, if the pipe is subjected to cubic motion–limiting constraints, nonlinear dynamic phenomena emerge. The route to chaos was shown to be reached through a sequence of period–doubling bifurcations. The problem of a block repeatedly sliding, jumping and being conveyed inside a vibrating spatial–curved tube was studied by Long et al [600]. Analytical models and the governing equations of motion were developed taking into account three types of motion regimes. These were relative sliding motion, flying motion and the bumping process. The study included the inclined friction with component velocities for oblique impact. The detailed motion of the entire response, manifested in the form of 'subperiodic motion', was revealed through numerical solutions.

6.4.3 Modeling and Response Analysis

The nonlinear equations of motion of pipes conveying fluid were developed using energy and Newtonian methods by Semler et al (1994). The derivations were made for cantilevered and clamped end pipes. It was shown that the origin of various terms and the structure of the equations of motion are distinctly different in these two cases. A critical assessment of derivations developed by others was presented. Some of the equations were found to be fully correct, while others were found to be inadequate, due to either the assumptions made or inconsistencies in the derivations; for pipes with both

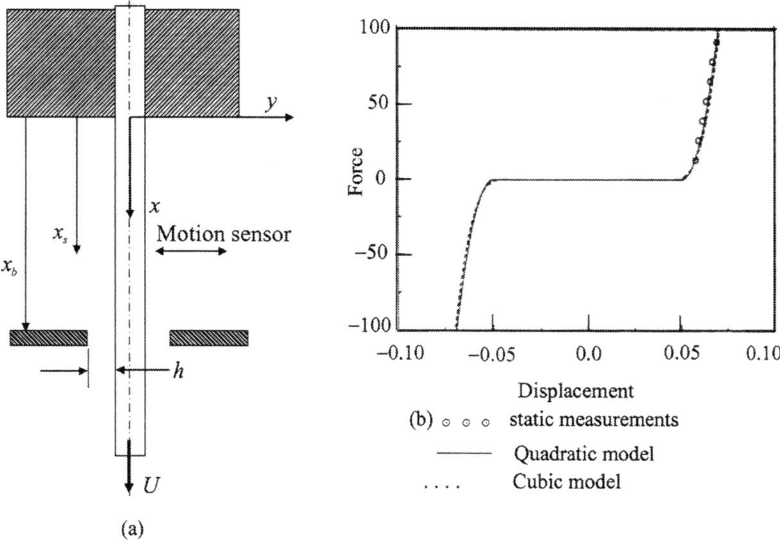

Fig. 6.7. Restraint pipe and restrained force characteristics: (a) schematic diagram of a constrained cantilevered pipe, (b) dependence of restrained force on the displacement, [770].

ends fixed. It was claimed that the equations presented by Semler et al [920] are complete and correct.

In modeling the dynamics of a constrained cantilevered pipe shown in Fig. 6.7(a), Païdoussis et al ([769], [770]) considered a tubular vertical beam of length L, internal cross–sectional area A, mass peer unit length m, flexural rigidity EI and coefficient of Kelvin–Voigt damping c, conveying a fluid of mass M per unit length with a constant axial velocity U. The equation of motion of the pipe in the presence of motion limiting restraints may be described by the Euler–Bernoulli beam theory in the non–dimensional form, ([769], [770], [772]),

$$\frac{\partial^2 \eta}{\partial \tau^2} + \left(1 + \alpha \frac{\partial}{\partial \tau}\right) \frac{\partial^4 \eta}{\partial \xi^4} + \left[u^2 - \gamma(1-\xi)\right] \frac{\partial^2 \eta}{\partial \xi^2}$$
$$+ 2u\sqrt{\mu} \frac{\partial^2 \eta}{\partial \xi \partial \tau} + \gamma \frac{\partial \eta}{\partial \xi} + f(\eta)\delta(\xi - \xi_b) = 0 \tag{6.12}$$

where the last term represents the high power nonlinear impact force due to the presence of restraint at $x = x_b$, and $\delta(\cdot)$ is the Dirac delta function. $\eta = w/L$, $\xi = x/L$, $u = UL\sqrt{M/EI}$, $\tau = t\sqrt{EI/(M+m)}/L$, $\mu = M/(M+m)$, $\gamma = (M+m)gL^3/EI$, $\alpha = c\sqrt{EI/(M+m)}/L$, and $f(\eta) = F(w)L^3/EI$.

The stiffness of the constraint was measured and confirmed the power law phenomenological modeling, $f(\eta) = \kappa_n \eta^{2n-1}$, as described in Chapter 1. With reference to Fig. 6.7(b), Païdoussis et al [769] adopted the impact cubic model:

$$f(\eta) = \kappa \eta^3 \tag{6.13}$$

where $\kappa = kL^5/EI$, k is the stiffness of the cubic spring. Alternatively, Païdoussis et al [769] and Païdoussis and Semler [772] adopted smoothened tri–linear spring model

$$f(\eta) = \kappa_n \left[\eta - \frac{1}{5} \left(|\eta + \eta_{nb}| - |\eta - \eta_{nb}| \right) \right]^n \tag{6.14}$$

This modeling enables one to represent adequately the free gap in which the constraints are zero and to smoothen the sharp discontinuity at $|\eta| = \eta_b$. Païdoussis and Semler [772] adopted the cubic, $n = 3$, tri–linear model with $\kappa_3 = 5.5 \times 106$, and $\eta_{b3} = 0.044$. The approximation of the cubic spring stiffness was found to be $\kappa = 10^5$, which fits the experimental measurements.

Païdoussis and Semler [772] obtained added inertia nonlinearity to equation (6.12) to account for large pipe deflection. Equation (6.12) was discretized through the modal expansion

$$\eta(\xi, \tau) = \sum_i \phi_i(\xi) q_i(\tau)$$

together with Galerkin's method for the first two modes. This resulted in the following equations in the matrix form

$$\left\{ \frac{d^2 q}{d\tau^2} \right\} + [C] \left\{ \frac{dq}{d\tau} \right\} + [K] \{q\} + \{f(q)\} = \{0\} \tag{6.15}$$

The elements of damping, C_{sr}, stiffness, K_{sr}, and the nonlinear stiffness, $f(q)$, matrices are $C_{sr} = \alpha \lambda_r^4 \delta_{sr} + 2ub_{sr}\sqrt{\mu}$, $K_{sr} = \lambda_r^4 \delta_{sr} + \left(u^2 - \gamma\right) c_{sr} + \gamma \left(d_{sr} + b_{sr}\right)$, $f_r = \kappa \left[\phi_{rb} q_r + \phi_{sb} q_r\right]^3 \phi_{rb}$. The constants b_{sr}, c_{sr} and d_{sr} are documented by Païdoussis and Moon [771]. λ_r is the non–dimensional eigenvalue of the cantilever pipe.

For mass ratio $\mu = 0.2$, weight to stiffness ratio $\gamma = 10$, damping parameter $\alpha = 5 \times 10^{-3}$, barrier location $\xi_b = 0.82$, and barrier stiffness coefficient $\kappa = 100$, a series of numerical simulations were performed to construct the bifurcation diagrams shown in Figs. 6.8(a–c). These diagrams represent the dependence of the pipe free–end displacement on the flow velocity parameter. The displacement was estimated using two–mode approximation of the free end displacement, i.e., $\eta(1, \tau) = \phi_1(1) a_{q1}(\tau) + \phi_2(1) q_2(\tau)$. The corresponding velocity was estimated and the displacement corresponding to zero–velocity was recorded to obtain the bifurcation diagrams of Fig. 6.8. The symmetry of the solutions was broken for certain values of the flow velocity, u. Païdoussis

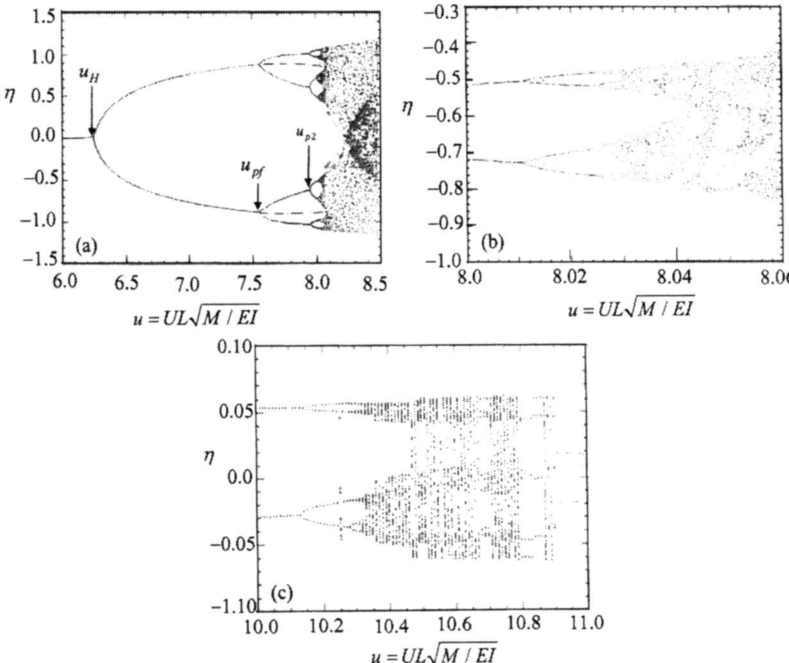

Fig. 6.8. Segments of bifurcation diagram over different values of flow speed parameter showing the dependence of the free-end displacement on the flow speed parameter, for $\mu = 0.2$, $\gamma = 10$, $\alpha = 5 \times 10^{-3}$, $\xi_b = 0.82$, and $\kappa = 100$: (a) lower speed region, (b) magnification over small range of flow speed, (c) large range of flow speed, ([769], [772]).

et al [769] found that chaotic motion occurs at the flow speed parameter $u = 8.03$ after a sequence of period–doubling bifurcations as shown in Figs. 6.8(a) and 6.8(b). Fig. 6.8(a) shows that Hopf bifurcation occurs at the critical flow speed $u_H \approx 6.22$ with two solution branches for $u > u_H$. At $u_{bf} \approx 7.52$, a pitchfork bifurcation occurs and the limit cycle oscillation loses its symmetry. The dashed–dot curves shown in Fig. 6.8(a) represent unstable solutions, and the upper and lower branches, in the positive and negative parts of the diagram, bifurcate into period doubling at u_{p2}. Fig. 6.8(b) is a magnification of the bifurcation diagram over a small range of flow speed ($u = 8.0$ to 8.06) and it shows a sequence of period–doubling bifurcations. The motion of the pipe was found to be narrow–band chaotic for $u > 8.03$, and then at $u \approx 8.2$ it becomes wide–band chaotic.

When the pipe inertia nonlinearity is included in the analytical modeling, the pipe does not experience any chaos, even for higher flow values of flow speed. In this case the inertia nonlinearity acts as a stabilizing mechanism. Another observation is that chaotic oscillations can occur for sufficiently large barrier location ξ_b. The flow velocity range in Fig. 6.8(c) was extended beyond the one given in Fig. 6.8(a), [771], and thus it reveals that the pipe regains stability after the region of chaos, around $u = 10.87$. This feature was observed experimentally where the pipe sticks permanently to one of the barriers. This implies that the system becomes unstable by divergence. The existence of fixed points was examined and it was indicated that a subcritical saddle–node bifurcation occurs at $u = 9.85$. This is followed by two fixed points one of them is stable and the other one unstable. Other bifurcation diagrams represented by the dependence of the pipe–end displacement on the location ξ_b of the barriers were also obtained for different values of barrier stiffness. It was shown that the pipe amplitude displacement decreases as the barrier location increases. Furthermore, as the barrier location increases different scenarios of response were reported. These include periodic solutions with symmetric limit–cycle oscillation, followed by a transcritical pitchfork bifurcation, then cascades of period–doubling bifurcations leading to chaotic motion.

6.5 Nuclear Reactors and Heat Exchangers

6.5.1 Fretting Wear Problem

Nuclear reactor components such as fuel rods and steam generator tubes are usually arranged with gaps at support points to allow for thermal expansion. In nuclear fuel systems, gaps may develop during service due to relaxation of support springs and creep down of cladding tubes. The rod cluster control assemblies in these plants can be damaged by impact–sliding wear due to flow–induced vibrations, which generate contacts with their guidance devices (guide tube) inside the reactor pressure vessel. Païdoussis [763] presented an excellent account of the practical experiences and state of knowledge of flow–induced vibrations in nuclear reactors and heat exchangers.

The occurrence of flow–induced vibration fretting wear in process equipment such as heat exchangers and steam generators accounts for the majority of failures due to vibration. Fretting (or fretting corrosion) refers to corrosion damage at the asperities of contact surfaces. This damage is induced under load and in the presence of repeated relative surface motion. Fretting wear prediction requires nonlinear computations of the tube dynamics in which proper modeling of the fluid forcing function plays an important role. Shin et al ([939], [940]) carried out a series of experimental tests to determine the effects of tube/support misalignment, tube/support–hole clearance, support thickness, exciting force amplitude, and support spacing on the vibrational characteristics (resonant frequencies, mode shapes, and damping) and displacement response amplitude of a heat exchanger tube. The test results were compared with analytical results based on a multi–span beam with simple intermittent supports.

Yetisir et al [1103] proposed a simple criterion to estimate fretting wear damage in heat exchanger tubes with clearance supports. The criterion was based on parameters such as vibration frequency, mid–span vibration amplitude, span length, tube mass, and an empirical wear coefficient. Fretting wear damage is known to be proportional to a parameter called "work rate." Work rate, defined later by equation 6(33), is a measure of the dynamic interaction between a vibrating tube and its supports. Work rate calculations for heat exchanger tubes require specialized nonlinear finite element codes. These codes will be described in the next subsection and they take into account contact models for various clearance support geometries. The proposed criterion was obtained from an extensive parametric study that was conducted using a nonlinear finite element program. It was shown that work rate can be estimated within a factor of two.

One of the parameters, which plays a vital role in the prediction of tube wear rate, is the impact force that occurs when the free displacements of the tube exceed the clearance in the support plates, resulting in a collision ([516], [916]). Sauve and Teper [916] developed an analytical approach to simulate the nonlinear dynamic–impact response of multi–supported tubes including U–bends and the effect of non–uniform gap clearances at the supports. The approach was incorporated into a computer code based on finite element and displacement methods using an unconditionally stable numerical integration scheme to solve the nonlinear equations of motion. The method simulated impacts between tubes and support plates in steam generators and heat exchangers in order to determine tube bundle susceptibility to fretting wear failure at the design stage or operational phase. The impact and/or sliding forces which occur at interfaces lead to progressive thinning of tube walls and ultimate failure in the extreme case. Fricker [345] presented a method for analyzing the impact behavior of a steam generator tube, which has one or more loose supports and is fluidelastically unstable. Fluidelastic instability under conditions of turbulence and nonlinearities in nuclear power plants was analyzed by Mureithi et al [699]. The results confirmed the existence of an attractor distinguishable from randomness.

Boucher and Taylor [131] conducted an experimental investigation to simulate two–phase cross flow in the region of U–bend tubes of heat exchangers. The effectiveness of U–bend tube restraints was studied by monitoring the tube vibration and the tube–to–support contact during single– and two–phase cross flow. Work–rate was taken as a measure of the contact criteria that combines both contact force and sliding distance. The work–rate is directly related to the wear rate between the tube and support. Effectiveness of the tube supports in the in–plane and out–of–plane directions was studied by examining tube response and the resultant work–rates for two different tube–to–support clearances. U–bend tube restraints limit the out–of–plane tube motion to the magnitude of the tube–to–support clearance while in–plane motion continues to increase with flow rate.

A robust feedback controller to suppress flutter–type chaotic vibrations in baffled heat exchanger tubes was presented by Bedout et al [101]. The vibrations are the result of the fluid dynamic forces on the tube, which behave as a negative damping element. These vibrations result in tubes impact with baffle plates. The heat exchanger tube and fluid dynamic forces acting on the tube were modeled with linear delayed differential equations. The feedback controller was realized using a frequency domain loop shaping approach. The control effector is a magnetic force transducer that acts on the heat exchanger tube. The feedback controller was shown to provide robust stability and performance over a large flow velocity regime.

6.5.2 Computational Methods

The relatively small inherent tube–to–baffle hole clearances associated with manufacturing tolerances in heat exchangers are known to affect the vibrational characteristics and the tube response. Various computational methods were employed to predict some aspects of vibro–impact dynamics. Rogers and Pick ([892], [893]) and Fisher et al [326] developed a dynamic finite element code VIBIC (Vibration of Beams with Intermittent Contacts) for predicting the motions and baffle contact forces of a single heat exchanger tube. The modal equations of motion were generated and numerically integrated. Essentially the numerical code VIBIC simulates the dynamic response of a heat exchanger tube as it impacts and rubs against its supports.

A simplified two–dimensional dynamic model describing the interaction of a heat exchanger tube and its support, with transitions to and from solid contact, was developed by Zhou and Rogers [1136]. A new friction model with six degrees of freedom was merged into the test code. Numerical analyses and simulations were performed. The simulation results for three typical tube motions were generated from the test code and one was from an improved VIBIC code. A WOrk–Rate Measuring Station (WORMS) was developed by Fisher et al [328] to measure the relative motion and contact forces between a vibrating fuel element and its support. These measurements confirmed numerical simulations of in–reactor interaction predicted earlier using the VIBIC code.

The fluidelastic vibration of a tube array caused by cross flow was examined by Nakamura and Fujita ([707], [708]) using a time–history simulation. It was possible to extract impact vibration from the time series. The computed results were compared with the measured impact force and were extended to account for a steam–water two–phase flow. Flow–induced vibration on a tube array caused by two–phase flow at high–pressure and high–temperature conditions was examined experimentally by Nakamura et al [709]. The experimental measurements included the turbulent buffeting force by air–water two–phase flow and by steam–water flow of extreme conditions up to pressure $5.8 MPa$ and temperature $272°C$. The main source of the buffeting force by two–phase flow in slug or froth flow pattern was recognized to be the impact force caused by the intermittently rising liquid slug. The slug speed and the fluid force acting on a tube were estimated, combined with the estimation of the intermittence of the occurrence of the liquid slug rising.

Numerical investigations utilizing the finite element method were conducted to simulate the nonlinear tube response. These studies considered tubes supported by loose baffle plates ([893], [916], [884]), broached hole [327], and flat–bar supports [1104]. Axisa et al [46] further elaborated on the method of Rogers and Pick ([892], [893]) to compute displacements and impact forces in tubular structures using a nonlinear tube–support contact force. The impact behavior of a periodically forced oscillator with limiting stops was studied by Nguyen et al ([726], [727]) and Axisa et al [44] for different values of excitation and system parameters. The results were utilized in related studies of noise and wear in mechanical systems with clearances. Later, Axisa and Izquierdo [47] utilized this method to assess experimental measurements on vibro–impact dynamics of loosely supported tubes under harmonic excitations. Sauve and Teper [916] analyzed the vibro–impact tube dynamics using finite element method and an unconditionally stable numerical scheme to solve the nonlinear equations of motion. They introduced Rayleigh proportional damping for the damping matrix in the equations of motion. Lin and Bapat [596] used three approaches to estimate clearances and impact forces for a simple undamped beam–stop system. The predictions of impact forces and clearances obtained using data from the mechanical experiments, were compared with measured impact forces and actual clearances, respectively, and were found in good agreement. Peterka ([817], [818]) examined the impact interaction of single and two heat exchanger tubes.

Hassan et al [396] presented numerical simulations of a loosely supported heat exchanger tube excited by turbulence. The effects of support clearance and flow orientation were examined for various support geometries and lattice–bar support offset. The finite element method was utilized to model the vibrations and the impact dynamics. Three different friction models were introduced to account for the tube/support friction forces. It was found that some flow orientations, support types, and support offsets provide favorable support conditions for higher tube sliding motion against the support. This results in potentially greater wear rates under service conditions. Furthermore, minimizing tube–to–support clearance was found beneficial in terms of reducing wear in heat exchangers. Offset of rhomboid–flat bar support was found to result in nonlinear coupling of tube modes and even greater complexity in tube dynamics. Increasing offset would result in increasing normal work rate since constraint of the tube by one support increases sliding contact with the adjacent support. Thus, wear should be minimized by minimizing support offset. Later, Hassan et al [397] and Hassan [392] proposed computational algorithms to examine the tube/support impact considering a finite support width. The tube/support contact was modeled by a distributed stiffness to account for the segment contact. The impact forces were distributed along the contact segment using the beam displacement interpolation function. The tube/support impact was shown to be a combination of edge (point) and

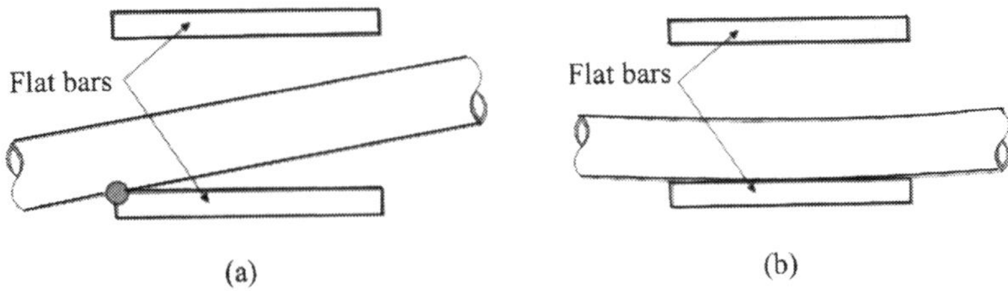

Fig. 6.9. Tube-contact modeling: (a) Point contact, (b) Segment Contact, [397].

segmental (line) contact. It was also found that smaller tube/support clearances tend to produce segmental contact and therefore larger contact areas as well as increased support stiffness against tube rotation.

Various support geometries currently used by manufacturers in heat exchanger U–bends were discussed by Weaver & Schneider [1066]. The geometry of the support affects the dynamics of the tube due to variances in the contact configurations. An accurate modeling of the nonlinear tube–support interaction is very important in order to predict its response. Note that the tube/support contact configuration may change with the heat exchanger operating conditions. Kim et al [503] and Hassan et al [392] classified the tube/support contact into two main categories shown in Fig. 6.9. The first is the point contact at the support edge and the second is the segment contact over a line. The two models were formulated in details by Hassan et al [397], and the following treatment is adapted from their derivation.

6.5.3 Point Contact Model

Fig. 6.10 shows a segment of the tube with the point contact model. The segment line AE connects the principal contact node A and the neighboring contact node E. Two side supports each modeled by a spring–dashpot system. Note the maximum penetration distance δ_B between the tube's surface B and the support plane is

$$\delta_B = d_n^B - C_r, \tag{6.16}$$

where C_r is the radial clearance and d_n^B is the normal displacement of point B.

The impact force, $\mathbf{F}_{imp}$, constitutes the spring force component, $\mathbf{F}_{pring}$, resulting from a concentrated spring at point B and the damping force, $\mathbf{F}_{damp}$, accounting for energy dissipation during impact, i.e.,

$$\mathbf{F}_{imp} = \mathbf{F}_{pring} + \mathbf{F}_{damp}, \tag{6.17}$$

where

$$\mathbf{F}_{pring} = (K_{spring}\delta_B) \cdot \mathbf{u}_{smn}, \tag{6.18a}$$

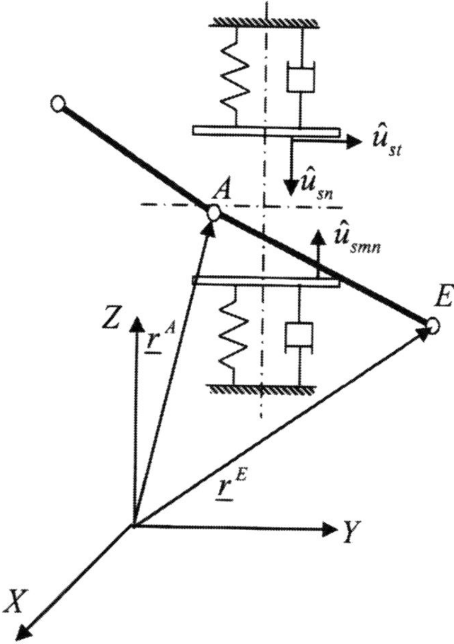

Fig. 6.10. Contact node set showing the principal contact node A and the neighboring contact node E, [397].

$$\mathbf{F}_{damp} = -sgn\left(\dot{\delta}_B\right)\left(\frac{3}{2}\zeta_{imp}|K_{spring}\delta_B|\right)\cdot\mathbf{u}_{smn}, \tag{6.18b}$$

$\mathbf{u}_{smn}$ is the support normal unit vector, ζ_{imp} is an impact damping parameter related to the coefficient of restitution [434]. Decomposing the impact force into tube local normal, F_{in}, and axial, F_{ia} components, gives

$$F_{in} = \mathbf{F}_{imp}\cdot\mathbf{u}_{tn}, \qquad F_{ia} = \mathbf{F}_{imp}\cdot\mathbf{u}_{ta} \tag{6.19a,b}$$

where $\mathbf{u}_{tn}$ and $\mathbf{u}_{ta}$ are the tube unit vectors along normal and axial directions.

The frictional force acting in the plane containing the tube tangential velocity and the axial unit vector requires the estimation of the tube tangential velocity in the friction plane. The resultant tangential velocity, V_{rt}^B, in the friction plane was determined by the vector sum of the tube velocity components in the directions of $\mathbf{u}_{ta}$ and $\mathbf{u}_{tt}$. Note that the velocity of point B must be obtained by interpolating the velocity with respect to the nodal velocities. Hassan et al [397] incorporated the following three friction models ([18], [976], [396]):

Velocity–limited friction model: This model was also adopted by Rogers and Pick [893], and Yetisir and Weaver [1104]. It employs a limiting velocity, V_0, to overcome the problem of discontinuity of the classical Coulomb friction model. The friction force, F_f, was modeled by the two expressions

$$|F_f| = \mu F_{in} \qquad if\ |V_t| > V_0, \tag{6.20a}$$

$$|F_f| = \frac{|V_t|}{V_0}\mu F_{in} \qquad if\ |V_t| \leq V_0, \tag{6.20b}$$

where μ is the kinetic coefficient of friction and F_{in} is the normal force.

The velocity–limited friction model was found to yield good tube response, impact force, contact ratio and work rate for a small preload and no substantial initial eccentricity. This model does not address the influence of sticking on wear. Under significant preload and initial contact, the next two models are recommended ([396], [394]).

Spring damper friction model: This model was modeled by Antunes et al [18] who introduced adherence stiffness, K_a, and an adherence damper, C_a, for the sticking force

$$|F_f| = \mu F_{in} \quad \text{for sliding}, \tag{6.21a}$$
$$|F_f| = K_a\,(d_c - d_0) + C_a V_t \quad \text{for sticking} \tag{6.21b}$$

where d_c and d_0 are the current and zero–velocity tangential displacements, respectively. Note that sticking is detected by a negative dot product of the present tangential velocity vector with its value at the previous time step.

Force balance friction model: This model was proposed by Tan and Rogers [976] who detected sticking whenever the absolute velocity is less than a small limiting velocity, V_0. During sticking the friction force was calculated such that it balances the net force, i.e.,

$$F_f = Ku - F_e \tag{6.22}$$

where Ku represents the internal tangential forces at the point of contact and F_e is the external force.

The condition $|F_f| < \mu_s F_{in}$ must be satisfied for the occurrence of sticking, where μ_s is the static friction coefficient.

Hassan et al [397] adopted the third modeling in developing a computational algorithm to describe tube/support impact interaction. The calculated resultant frictional forces and moments were translated to the tube local coordinates and the following results were obtained

$$\begin{aligned} F_{fa} &= \mathbf{F}_f \cdot \mathbf{u}_{ta}, \qquad & F_{ft} &= \mathbf{F}_f \cdot \mathbf{u}_{tt}, \\ M_{fa} &= \mathbf{M}_f \cdot \mathbf{u}_{ta}, \qquad & M_{ft} &= \mathbf{M}_f \cdot \mathbf{u}_{tt} \end{aligned} \tag{6.23}$$

The total contact forces and moments in the local tube directions are

$$\begin{aligned} F_{ca} &= F_{ia} + F_{fa}, \qquad & F_{cn} &= F_{in}, \qquad & F_{ct} &= F_{ft} \\ M_{ca} &= M_{fa}, \qquad & M_{cn} &= 0, \qquad & M_{ct} &= M_{ft} \end{aligned} \tag{6.24}$$

These components of forces and moments are defined at the impact point B. Since in the finite element algorithm loads are applied at the nodes, these components have to be expressed as concentrated loads and moments applied at the nodes. In this case the work done by impact forces is equivalent to the work done by equivalent concentrated forces at the nodes. Thus the work done, W, by the impact and frictional forces and moments is

$$W = F_{ca} d_a^B(\eta) + F_{cn} d_n^B(\eta) + F_{ct} d_t^B(\eta) + M_{ca} \theta_a^B(\eta) + M_{ct} \theta_t^B(\eta) \quad (6.25)$$

where $\eta = b/L_e$, b is the distance between point B and the node A, L_e is the length of the element of the tube AE, d_a^B, d_n^B, and d_t^B are the axial, normal, and tangential components of the displacement vector of the node, respectively, θ_a^B and θ_t^B are the rotational displacement components.

6.5.4 Segment Contact Model

When the displacements of the principal contact node A or the pair of points A and E exceeds the support gap, an overlap segment takes place as shown in Fig. 6.11. The strain energy, U_g, of the deformed distributed stiffness is

$$U_g = \frac{1}{2} K_g L_e \int \left[d_n(\xi) - C_r \right]^2 d\xi \quad (6.26)$$

where K_g is the gap stiffness, assumed constant over the overlap segment η. $d_n(\xi)$ is an interpolated displacement using the beam displacement shape

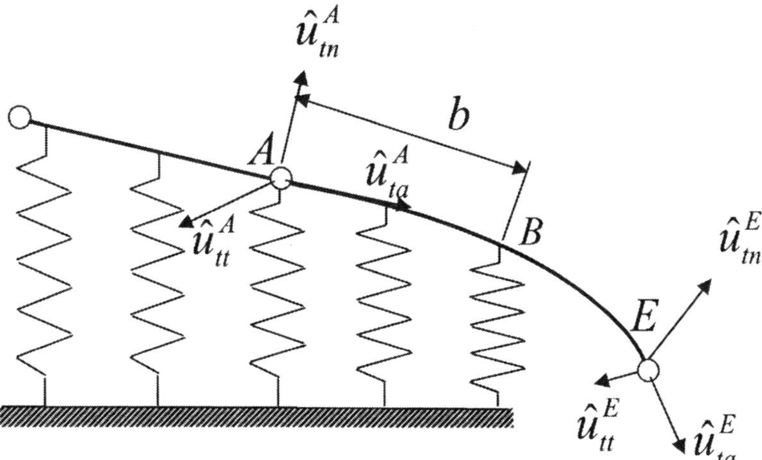

Fig. 6.11. Schematic diagram of the segment contact model showing a pair of nodes A and E, [397].

function. Upon integrating equation (6.26) the nodal forces and moments can be obtained by differentiating the strain energy with respect to the nodal degrees of freedom with the result

$$\begin{Bmatrix} F_n^A \\ F_t^A \\ M_n^A \\ M_t^A \end{Bmatrix} = \frac{1}{2} K_g L_e \begin{bmatrix} 2H_1 & L_e H_2 & H_3 & L_e H_4 & H_5 \\ L_e H_2 & L_e^2 H_6 & L_e H_7 & L_e^2 H_8 & L_e H_9 \\ H_3 & L_e H_7 & 2H_{10} & L_e H_{11} & H_{12} \\ L_e H_4 & L_e^2 H_8 & L_e H_{11} & 2L_e^2 H_{13} & L_e H_{14} \end{bmatrix} \begin{Bmatrix} d_n^A \\ \theta_n^A \\ d_n^E \\ \theta_n^E \end{Bmatrix} \quad (6.27)$$

where H_i, $i = 1, 2, 3, ..45$, are polynomials of the dimensionless contact segment η and are documented in reference [397]. The work done by the sliding friction forces, W_{fa}, in the axial direction is

$$W_{fa} = -\mu_d \left(\mathbf{u}_{tr} \cdot \mathbf{u}a \right) K_g L_e \left\{ d_a^A \; d_a^E \right\} \begin{bmatrix} H_{11} & H_{12} & H_{13} & H_{14} & H_{15} \\ H_{21} & H_{22} & H_{23} & H_{24} & H_{25} \end{bmatrix} \begin{Bmatrix} d_n^A \\ L_e \theta_t^A \\ d_a^E \\ L_e \theta_t^E \\ C_r \end{Bmatrix} \quad (6.28)$$

The consistent load vector describing the sliding tangential friction force is

$$\{F_{fa}\} = -\mu_d \left(\mathbf{u}_{tr} \cdot \mathbf{u}a \right) K_g L_e \begin{bmatrix} H_{11} & H_{12} & H_{13} & H_{14} & H_{15} \\ H_{21} & H_{22} & H_{23} & H_{24} & H_{25} \\ H_{31} & H_{32} & H_{33} & H_{34} & H_{35} \\ H_{41} & H_{42} & H_{43} & H_{44} & H_{45} \end{bmatrix} \begin{Bmatrix} d_n^A \\ L_e \theta_t^A \\ d_a^E \\ L_e \theta_t^E \\ C_r \end{Bmatrix} \quad (6.29)$$

where μ_d is the dynamic friction coefficient (projected on the friction plane).

The time history of the tube response was computed using the standard finite element solution of the beam equation

$$[M] \left\{ \ddot{d} \right\} + [C] \left\{ \dot{d} \right\} + [K] \{d\} = \{F_e(t)\} + \left\{ F_{imp} \left(d, \dot{d} \right) \right\} \quad (6.30)$$

The total power induced by turbulence, W_{tur}, and absorbed by the tube of length L and mass per unit length m may be expressed by the formula [1103]

$$W_{tur} = \sum \frac{S_{FF}(f_i) J_i^2 L}{2m} \quad (6.31)$$

where $S_{FF}(f_i)$ is the power spectral density of the local force per unit length in the ith mode and J_i^2 is the joint acceptance given by the expression ([40], [41])

$$J_i^2 = \frac{1}{L} \int_0^L \int_0^L \Phi_i(x') \Gamma(x', x'', \omega) \Phi_i(x'') dx' dx'' \quad (6.32)$$

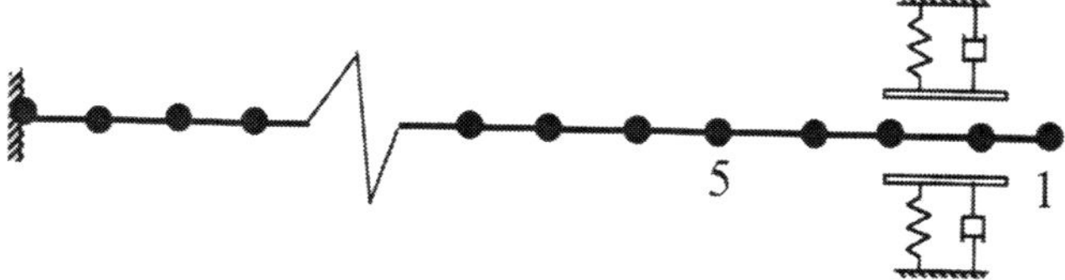

Fig. 6.12. Schematic diagram of the tube/support finite element model studied by Hassan et al, [397].

where $\Phi_i(x')$ is the mode shape function, $\Gamma = |S_{px'x''}(f)|^2 / [S_{px'}(f)S_{px''}(f)]$ is the coherence function, $S_{px'x''}(f)$ is the pressure cross–spectral density, $S_{px'}(f)$ is the pressure power spectral density. The theoretical development of the acceptance integral method used to estimate the random vibration of structures subjected to turbulent flow was critically reviewed by Au–Yan ([40], [41]). Based on the finite element of tube/stop interaction model shown in Fig. 6.12, Hassan et al [397] estimated the dependence of dimensionless root-mean square (*rms*) impact force, $F_{imp}/(F_{tur}L)$, where F_{tur} is the distributed turbulence force per unit length on the dimensionless clearance ratio, C_r/d_{rsi}, where d_{rsi} is the tube response at the support location obtained by applying the same excitation on the linear unconstrained tube. The results of their numerical algorithm are shown in Figs. 6.13(a) and 6.13(b) for the segment impact force and point impact force respectively and different values of support width ratio, w/D, where w is the support width and D is the tube diameter. For the case of segment impact, Fig. 6.13(a) shows that the *rms* impact force decreases as the width ratio increases. Furthermore, the impact

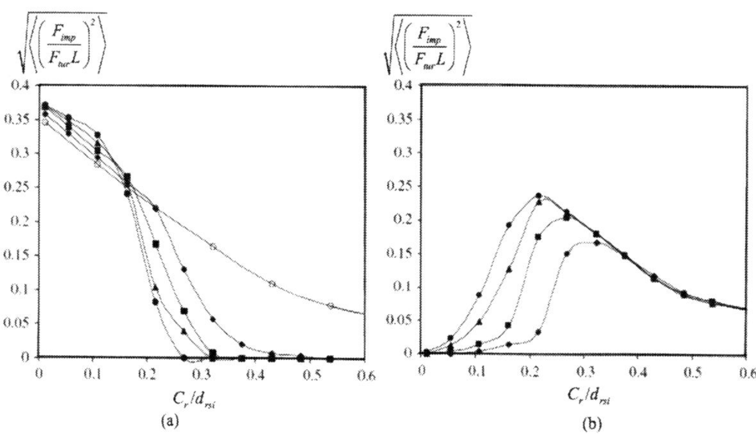

Fig. 6.13. Dependence of the *rms* impact force on the clearance size and for different values of width ratio w/D : ♦ 0.63, ■ 0.94, ▲ 1.26, · 1.56, ○ single point contact model: (a) Segment impact, and (b) point impact, [397].

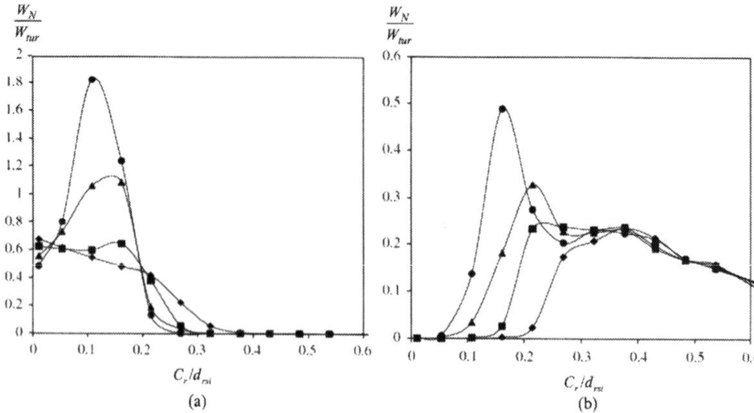

Fig. 6.14. Dependence of the normal work done ratio on the clearance size and for different values of width ratio w/D : ♦ 0.63, ■ 0.94, ▲ 1.26, · 1.56, ○: (a) Segment impact, and (b) point impact, [397].

force predicted by the single point contact model is very close to all other tested support width ratio over the lower region of clearance ratio up to 0.2 above which the impact force of single point model exceeds those values due to other support widths. Note that for all support width values, the tube has 100% segment contact ratio, $\eta = b/l_e$, with the support at zero clearance.

For the case of point impact force, the impact force is zero for zero clearance, see Fig. 6.13(b), in contrast with case of segment impact force. As the clearance ratio increases the impact force increases for all values of support width ratio up to a clearance ratio $C_r/d_{rsi} > 0.2$ depending on the support width ratio. Beyond the peak value of the impact force, the *rms* of impact force begins to decrease with the support width ratio and eventually all curves merge in one curve as shown in Fig. 6.13(b). As the clearance ratio increases, the edge contact ratio increases dramatically up to a clearance ratio $C_r/d_{rsi} >$ 0.2 depending on the support width ratio, then follows the same trend of the impact force.

Tube fretting wear is usually estimated by correlating the experimental wear with the computed work rate. Work rate, W_N, is estimated by averaging the product of the impact forces and tube sliding displacement, i.e.,

$$W_N = \frac{1}{N}\int_0^T F_N ds \tag{6.33}$$

where F_N is the normal impact force and ds is the displacement. The dependence of the normal work rate ratio W_N/W_{tur} on the clearance ratio for different values of width ratio is shown in Figs. 6.14(a) and 6.14(b) for segment contact and edge contact, respectively. It is seen that the normal

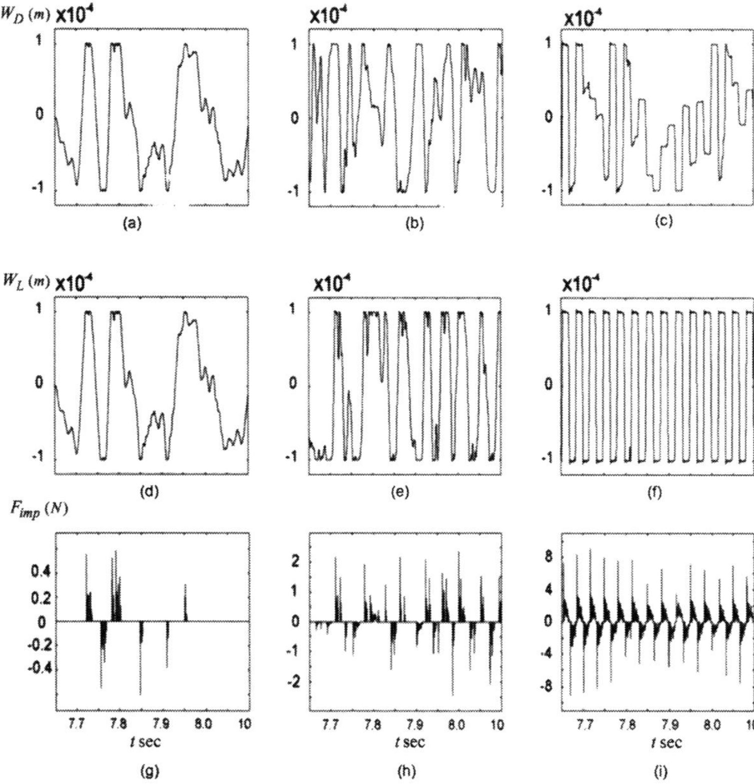

Fig. 6.15. Time history records of the pipe response (a-c) along the flow, (d-f) along the lift, and (g-i) impact force, for clearance $C_r = 0.1mm$, and three different values of flow velocity parameter $U_r = 31$, 50, and 92, [393].

work rate curves of the edge contact case are lower than those of the segment contact for all support contact width ratios.

Hassan and Hayder [393] extended the work of Hassan et al [395] and developed a time–domain model for fluidelastic instability forces of tubes with loose–supports. In their model the fluidelastic force, which is dependent on the flow velocity and array geometry, was superimposed on the turbulent forcing function. Fig. 6.15 shows time history records of the tube tip displacement along the flow, along the lift, and the time history of the impact force for three different values of the flow velocity parameter, $U_r = U/fD$, where U is the undisturbed fluid flow velocity across tubes bundle, and f is the fundamental tube velocity. These velocity parameters exceed the critical flow velocity. Below that flow velocity the tube was found to oscillate periodically without impact.

6.5.5 Experimental Results

The vibro–impact of an instrument tube against adjacent fuel channel boxes in a boiling water reactor (BWR) has been the subject of many studies (see, e.g., [103], [698], [665]). It was proposed that the magnitude of the normalized cross–spectral density between the signals of two detectors in the instrument tube in the frequency range 3.5 Hz to 6.0 Hz to be used as an indicator of impact. Axisa and Antunes [45] reported some results of experimental tests dealing with the vibro–impact motion of linearly unstable multi–supported tubes. A simple method was described for simulating the fluidelastic forces, using a feedback velocity loop. Experiments were performed for different values of the instability growth rate and for several initial motion conditions of the system. The results revealed that various stationary tube responses are possible depending on these parameters. Jacquart and Gay [461] described the Electricite de France (EDF) numerical development with the Aster mechanics computer code to calculate the nonlinear dynamics of tubular structures with loose supports. Both numerical and experimental validations of this computer code were presented. They also reported the research activities of nuclear power plant components of EDF and the design requirements to prevent various damaging processes including flow–induced vibration and wear mechanisms. The vibration and impacting of an instrument tube in a (BWR) were studied by Laggiard et al [561] using one–dimensional bimodal model. Four modal nonlinear boundary conditions were applied and a set of coupled nonlinear equations describing the temporal evolution of two continuous modal amplitudes were obtained. These equations were numerically solved by means of a generalized Runge–Kutta algorithm.

Due to tube–support gaps in heat–exchangers, low–frequency modes may develop and become unstable at comparatively low flow velocities. This kind of linear fluidelastic instability results in a negative value of the modal damping, which is a function of the flow velocity. The response amplitude of the unstable tubes was found to increase steadily until tube–support impact becomes unavoidable. Antunes et al [15] conducted a series of experimental tests to validate nonlinear predictions on vibro–impact dynamics of heat exchanger tube bundles under fluidelastic instability. In particular the actual behavior of the U–band portion of heat exchanger tube bundles was the main focus of the study. The results showed that several steady motion regimes may arise, depending on the system parameters and initial conditions of the motion. In another study, Antunes et al [16] reported some results on a series of laboratory experiments with the purpose of validating numerical predictions of vibro–impact dynamics of heat exchanger tube bundles under fluidelastic instability. The test model was designed for unidirectional motion. The system instability was generated by a velocity feedback loop. This method presents significant advantages due to simplicity of the setup and the controllability of the system parameters, in particular concerning the negative damping ratio of the unstable model. A comparison of experimental and computed system response for several values of instability growth rate and different initial conditions revealed several steady state motion regimes. Furthermore, a satisfactory qualitative and quantitative agreement was obtained between theoretical predictions and test data. Antunes et al [17] developed a theoretical analysis of a simplified model with reduced dimension to predict some essential aspects of the system nonlinear dynamics. The system was found to develop a number of possible nonlinear periodic motions, which are essentially controlled by the modal parameters of the first and second unconstrained modes.

Whiston [1072] and Jordan and Whiston [485] proposed a technique for remote impact analysis applicable to on–line vibration wear assessment. Comparison was made between predicted and measured Timoshenko transfer functions between the remote acceleration transform and the impact force–time history transform. The inversion process was applied to experimental impacting data and good representations of impact force–time histories were obtained. De Aroujo et al [211] developed experimental identification of the wave path propagation parameters and impact forces by using tube response measurements at remote locations. Experiments performed on a long steel beam were described and a simple method was developed to deal with the boundary reflections of a wave generated by a single impact. Experimental identification of the wave–path properties of isolated impact forces and impact locations was performed. This study was extended by Antunes et al [19] who found that the loosely supported tubes display very complex rattling motions. The rattling was associated with the impact–generated primary waves completely immersed in countless wave reflections traveling between the tube boundaries. As a consequence, multiple–impact patterns of tube–support interaction were found to be much more difficult to identify than isolated force spikes. However, Antunes et al [19] considered the identification of impacts for realistic tube vibrations by using a signal–processing technique for separating the multiple wave sources. The technique uses the information provided by a limited number of vibratory transducers.

The remote identification of impact forces on loosely supported tubes was performed by Paulino et al [795]. Experiments were performed on a long beam with three clearance supports, excited by random forces. Antunes et al [20] extended their previous work by including several simultaneous responses. From numerical simulations and experiments, it was shown that the robustness to noise contamination is increased by using multiple response data.

6.5.6 Case Study

Knudsen et al [515] and Knudsen and Massih ([512], [513], [514]) considered the dynamic response of a cantilever beam having loose supports with a prescribed clearance subjected to a harmonic excitation as shown in Fig. 6.16(a). The results were compared with those measured experimentally. The dynamic characteristics of vibro–impacts were studied by evaluating the

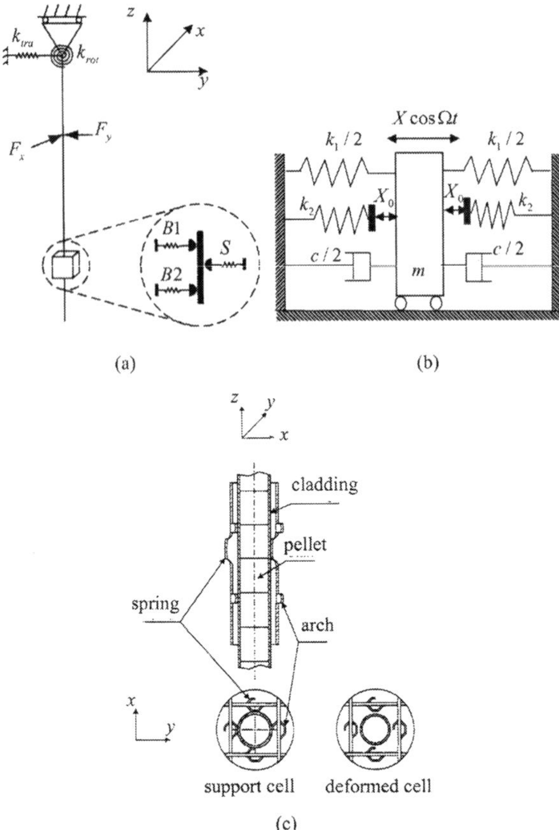

Fig. 6.16. Schematic diagrams of (a) the cantilever beam with loose supports at one end subject to a time-varying force, (b) single-degree-of-freedom oscillator with two-sided constraints subjected to a harmonic load, and (c) the rod-support cell system, showing also the deformed support cell with gaps, [513].

contact velocity as a function of excitation frequency for two kinds of structures, namely a two–sided single–degree–of–freedom impact oscillator, and a two–dimensional cantilever beam with a double–sided support clearance. It was found that neglecting friction at contact would lead to an overestimation of the mean wear work rate.

Figs. 6.17(a) and 6.17(b) show the impact forces versus time in the $x-$ and $y-$directions, respectively. The coordinate directions are defined in Fig. 6.16(c), i.e., a positive force in the $x-$direction represents the force acting on the arches, and a negative force in the $x-$ direction is the force acting on the opposing soft spring. In the $y-$direction, the situation is reversed. The magnitude of the impact forces acting on the arches are overestimated, especially in the $y-$direction. However, the forces acting on the soft springs were found in good agreement with the measured values.

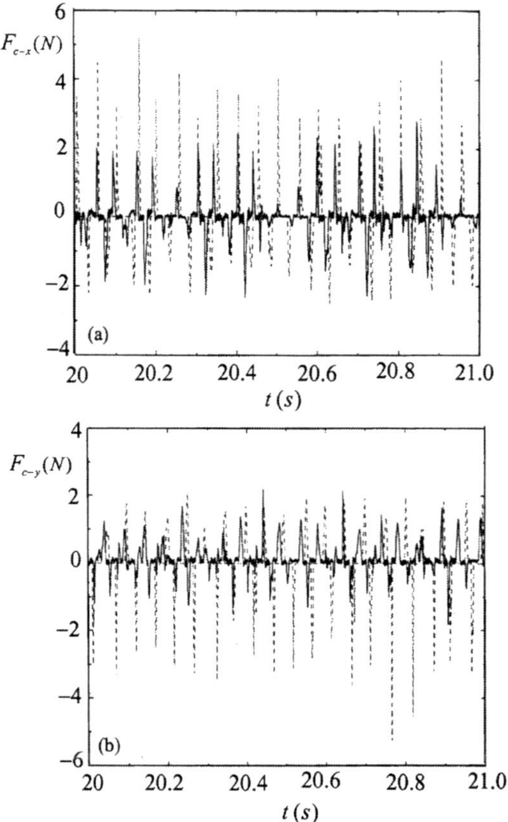

Fig. 6.17. Vibro-impact dynamic behavior of a loosely supported rod under harmonic excitations with a clearance of $0.25mm$ and 0.20 mm in x and y directions, respectively. The driving frequency is $20Hz$. Computations utilize the simple contact algorithm, neglecting friction: (a) Impact force in x direction; (b) force in y direction. The solid line shows measured values while the broken line indicates calculated values, [513].

Knudsen and Massih [513] tried to make the single–degree–of–freedom oscillator shown in Fig. 6.16(b) and the cantilever beam dynamically comparable. The extent of the agreement between the behavior of the two systems can be seen in Figs. 6.17(a) and (b). For low frequencies, the two systems show similar features, albeit the velocities are somewhat higher in the beam case. The difference increases with increasing the applied frequency as the higher modes of the beam are excited. These differences were found to exist and cannot be avoided when comparing one freedom system and continuous beam solutions. There are, however, differences in modeling the two cases. These differences are mainly due to initial conditions and damping parameters.

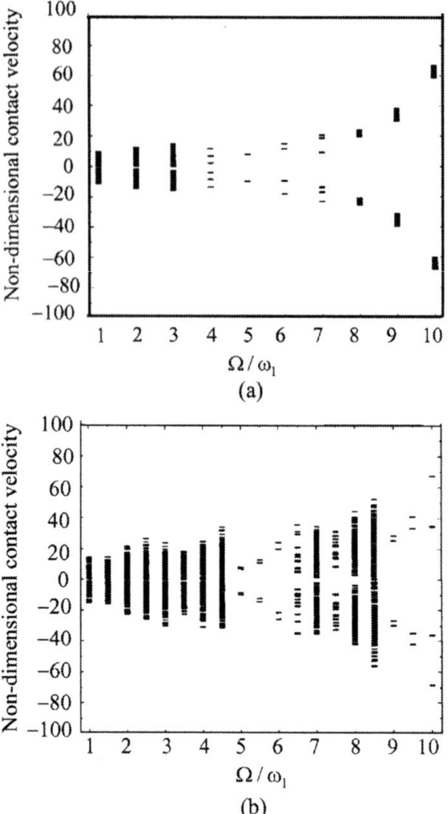

Fig. 6.18. Dependence of non-dimensional contact velocity on excitation frequency ratio for (a) single–degree–of–freedom two–sided impact oscillator and (b) cantilever beam with two–sided supports subjected, [513].

Fig. 6.18 shows the dependence of the non–dimensional contact velocity on the excitation frequency ratio Ω/ω_1, where Ω is the excitation frequency and ω_1 is the natural frequency of the system without impact. It shows the impact and release velocities at 200 consecutive impacts for each value of the forcing frequency. The figure reveals that the system experiences a multitude of complex impact patterns including chaotic behavior. For frequency ratios $\Omega/\omega_1 = 1, 2,$ and 3 the system displays aperiodic behavior. Over the interval $4 \leq \Omega/\omega_1 \leq 7$, the system assumes periodic oscillations with one or more impacts per support per forcing cycle. For a frequency ratio exceeding the value 8, the contact velocity displays an unsettled periodic motion. The overall increase in contact velocity associated with increasing the forcing frequency corresponds to the response of a single–degree–of–freedom oscillator with spring stiffness $k = k_1 + k_2$ subjected to harmonic excitation approaching resonance.

The dependence of the non–dimensional contact velocity on the excitation frequency ratio Ω/ω_1 for the beam of Fig. 6.16(a) is shown in Fig. 6.18(b).

Impact occurrence was identified when $|F_n(j\Delta t)| > 0$ and $F_n\left[(j-1)\Delta t\right] = 0$, where j is an integer identifying the current iteration step and F_n is the normal component of the contact force. Calculations were made for forcing frequencies corresponding to integer multiples of the first free beam eigenfrequency (≈ 7.5 Hz) and halfway between these, i.e., $\Omega/\omega_1 \in \{1, 1.5, 2.0, ..., 10\}$. At forcing frequencies of $\Omega/\omega_1 < 4$ and $7 \leq \Omega/\omega_1 \leq 8$, the beam experiences aperiodic motion. Periodic solutions with one impact per contact site per forcing cycle, were found between these intervals.

6.6 Plates

Switching devices usually contain units consisting of plates activated by a magnetic field generated by a control coil. When the control coil windings are connected to a power source, a magnetic field is generated which magnetizes the plates. As a result, these plates move towards each other and collide. The collision of plates is accompanied by their rebounding and vibrate in several modes. Computing the dynamic characteristics of such systems requires knowledge of the distribution of the magnetic flux. This flux generates the magnetic forces acting on the elastic plates. The behavior of the plates under the effect of the magnetic flux was studied using the finite element method by Ostasavichyus et al ([755], [756], [757]) and Gaidis et al [355].

Noise generation by the impact of thin plates is an example of energy transfer from the low frequency vibration of the fundamental modes to high frequency vibration modes. The noise radiation from the impact is determined by the impact dynamics. Qiu and Feng [878] extracted low–dimensional models for the impact dynamics based on a single–mode impact modeling. It was found that many periodic solutions bifurcate from the grazing bifurcation point. Most of these periodic orbits terminate at secondary grazing bifurcations due to an additional impact.

Fig. 6.19 shows a bifurcation diagram in the space of excitation force amplitude parameter, f/h, and the excitation frequency parameter, ν, where f and ν are non–dimensional excitation amplitude and frequency, respectively, and h is a non–dimensional gap between the plate mid–point and the barrier. Fig. 6.19 was generated for plate damping factor $\zeta = 0.05$, and coefficient of restitution $e = 0.6$. For one impact per $n-$forcing cycles $(1:n)$, Fig. 6.19 reveals four stable regions of $1:1, 1:2, 1:3$, and $1:4$ periodic orbits. Curve G corresponds to the primary grazing bifurcation. Below this curve the linear analysis predicts non–impact solution. However, the nonlinear analysis reveals the existence of impact solutions below curve G. There is a region between G and H where stable $1:1$ impact solution coexists with the stable non–impact solution (see also [121]).

Figs. 6.20(a)–(c) show the dependence of three kinematic parameters on the excitation frequency parameter, ν. These parameters are the impact velocity, V_{im}, the positive velocity crossing the line of static equilibrium position, $V_{x=0}$, and the average impact velocity per forcing cycle, V_{av}. As the

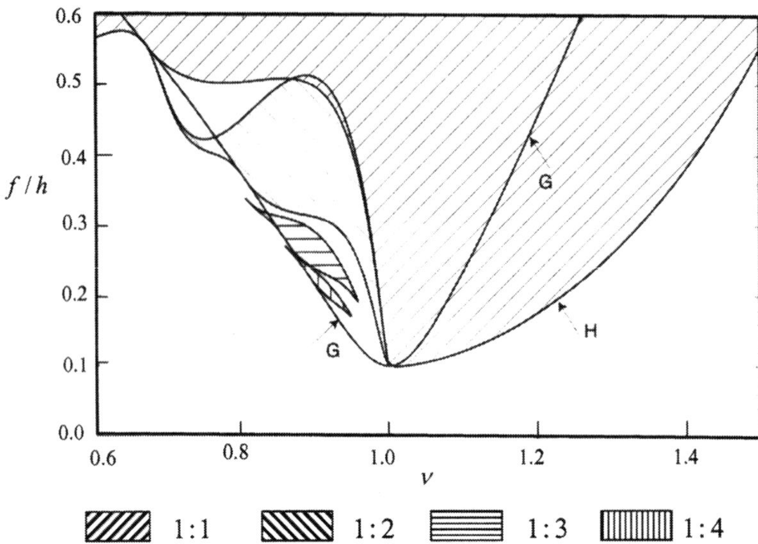

Fig. 6.19. Bifurcation diagram shows stable regions of periodic solutions of $1/n$ type for $n = 1, 2, 3, 4$. Curve G belongs to grazing bifurcation as predicted by linear analysis. The region between G and H corresponds to two stable steady solutions: one with impact of $1:1$ type and the other without impact, [878].

excitation frequency slowly increases, the non–impact solution ceases to exist at the grazing bifurcation point $\nu = 0.907$ as predicted by the linear theory. After grazing bifurcation, $1:5$ periodic orbit emerges and terminates at a secondary grazing bifurcation at $\nu = 0.913$. This is followed by $1:4$ orbit until $\nu = 0.928$ at which the plate experiences chaotic motion. At $\nu = 0.95$ a stable $1:3$ orbit emerges, which terminates at $\nu = 0.953$ through a secondary grazing bifurcation. This is again followed by chaotic dynamics and at $\nu = 0.973$ another bifurcation occurs to stable $1:2$ orbits. These orbits terminate at $\nu = 0.985$ through a secondary grazing bifurcation with $1:1$ orbits. They remain in effect up to an excitation frequency of $\nu = 1.23$ above which unstable $1:1$ orbits emerge through a saddle–node bifurcation and then disappear.

Experimental investigations on a thin plate impacting a striker revealed significant differences of plate impact depending on whether the forcing is above or below the resonance frequency. The hysteresis above the resonance frequency was found much larger than those below resonance. In addition the impact above resonance frequency was noisier than that below resonance.

Fegelman and Grosh [311] analyzed the fundamental mechanics of rattling plates. A flexible beam model was found to capture more of the high–frequency response, but the rigid model captured the qualitative behavior of the rattling beam. These models were used to predict stable and chaotic ranges of rattle motion as a function of excitation level and frequency. Later,

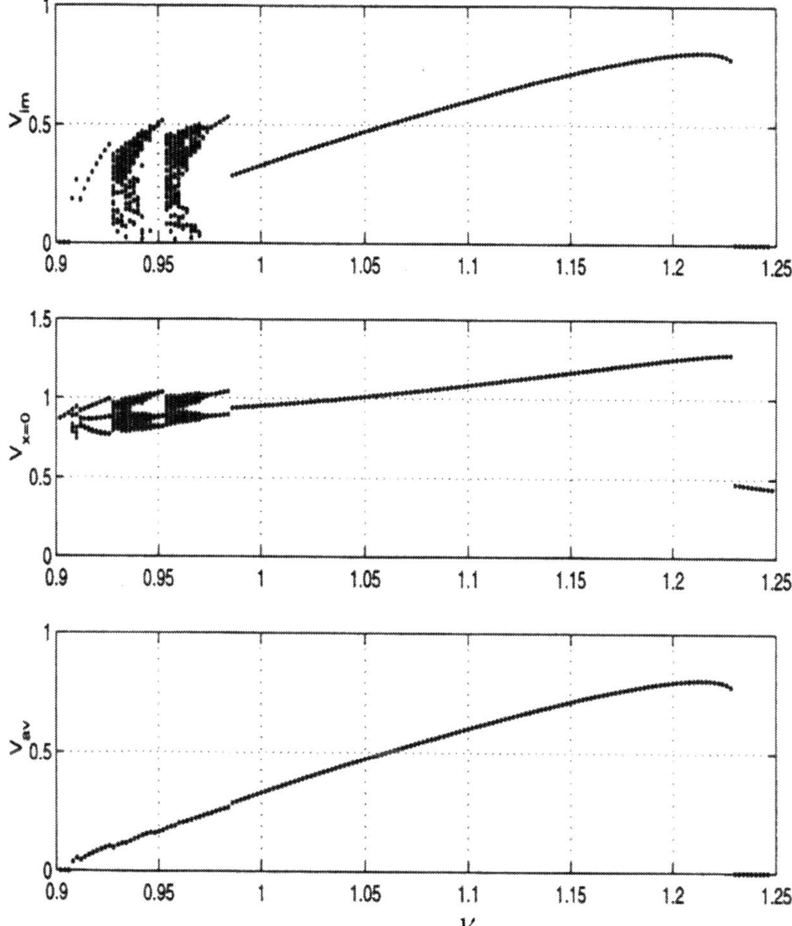

Fig. 6.20. Dependence of the impact velocity, V_{im}, Velocity crossing, $V_{x=0}$, and average velocity, V_{av}, on excitation frequency upsweep at fixed forcing amplitude $F = 0.2$, and $e = 0.6$ non-dimensional gap parameter $h = 1$, [878].

Oppenheimer and Dubowsky [752] presented a methodology for predicting noise and vibration of machines and their support structures. The methodology was implemented using highly idealized closed form and more elaborate numerical descriptions. The predicted results were verified by vibration and sound measurements of a plate subjected to periodic impacts by balls and a beam that rattles within a clearance bearing. The energy–based criterion was found to indicate situations in which mechanism–support coupling affects noise radiation. In some cases the coupling was observed to significantly affect vibration and noise radiation of the support structure, while having a relatively minor effect on the response mechanism.

6.7 Slamming of Ocean Waves

The design of floating systems, liquid storage tanks and off–loading platforms, and floating storage units depends on the values the forces acting on these systems. The impact between sea water waves and ships, known as slamming, can cause important global and local effects. The global loads are induced by the unsteady hydrodynamic pressure due to fluid oscillatory motions surrounding the ship hull. The local component represents the local or secondary loads, such as slamming and whipping, are due to wave impacts. Ochi and Motter [744] and Kawakami et al [500] developed some simple formulae for estimating the slamming loads based on experimental results. Belik et al [104] and Belik and Price [105] divided the bottom slamming into impact and momentum slamming components. They found that the predicted magnitudes of the responses after a slam depend very much on the mathematical model adopted in estimating slamming loads. Molin et al [687] found that the values of impact force as determined by conservation of momentum are almost identical to those predicted by conservation of kinetic energy.

Cooker and Peregrine [190] considered a mathematical model of the large, short–lived pressures brought about by waves breaking against coastal structures. They used the pressure impulse to simplify the equations of ideal incompressible fluid motion and solved analytically a two–dimensional boundary–value problem, which models an idealized wave striking a vertical wall. Expressions were derived for the impulse on the wall, the peak pressure distribution, and the change in fluid velocity due to impact. The results were found insensitive to the shape of the wave far from the wall. Later, Lugni et al [604] analyzed the impact of waves on a vertical, rigid wall during sloshing with reference to the modes that lead to the generation of a flip–through. Experimental measurements were used to characterize the details of the flip–through dynamics while wave loads were computed by integrating the experimental pressure distribution.

The hydrodynamic elastic response of ship hulls considering slamming impact loads due to navigation in rough seas was analyzed by Park et al [789]. Ship hull structures were modeled as elastic body based on Timoshenko's beam theory. The momentum slamming theory was used to derive nonlinear hydrodynamic forces considering the intersection between free surface and ship hull surface. The results provided various design information such as time history records of relative displacement, velocity, acceleration, vertical shear force and vertical bending moment of all sections of ship involving the effects of slamming.

Greco et al ([375], [376]) experimentally and numerically studied the shipping of water on the deck of a vessel in head–sea conditions and zero–forward speed. The fluid–structure interaction was formulated by coupling the non-linear potential flow model with a linear Euler beam to represent a portion of the deck house under the action of the shipped water. The loading conditions related to violent fluid impacts and air–cushion effects were considered. For realistic system parameters, Greco et al [376] were able to define the occurrence of critical conditions for structural safety.

Faltinsen and Chezhian [306] numerically estimated the water entry loads on three–dimensional bodies. Their experimental results for vertical force revealed a strong oscillatory nature, which was analyzed using a simplified hydroelastic model. The hydroelastic model was found to provide reasonable representation of the dynamic oscillations found in the vertical force. Sun and Faltinsen [971] developed a two–dimensional boundary element method to simulate the water flow during the water impact of a horizontal circular cylinder. They solved for the water impact problem of an elastic cylindrical shell in which water–structure interaction was considered. Craig and Kingselev [193] presented a multidisciplinary design and optimization approach for the design process of partially filled liquid containers involving sloshing and impact.

6.8 Closing Remarks

The study of vibro–impact dynamics of continuous systems plays an important role in the design and safety of mechanical systems. In particular, the design and safe operation of nuclear reactor components such as fuel rods and steam generator tubes take into account the gaps developed during service due relaxation of support springs and creep down of cladding tubes. This problem motivated design engineers to develop computational algorithms and computer codes based on finite element and boundary element methods. This chapter addressed classical continuous elements such as strings, beams, tubes conveying fluid and plates in the presence of motion restraints. Slamming of ocean waves acting on navigating ships was also considered in few studies. However, there are many issues have not been studied such as the influence of ocean wave impact on the aging problem and joint relaxation of ship outer structure under severe environmental conditions.

This chapter did not address the impact of continuous systems with a high velocity that may result in fracture of the stricken structure. Examples of this problem include the deformation of continuous system from projectile impact and the damage of an airplane with the exterior columns of buildings as occurred on September 11, 2001 terrorist attack. This type of impact is not associated with vibration but with damage and the reader may refer to Jones [483], Teng and Wierzbicki ([988], [989]), and the references cited.

Chapter 7
Stochastic Vibro–Impact Dynamics

7.1 Introduction

The analysis of random excitation of vibro–impact systems is not a simple task. The difficulty arises due to the fact that impact loading introduces strong nonlinearity. The theory of nonlinear random vibration encompasses analytical techniques that can handle developed for weakly nonlinear systems. Thus it is imperative to recast the vibro–impact system into a form amenable for the traditional analytical techniques. A nice overview of vibro–impact dynamics under random excitation has been presented by Dimentberg and Iourtchenko [251]. The article addressed analytical approaches and some results pertaining to random excitation of systems with lumped parameters and "classical" impacts. Emphasis was given to special piecewise–linear transformation of state variables using Zhuravlev transformation. Exact analyses for stationary probability densities of the response to white–noise excitation were found in few cases, whereas the stochastic averaging method was applied in some other cases. The method of direct energy balance was also illustrated based on direct application of the stochastic differential calculus between impacts. The problem of random excitation of vibro-impact systems attracted the attention of several researchers in the former Soviet Union (see, e.g., [56], [57], [58], [253], [254], [243], [244], [245], [247], [248], [252], [249], [250], [133], [448], [526], [527], [528], [529], [530], [61]). These studies reduced the modeling by using the stochastic averaging method and thus it was possible to estimate the statistical response characteristics of the vibro–impact motion. The response statistics revealed how the energy is transferred from the impacting mass to the secondary structure.

Iourtchenko and Song [449] estimated the response probability density function of vibro–impact systems with one or two rigid barriers. The system was subjected to an additive Gaussian white noise and the response statistics were estimated using Monte Carlo simulation. It was shown that the response probability density functions possess peaks near the barrier, which increase as the value of the restitution coefficient decreases. In another study, Song and Iourtchenko (2006) used the energy balance method to estimate the average energy of stochastic vibro–impact systems with inelastic impacts.

R.A. Ibrahim: Vibro-Impact Dynamics: Model., Map. & Appl., LNACM 43, pp. 193–216.
springerlink.com

Metrikyn [671] established a theoretical analysis of vibro–impact devices with randomly varying parameters. The bouncing ball under a Gaussian random process was studied by Wood and Byrne ([1082], [1083]) who obtained a stationary distribution for the impact velocity. Krupenin [548] studied random oscillations of a periodic structure near a plane travel limiter installed parallel to the plane of structure static equilibrium. Namachchivaya and Park [710] developed an averaging approach to study the dynamics of a vibro–impact system excited by random perturbations and obtained a model–reduction through stochastic averaging. They were able to estimate the mean exit time, probability density functions and stochastic bifurcations.

Yokomichi et al [1108] studied the random oscillation of engine mounting systems with motion–limiting stops. Lee and Byrne [569] studied the stochastic rattling using a small mass constrained to move along a slot of fixed length in a large mass, which is vibrating randomly in the direction of the slot. The statistics of impacts between the rattling ball and the large randomly driven mass were numerically determined. Rattling in change–over gears of road vehicles due to backlash of their teeth was modeled by an impulsive system consisting of some unloading gears ([554], [555], [839], [320], [1071]). Probability density functions of the unperturbed system and the perturbed system due to an added random noise were obtained using the stationary Fokker–Planck equation of the system response. A discrete stochastic model described by a mean map was developed using the non–Gaussian closure technique. The analysis revealed stochastic chaotic behavior.

A combined tuned absorber and pendulum impact damper under random excitation was studied numerically and experimentally by Collette [185]. The optimal mass ratio between the impact damper and tuned absorber was found to be 25%. The clearance was found to be directly proportional to the excitation level provided that the primary system is linearly elastic.

Vibro–impact devices are commonly used in automated assembly lines. The existence of nonlinear phenomena such as cascade of bifurcations and chaotic solutions were examined by Hongler [424]. A micro–electro–mechanical system (MEMS) inertial switch can be modeled as a classical vibro–impact dynamic system in the form of a single degree–of–freedom oscillator impacting against one-sided rigid barrier. Field and Epp [324] considered the random excitation of a MEMS device that causes repetitive impacts with the barrier.

This Chapter considers the random excitation of structural elements such as beams, with one– and two–sided barriers. The modeling of the barriers will be either a stiff spring or a Hertzian contact stiffness. The stochastic equivalent linearization, stochastic averaging method, Fokker–Plank–Kolmogorov equation, and Monte Carlo simulation are used to estimate the response statistics. The roll dynamics of ships subjected to beam random impulsive loading, modeled as a Poisson process will also be considered using the path integral approach.

7.2 Beam–Stop under Random Excitation

Chapter 6 outlined the dynamic behavior of a beam–stop system used in nuclear systems where flow–induced vibrations or seismic excitations can result in impact interactions in pipe–baffle interfaces. Under random excitation, the analysis of such systems needs a special treatment. The random excitation of a beam impacting with elastic barrier (two–sided spring) was considered by Davies [208] who employed an equivalent linearization analysis. The spring barrier was found to introduce coupling between the beam normal modes. Bellizzi and Bouc [106], Bouc and Defilippi ([129], [130]) and Falsone and Muscolino [305] employed an equivalent non-linearization approach to estimate the response statistics of a beam–stop system subjected to a boundary random excitation. The original bilinear impact force was replaced by a cubic term whose coefficient is evaluated by minimizing the difference between the original and the equivalent systems. Fogli et al [337] and Fogli and Bressolette [336] determined the response spectra of vibro–impact systems. For example, the perturbation method and stochastic averaging were adopted by Fogli and Bressolette [336] to estimate the response spectrum of an oscillator with elastic impacts under a Gaussian white noise excitation. They replaced the initial system by a regular system obtained by approximating the nonlinear restoring force using a Chebyshev polynomial. Two approximations were included: one for the flow and one for the stationary distribution of the response amplitude.

The response of strongly nonlinear dynamic systems to stochastic excitation exhibits many interesting characteristics in the frequency domain. De Kraker et al [212] and Van de Wouw et al [1025] performed numerical simulations and experimental tests to examine the random response of an impacting beam system under Gaussian wide– and narrow–band random excitation. It was shown that in modeling the linear system with a local nonlinearity, the linear part can be effectively reduced to a description based on several modes. Combining this reduced linear part with the local nonlinearity in a reduced nonlinear model was shown to result in a nonlinear model. The nonlinear model can be used to accurately predict the stochastic response characteristics of the original system. By including more modes one would obtain results that differ significantly from that of a single–degree–of–freedom model. One also may obtain better correspondence with experimental results. The results revealed the occurrence of multiple resonance frequencies and stochastic equivalents of harmonic and subharmonic solutions. Sampaio and Soize [912] studied the transient dynamics of a Timoshenko beam with elastic barriers under a deterministic transient force whose Fourier transform is limited to a narrow frequency band. The mechanical energy transferred outside the frequency band of excitation was found to be a source of excitation for other subsystems. The influence of random uncertainties was also considered.

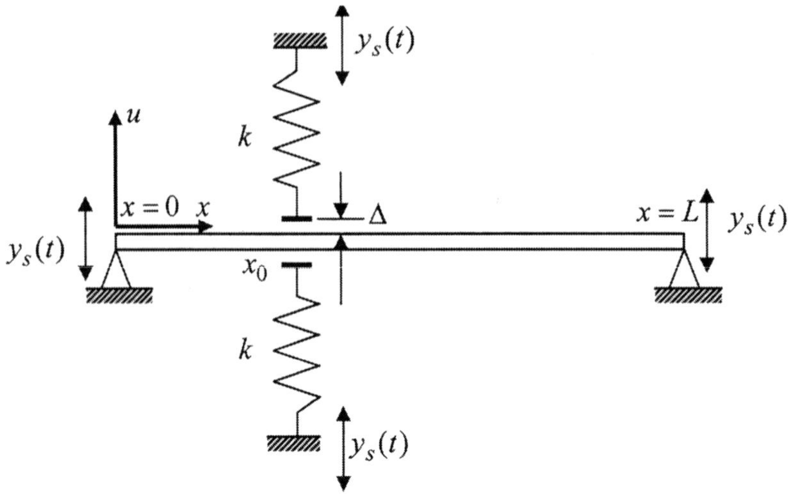

Fig. 7.1. Schematic diagram of a beam with elastic stop.

Bouc and Defilippi [129] estimated response spectra of a randomly excited beam with elastic stop of springs of stiffness k and clearance Δ as shown in Fig. 7.1. They described the motion of the beam by the partial differential equation

$$\rho A\frac{\partial^2 u(x,t)}{\partial t^2}+\frac{\partial^2}{\partial x^2}\left[EI\frac{\partial^2 u(x,t)}{\partial x^2}\right]+F(u_0(t))\delta(x-x_0)=-\rho A\frac{d^2 y_s(t)}{dt^2}, \quad (7.1)$$

where $u(x,t)$ is the beam deflection, ρ, A, E and I are the mass density, area of the beam cross–section, modulus of elasticity, and area moment of inertia of the beam, respectively. $y_s(t)$ is the support displacement excitation, which was assumed stationary wide–band random process with zero–mean (not necessarily Gaussian). $\delta(x-x_0)$ is the Dirac delta function and $F(u_0(t))$ is the impact force at x_0 described by the three segments

$$\begin{aligned} F(u) &= k(u+\Delta) && if\ -\infty < u \le -\Delta \\ F(u) &= 0 && if\ -\Delta < u < +\Delta \\ F(u) &= k(u-\Delta) && if\ \Delta \le u \le +\infty. \end{aligned} \quad (7.2)$$

According to the stochastic equivalent linearization, equation (7.1) may be replaced by the linear equation

$$\rho A\frac{\partial^2 u(x,t)}{\partial t^2}+\frac{\partial^2}{\partial x^2}\left[EI\frac{\partial^2 u(x,t)}{\partial x^2}\right]+K_{eq}u_0(t)\delta(x-x_0)=-\rho A\frac{d^2 y_s(t)}{dt^2}, \quad (7.3)$$

where the equivalent spring stiffness is

$$K_{eq} = \arg\min_{K} E\left[(F(u_0) - Ku_0)^2\right], \tag{7.4}$$

such that

$$K_{eq} = \frac{E\left[F(u_0)u_0\right]}{E\left[u_0^2\right]}, \tag{7.5}$$

where $E[\cdot]$ denotes expectation of the expression in the brackets. For the case of a simply–simply support beam whose natural frequency $\omega_n = (\pi m/L)^2 \sqrt{EI/\rho A}$, and mode shapes $\Phi_m(x) = \sqrt{2}\sin(m\pi x/L)$, $m = 1, 2, 3$, the solution may be expressed by the modal summation

$$u(x,t) = \sum_{m=1}^{3} \Phi_m(x) q_m(t),$$

where q_m are the generalized coordinates. The equivalent stiffness, K_{eq}, was expressed in terms of the amplitude a of the beam deflection at the impact location. The deflection at x_0 was written in the form $u_0(t) = \Phi_1(x_0, t)q_1(t)$, where $q_1(t) = a\cos\phi$, $a \geq 0$, $\phi(t) = \omega_1(a)t + \vartheta$. The expectations in equations (7.4) and (7.5) were evaluated for a given value of the amplitude a with respect to the phase ϕ, which was assumed uniformly distributed over $(0, 2\pi)$. Bouc and Defilippi [129] obtained the following result:

$$K_{eq} = \frac{1}{\pi\left[\Phi_1(x_0,a)a\right]^2} \int_0^{2\pi} F\left(\Phi_1(x_0,a)a\cos\phi\right)\Phi_1(x_0,a)a\cos\phi d\phi. \tag{7.6}$$

Setting $U_1^0(a) = |\Phi_1(x_0,a)a|$, equation (7.6) gives

$$K_{eq} = 0, \; if \; U_1^0(a) \leq \Delta, \tag{7.7}$$

$$K_{eq} = \frac{2m}{\pi}\left[\arccos\frac{\Delta}{U_1^0(a)} - \frac{\Delta}{U_1^0(a)}\sqrt{1 - \left(\frac{\Delta}{U_1^0(a)}\right)^2}\right], \tag{7.8}$$

$$if \; U_1^0(a) > \Delta.$$

Thus $\Phi_k(x,a) = \Phi_k(a)$ and $\omega_m(a) = \omega_m$ for $U_1^0(a) < \Delta$. The modal analysis with the equivalent linear system given by equation (7.3), gives

$$\Phi_m(x_0) = \frac{EI}{K_{eq}}\left[\frac{1}{L}\int_0^L G_0^2\left(x, x_0; \omega_m\right)dx\right]^{-1/2}, \tag{7.9}$$

where the Green function, G_0, is

$$G_0\left(x, x_0; \omega_m\right) = \frac{EI}{\rho AL} \sum_{m=1}^{3} \frac{\Phi_m(x)\Phi_m(x_0)}{\omega_m^2 - \omega^2(a)}. \tag{7.10}$$

The modal equations of motion after adding linear viscous damping take the form

$$\ddot{q}_m + 2\zeta_m \omega_m \dot{q}_m + \omega_m^2(a) q_m = -\nu_m \ddot{y}(t), \tag{7.11}$$

where $\nu_m = \frac{1}{L} \int\limits_0^L \Phi_m(x)dx$.

Expressions for the response power spectra and cross spectra are

$$S_u(\omega; x) = \sum_{m,n} \Phi_m(x) H_m(\omega) S_{\ddot{y}}(\omega) H_n^*(\omega) \Phi_n(x), \tag{7.12}$$

$$S_u(\omega; x, y) = \sum_{m,n} \Phi_m(x) H_m(\omega) S_{\ddot{y}}(\omega) H_n^*(\omega) \Phi_n(y), \tag{7.13}$$

where

$$H_m(\omega) = \frac{\nu_m}{\omega_m^2(a) - \omega^2 + 2i\zeta_m \omega_m(a)\omega}.$$

Fig. 7.2 shows the dependence of the first two modal natural frequencies on the amplitude of the first mode. It is seen that as the beam amplitude increases the first mode natural frequency does not exhibit any appreciable increase until the beam deflection approaches the gap clearance, $a = \Delta = 0.0019\ m$ above which the natural frequency experiences significant increase.

Bouc and Defilippi [130] employed the stochastic averaging method and solved for the stationary solution of the Fokker–Planck equation for the first

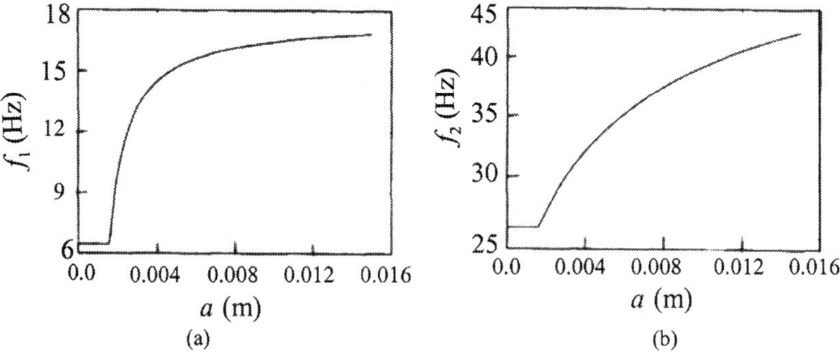

Fig. 7.2. Dependence of the first two modes natural frequency on the vibration amplitude for gap clearance $\Delta = 0.0019m$, beam length $L = 4.2m$, spring stop stiffness $k = 0.85 \times 10^6$ at $x_0 = 2.0m$: (a) first mode ($f_1 = 6.5Hz$ in the absence of spring stop), (b) second mode ($f_2 = 26Hz$ in the absence of spring stop), [130].

mode amplitude response. They obtained the following closed form solution for the response probability density function

$$p(a) = p_0 a \omega_1(a) \overline{D}(a) \exp\left[-\int_0^a \frac{4\zeta_1 \omega_1 \omega_1^2(a) \overline{D}(a)}{\nu_1^2(x) S_{\ddot{y}}(\omega_1(x))} dx \right], \quad a \geq 0, \tag{7.14}$$

where p_0 is the normalization constant, ζ_1 is the first mode damping factor (introduced in the modal equation of motion), ω_1 is the beam first mode natural frequency in the absence of spring stop, $\omega_1(a)$ is the first mode natural frequency that depends on the amplitude response as shown in Fig. 7.2,

$$\overline{D}(a) = (1/2\pi) \int_0^{2\pi} D(a,\phi) d\phi, \qquad D(a,\phi) = 1 + \frac{(\partial \omega_1(a)/\partial a)\, a}{\omega_1(a)} \sin^2 \phi,$$

$S_{\ddot{y}}(\omega_1(x))$ is the power spectral density of the base acceleration, and $\nu_1(a) = \frac{1}{L} \int_0^L \Phi_1(x,a) dx$.

Fig. 7.3 shows the response probability density function, which is essentially non–Gaussian with non–zero mean. The response power spectral density of the beam–spring system was obtained by averaging the conditional

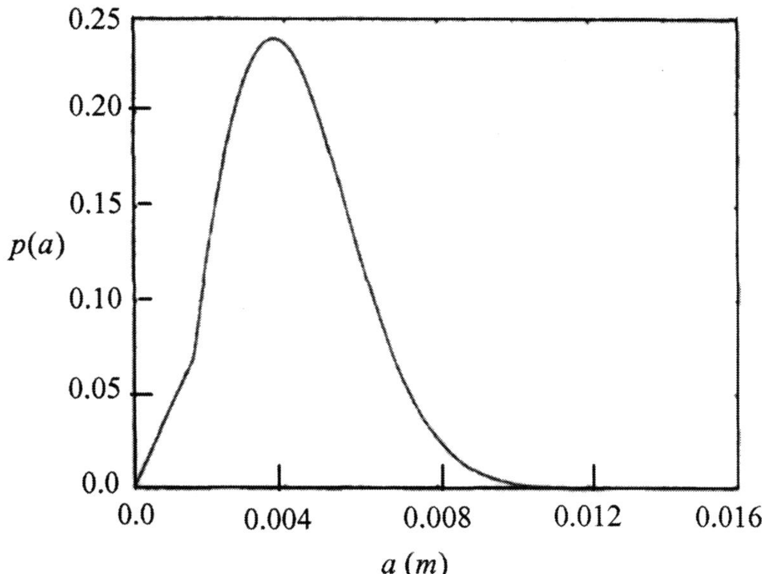

Fig. 7.3. Response probability density function pf the beam amplitude, [130].

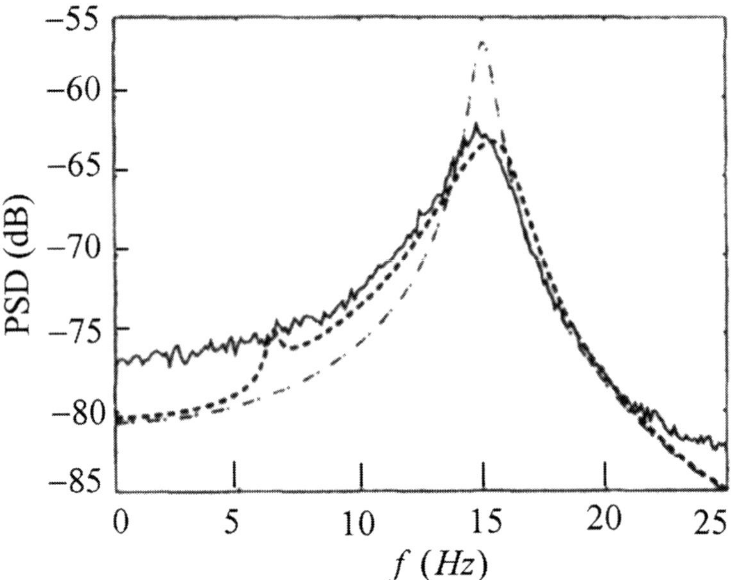

Fig. 7.4. Response power spectral density function as estimated by: —— numerical simulation, —· — stochastic Gaussian linearization, - - - proposed method by Bouc and Defilippi, [130].

power spectral density with respect to the response probability density function given by equation (7.14). For example, the power spectral density function of the beam deflection, $u(x,t)$, was given by the expression

$$S_u(\omega) = \int_0^\infty S_u(\omega; x|a) B^2(a) p(a) da$$

where

$$B^2(a) = \frac{2\zeta_1 \omega_1 a^2 \omega_1^2(a)}{\nu_1^2 S_{\ddot{y}}(\omega_1(a))}.$$

Fig. 7.4 shows the mean response power spectral density function over the entire length of the beam defined by

$$\nu_m = \frac{1}{L} \int_0^L S_u(x,\omega) dx.$$

Fig. 7.4 also shows stochastic Gaussian linearization result indicated by the dash–dot curve $(- \cdot -)$. The stochastic Gaussian linearization result yields correct resonance frequency but overestimates associated level and underestimates spectral band–width.

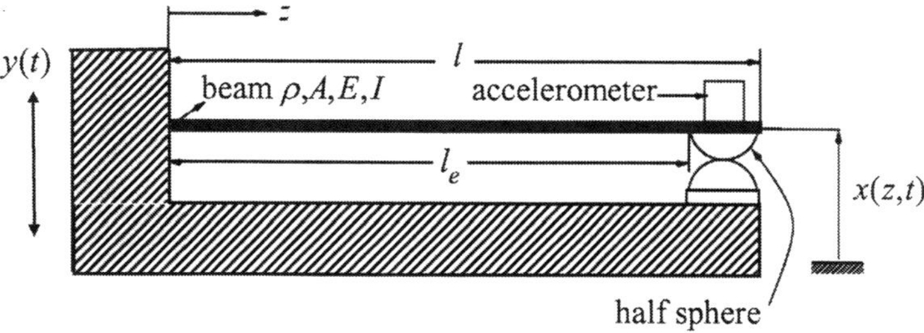

Fig. 7.5. Schematic diagram of a cantilever beam with its free constrained by a half spherical barrier, and shows the experimental set up, [1025].

7.3 Cantilever Beam with One–Sided Barrier

In an effort to understand the dynamic response of a cantilever beam whose free end is restrained by one-sided barrier Van de Wouw et al [1024] considered the first mode response under support Gaussian band limited random excitation. The beam was replaced by an equivalent mass–spring–dashpot system and the contact force was represented by the Hertzian contact with hysteretic energy dissipation. Nonlinear stochastic phenomena like multiple resonance peaks and high–energy low–frequency response content were observed under broad–band excitations. The origin of the multiple resonance frequencies was demonstrated by investigating the system's response to narrow–band excitations covering harmonic and subharmonic resonance regions. Stochastic equivalents of harmonic and subharmonic solutions were reported. Numerical simulations and experimental results were found in good agreement. The observed phenomena can also be found in systems with other one–sided nonlinearities.

The normal modal interaction of a cantilever beam with its free end constrained against half spherical barrier was considered by Van de Wouw et al [1025] under Gaussian random excitation of its support. Fig. 7.5 shows a schematic diagram of their experimental system. The contact force due to the beam impact with the elastic barrier may be represented by the Hertzian model with hysteretic energy dissipation as described in Chapter 1. The beam was modeled as an Euler beam and its equation of motion was discretized through Galerkin's method in terms of the first two modes. The estimated first two modal natural frequencies of the beam model are $\omega_1 = 109.1\ rad/s$ and $\omega_2 = 790.7\ rad/s$. The experimental measured values were found to be $\omega_1 = 101.5\ rad/s$ and $\omega_2 = 781.6\ rad/s$. The corresponding modal damping factors were 1.5% and 0.5%, respectively. The modal equations of motion may be written in the matrix form

$$[M]\{\ddot{q}\}+[C]\{\dot{q}\}+[K]\{q\}+[K_H]\varepsilon(\delta)\left\{\delta^{3/2}\right\}\left[1-\frac{3\dot{\delta}}{4\dot{\bar{\delta}}}\left(1-e^2\right)\right]=\{m_0\}\ddot{y}(t), \tag{7.16}$$

where

$$\varepsilon(\delta)=\begin{cases}1 \ for \ \delta>0\\ 0 \ for \ \delta\leq 0\end{cases},$$

a dot denotes differentiation with respect to time, e is the coefficient of restitution, $\{q\}$ is the vector of the natural coordinates of the first two modes, $[M]$, $[C]$, and $[K]$ are the mass, damping, and stiffness matrices, respectively. $[K_H]$ is a coefficient matrix of contact stiffness, $\{\delta\}=\{y-x(z=l+l_e/2),$ $y-x(z(l/2)\}^T$, $\dot{\bar{\delta}}$ is the velocity difference of the two colliding bodies at the beginning of impact, $x(\cdot)$ is the absolute displacement of a point on the beam at the designated location in the parentheses, $\{m_0\}$ is the mass vector associated with the random support motion $y(t)$, and T denotes transpose. Equations (7.16) were solved numerically utilizing Hénon's method to determine the time of impact. The system of equations (7.16) was written first in the space vector form, then rearranged without nonlinearities such that δ becomes the independent variable whereas the time t becomes one of the dependent variables. In this case, the nonlinear part becomes superfluous since the last time interval before impact was obtained. At the time step before impact the rearranged equations were integrated until $\delta=0$. This integration

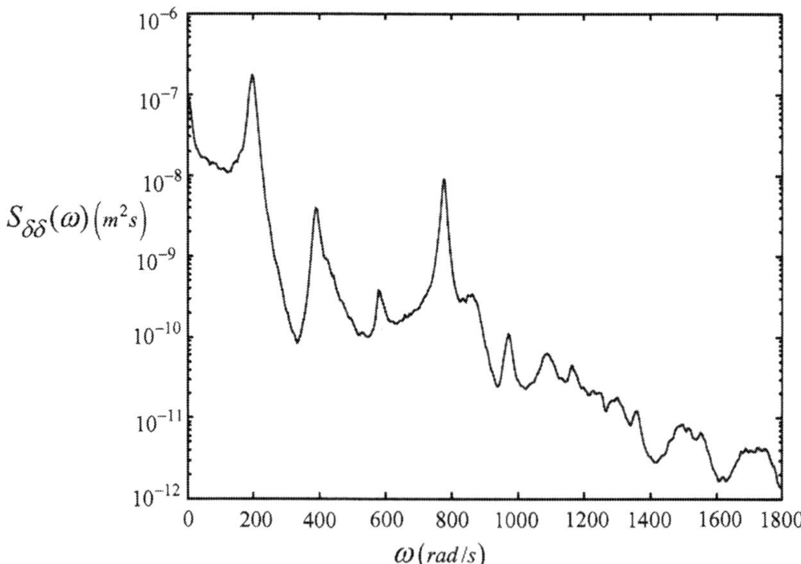

Fig. 7.6. Beam response power spectral density as estimated numerically by Van de Wouw et al, [1025].

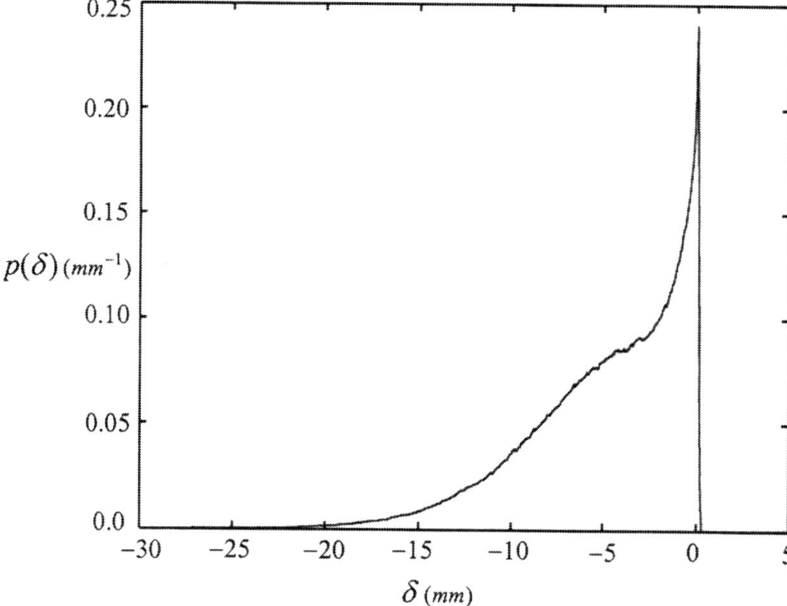

Fig. 7.7. Beam response probability density function for one-sided barrier as estimated by Van de Wouw et al, [1025].

step results in the variables $t_{contact}$, and $\{x\}_{contact}$, for which $\delta(t_{contact}) = 0$ holds. The process continues by switching to a small integration step size to solve the state vector equations. Upon leaving contact, the integration routine switches back to the large integration time step.

The numerical simulation was performed for a Gaussian wide band random excitation of the base with a constant power spectral density $S_{yy}(\omega) \approx 0.5 \times 10^{-10} m^2 s$ over the frequency band ω_{band} = (0.0 to 1,222.6 rad/s). Fig. 7.6 shows the beam response power spectral density and reveals the first eigenfrequency at $\omega_1 \approx 195\ rad/s$ while the second mode at $\omega_2 \approx 780\ rad/s$. It is known that for a piecewise linear system [930] the nonlinear resonance frequency is almost twice the linear eigenvalues of the same system without barriers. This feature is only manifested in the first mode since the impact plays a less important role in the nonlinear response of the second mode. Fig. 7.7 shows the beam response probability density function of the relative end displacement δ. The response probability plot reveals asymmetry and non–Gaussian distribution due to the presence of one–sided barrier.

Van de Wouw et al [1025] conducted an experimental investigation of the same model. However, the measured excitation power spectral density function was not uniform as it was assumed in the numerical simulation. Fig. 7.8 shows a typical plot of the measured excitation power spectral plot which was also used in the numerical simulation to verify the experimental

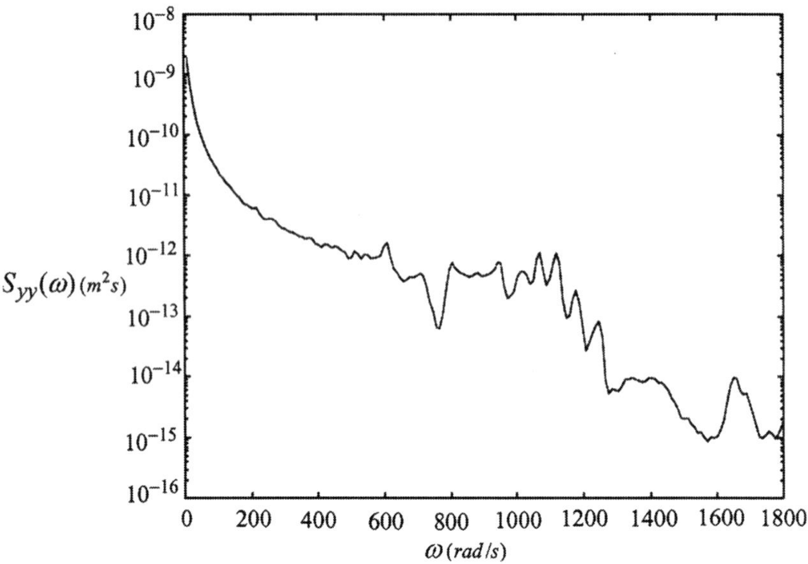

Fig. 7.8. Measured excitation power spectral density of the beam support, [1025].

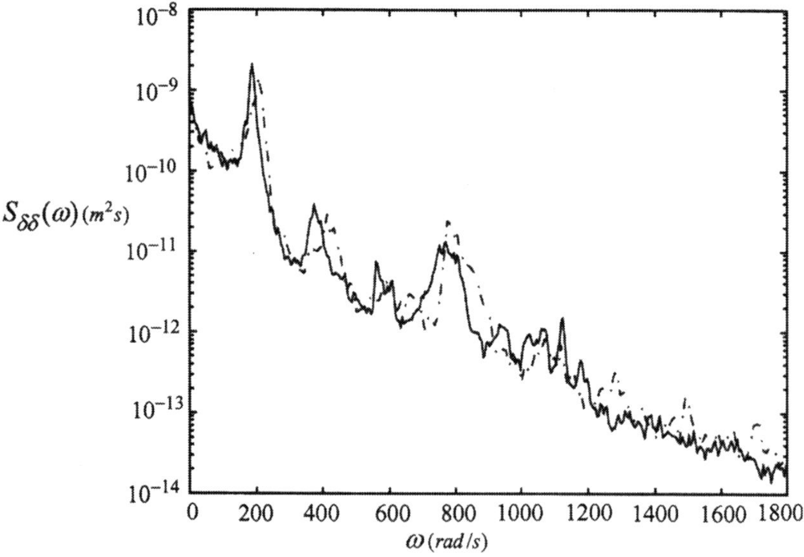

Fig. 7.9. Measured ——, and estimated -.-.- response power spectral density of the beam free end, [1025].

measurement of the response power spectral density. Fig. 7.9 shows a comparison between the response power spectra as measured (shown by solid curve) and estimated (shown by dashed–dot curve). The two results exhibit the second mode harmonic at $\omega_2 \approx 780\ rad/s$. Other experimental tests were conducted under narrow–band random excitation. Two excitation frequency bands were selected. The first was (144.5 to 270.2) rad/s, which covers the first mode nonlinear resonance frequency. The second was (351.9 to 477.5) rad/s, which is away from the first mode and is close to half the second mode frequency. In the second case the response displayed stochastic half subharmonic solution. In both cases, the results confirmed the importance of the inclusion of the second mode in the analytical modeling.

7.4 Pre–Loaded Vibro–Impact Hertzian Contact

Huang et al ([429], [430]) studied single and multi–degree–of–freedom vibro–impact systems under white noise excitations as a dissipated Hamiltonian system. The constraints were modeled as nonlinear springs according to the Hertzian contact law. Approximate stationary solutions of the system were obtained using the stochastic averaging method for quasi–Hamiltonian systems. It was shown that the stochastic averaging method is applicable if the nonlinear forces according to the Hertzian contact law take an important role in the response of the system. On the other hand, the stochastic averaging method for quasi–integrable–Hamiltonian systems is applicable if the nonlinear forces are neglected.

Rigaud and Perret–Liaudet [891] and Perret–Liaudet and Rigaud ([807], [808]) experimentally measured the nonlinear dynamic response of a normally excited pre–loaded Hertzian contact (including possible contact losses). They considered a system consisting of a double sphere–plane contact loaded by the weight of a rigid moving mass. Contact vibrations were generated by an external Gaussian white noise and exhibited vibro–impact responses when the input level is sufficiently high. Theoretical response characteristics were also predicted using the stationary Fokker–Planck equation and Monte Carlo simulations. When contact loss occasionally occurred, numerical results revealed a good agreement with experimental measurements. However, poor agreement was reported for the case of the occurrence of vibro–impacts. The contact loss nonlinearity was found to be rather strong compared to the Hertzian nonlinearity. It results in widening the spectral contents of the response. Perret–Liaudet and Rigaud [807] considered the random excitation of the system shown in Fig. 7.10. The system was described by the equation of motion

$$m\ddot{x} + c\dot{x} + \kappa\left[xH(x)\right]^{3/2} = F_0\left[1 + \sqrt{D}W(t)\right], \tag{7.17}$$

where x is the displacement of the system mass m, measured such that $x < 0$ corresponds to loss of contact. c is a linear viscous coefficient, κ is the contact stiffness coefficient, which is a function of the elastic properties and geometries of the contact bodies, and $H(x)$ is the Heaviside step function. F_0 is the static external load component, $W(t)$ is a stationary zero–mean

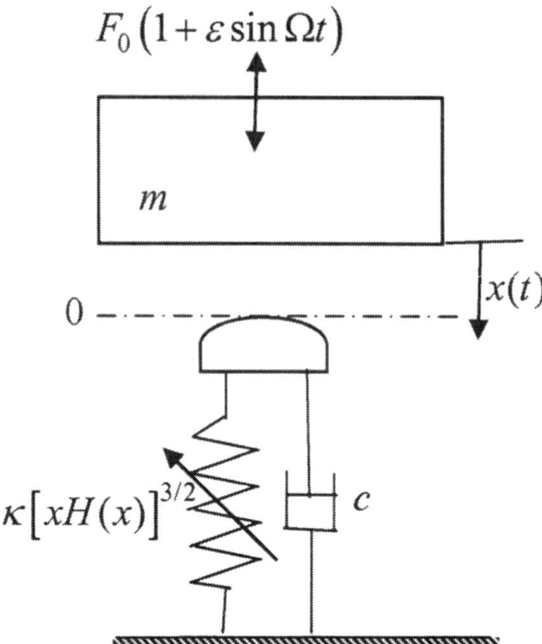

Fig. 7.10. Schematic diagram of a nonlinear oscillator with Hertzian contact.

Gaussian white noise, and D is the intensity of the random normal force. Defining the static contact compression and the linearized natural frequency as given in Chapter 1, i.e., $x_s = (F_0/\kappa)^{2/3}$, and $\omega_n^2 = \left(\frac{3\kappa}{2m}\right)\sqrt{x_s}$, respectively, equation (7.17) can be written in the non–dimensional form

$$q^{''} + 2\zeta q^{'} + \left[(1 + 2q/3)\, H\,(1 + 2q/3)\right]^{3/2} = 1 + \sqrt{D}w(\tau), \tag{7.18}$$

where $q = \frac{3}{2}\left(\frac{x}{x_s} - 1\right)$, $\tau = \omega_n t$, and $\zeta = \frac{c}{2}\frac{\omega_n}{m}$.

Note that $w(\tau)$ is selected such that it possesses a unit power spectral density, i.e., $S_w(\nu) = 1$, where $\nu = \omega/\omega_n$ is a non–dimensional frequency parameter. Perret–Liaudet and Rigaud [807] wrote equation (7.18) in the general form

$$q^{''} + 2\zeta q^{'} + G(q) = f(\tau), \tag{7.19}$$

where $f(\tau)$ is a zero–mean stationary Gaussian white–noise with auto–correlation function $R_f(\overline{\tau}) = 2\pi D\delta(\overline{\tau})$, and a prime denotes differentiation with respect to the non–dimensional time parameter τ. In this case the power spectral density $S_f(\nu) = DG(q)$ represents the nonlinear restoring force including the nonlinearity due to contact loss. The Fokker–Planck equation which governs the evolution of the response transitional probability density function $p\left(q, q^{'}|q_0, q_0^{'}\right)$ may be written in the form

$$\frac{\partial p}{\partial \tau} + q^{'}\frac{\partial p}{\partial q} = \frac{\partial p}{\partial q^{'}}\left[2\zeta q^{'} + G(q)\right]p + \pi D\frac{\partial^2 p}{\partial q^{'2}}. \tag{7.20}$$

The stationary joint probability function may be written in the form

$$p_s\left(q, q^{'}\right) = p_0 \times \exp\left(\frac{-2\zeta}{\pi D}\frac{q^{'2}}{2}\right) \times \exp\left[\frac{-2\zeta}{\pi D}\int_0^q G(y)dy\right], \tag{7.21}$$

where p_0 is the normalization constant. The marginal probability densities for the displacement and velocity are independent and a closed form of the probability density of the response displacement is

$$p(q) = \begin{cases} \overline{p} \times \exp\frac{-2\zeta}{\pi D}\left[\frac{3}{5}\left(1+\frac{2}{3}q\right)^{5/2} - q - \frac{3}{5}\right] & for\ q > -\frac{3}{2} \\ \overline{p} \times \exp\frac{-2\zeta}{\pi D}\left[-q - \frac{3}{5}\right] & for\ q \leq -\frac{3}{2} \end{cases}, \tag{7.22}$$

where $\overline{p}$ is the normalization constant. The corresponding probability density function of the response velocity is

$$p(q^{'}) = \widehat{p} \times \exp\frac{-2\zeta}{\pi D}\frac{q^{'2}}{2}. \tag{7.23}$$

The n–th statistical moment of the response displacement is

$$E\left[q^n\right] = \int_{-\infty}^{\infty} q^n p(q)dq. \tag{7.24}$$

The normal load, $N = G(q)$, may be defined by the expressions

$$N = \begin{cases} \left(1+\frac{2}{3}q^{3/2}\right) - 1 & for\ q > -\frac{2}{3} \\ -1 & for\ q \leq -\frac{2}{3} \end{cases}. \tag{7.25}$$

The roots of this equation were obtained numerically and are designated by q_i, $i = 1, ..., s$. In this case the probability density function of the elastic restoring force may be written in the form

$$p(N) = \sum \frac{p(q_i)}{\left|\frac{dN(q_i)}{dq}\right|}$$
$$= \begin{cases} \left(1+N\right)^{-1/3} p\left[\frac{3}{2}\left(1+N\right)^{2/3} - \frac{2}{3}\right] & for\ N > -1 \\ \delta(-1)\int_{-\infty}^{-3/2} p(y)dy & for\ N = -1 \\ 0 & for\ N < -1 \end{cases}, \tag{7.26}$$

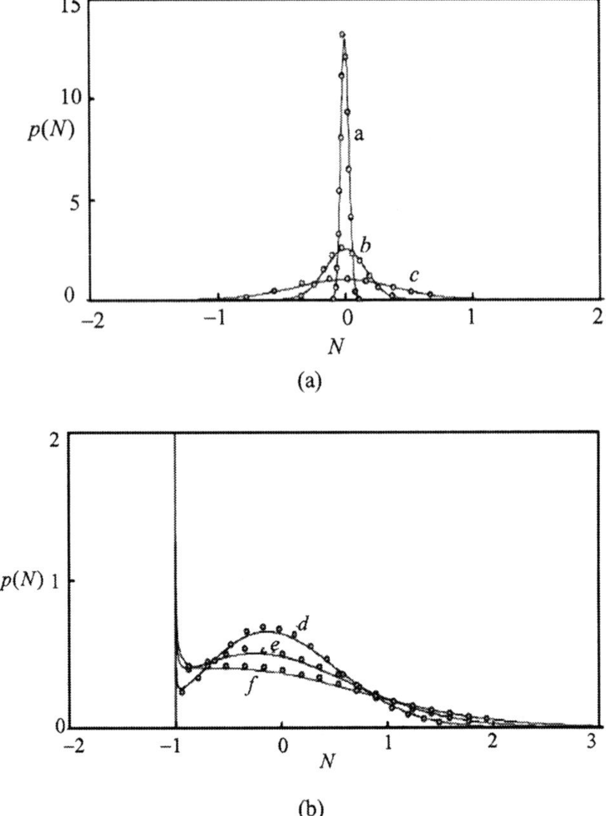

Fig. 7.11. Response probability density function of the elastic restoring force for different values of excitation level, D: $D_a = 3 \times 10^{-6}$, $D_b = 8 \times 10^{-5}$, $D_c = 5 \times 10^{-4}$, $D_d = 1.2 \times 10^{-3}$, $D_e = 2 \times 10^{-3}$, and $D_f = 3.2 \times 10^{-3}$. (——) Analytical results, (○) Monte Carlo simulation, [807].

where $\delta(\cdot)$ is the Dirac delta function. For $N < -1$, $p(N)$ contains an impulse of an area equivalent to the probability of loss of contact and $\lim\limits_{\substack{N \to -1 \\ N > -1}} p(N) = +\infty$. The n–th moment of the normal force is

$$E\left[N^n\right] = \int\limits_{-\infty}^{\infty} N^n p(N) dN. \tag{7.27}$$

Fig. 7.11 shows the stationary probability response of the restoring force for two different sets of excitation level and for damping ratio of 0.005. The analytical results are shown by solid curves while the results generated by Monte Carlo simulation are indicated by empty circles (◯). It is seen that as the

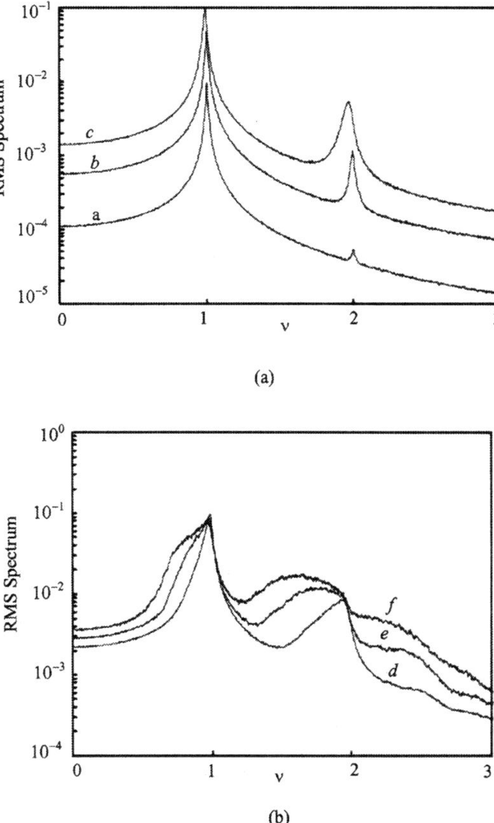

Fig. 7.12. Response spectra of the elastic restoring force for different values of excitation level, D: $D_a = 3 \times 10^{-6}$, $D_b = 8 \times 10^{-5}$, $D_c = 5 \times 10^{-4}$, $D_d = 1.2 \times 10^{-3}$, $D_e = 2 \times 10^{-3}$, and $D_f = 3.2 \times 10^{-3}$, [807].

excitation level increases the response probability density function becomes more flat for the first three levels (a, b, and d) as shown in Fig. 7.11(a). As the excitation level increases the analytical results reveal a spike at $N = -1$. Fig. 7.12 shows the corresponding response spectra and it demonstrates the broadening resonance associated with increasing the excitation level. This effect was verified experimentally as shown in Fig. 7.13. The measured results reveal a shift of resonance peaks as the excitation level increases. These features are well known for nonlinear systems under random excitation as documented by Ibrahim [435]. Depending on the type of nonlinearity, the shift moves to the left for soft nonlinearity, and to the right for hard nonlinearity as the excitation level increases. Note that as the excitation level increases there is a possibility of an intermittent contact loss as the normal force reaches the static value.

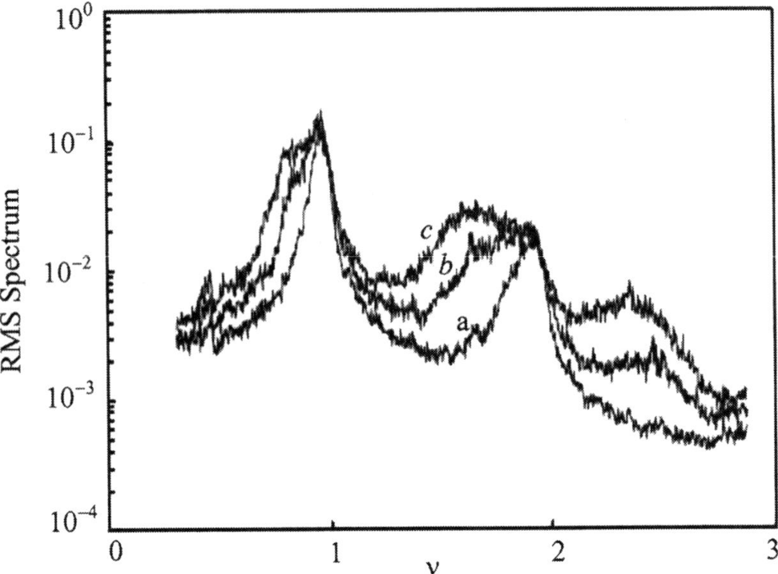

Fig. 7.13. Experimental response spectra of the elastic restoring force for different values of excitation level, $D_a = 6 \times 10^{-4}$, $D_b = 1.7 \times 10^{-3}$, $D_c = 2.5 \times 10^{-3}$, [807].

7.5 Random Impulsive Excitation

Random impulsive events are characterized by signals of short duration but with large magnitude and occur randomly. Examples of such loads include the hail–ice impact on structures [504] and impulsive loads of floating ice on ocean structures [438]. The impact was classified into the low–velocity and high–velocity regimes. Jackson and Poe [460] analytically studied the transition between high and low velocity impacts and highlighted the dramatic differences between the two velocity regimes. It was concluded that the peak force developed during impact is a key parameter in evaluating delamination formation of composite structures.

Random impulsive loads of floating ice interacting with ocean systems has been modeled by a Poisson arrival process of loading events, not necessarily with a constant arrival rate. Jordaan [484] adopted an expression for the distribution, $\Pi_Z(z)$, of extreme load, Z, which can be derived by writing the expression for zero arrivals in the process, with rate $N[1 - \Pi_X(Z)]$, where N is the number of arrivals in a year and X is the force or pressure under consideration. The probability of exceedance $\Pi_e = 1 - \Pi_X(x)$ is

$$\Pi_e = \exp\left[-(x - x_0)/\alpha\right] \tag{7.28}$$

where x_0 and α are constants.

The randomness of ice–induced loads was characterized in terms of probability distributions that were fitted to measured ice loads and stress levels. It was assumed that the number of load impacts on the bow of a ship navigating in solid ice, as a function of time, is distributed according to a Poisson law [502]. The measured stress amplitudes on a frame at the bow were found to follow an exponential distribution. Extreme value statistics were applied to the short–term maximum ice–induced pressures measured onboard the MS Igrim [522], and to daily maximum ice–induced pressures and stresses on plating and frames measured during winter 1978 at the bow of IB Sisu [1042].

Fuglem et al [350] indicated that the design load depends on the number of interactions per unit of time. The greater the number of interactions per unit of time, the further one has to approach the tail of the parent distribution, because the design naturally concentrates on extreme values. In formulating the design values recommended by Fuglem et al ([349], [351]), it is important to distinguish the factors contributing to the number of interactions as opposed to those affecting the failure load itself.

In some cases, ice loads are of impact type and have been assumed as a Poisson arrival process of loading events. Sample functions of Poisson white noise process $W_P(t)$ may be written in the form

$$W_P(t) = \sum_{i=1}^{N(t)} R_i \delta(t - T_i) \tag{7.29}$$

where $\delta(\cdot)$ is the Dirac's Delta function, R_i is the ith realization of the random variable R with assigned probability density function $p_R(r)$, T_i is the ith realization of the random variable t independent of R and distributed in time according to the Poisson law, and $N(t)$ is the so–called *counting process* giving the total number of impulse occurrences in $[0, t)$. The whole process defined by equation (7.29) is fully described in probabilistic setting by the cumulants, K_n:

$$K_n \left[W_P(t_1) W_P(t_2) ... W_P(t_n) \right] = \lambda E \left[R^n \right] \delta(t_1 - t_2) ... \delta(t_1 - t_n) \tag{7.30}$$

where $E[\cdot]$ denotes the expectation, and $\lambda > 0$ is the mean number of impulse occurrences per unit time. When λ approaches infinity and at the same time $\lambda E\left[R^2\right]$ remains constant, the Poisson white noise approaches the normal white noise. The treatment of dynamical systems under Poisson random processes has been considered in many references (see, e.g., [458], [459], [532], [258], [869]).

The response of ship roll oscillation under random ice impulsive loads described by a Poisson arrival process is very important in studying the safety of ships navigation in cold regions. Under both external and parametric random excitations the evolution of the roll response in terms of its probability density function was evaluated using path integral solution [191]. The path integral method relies on the Chapman–Kolmogorov equation which governs the response transition probability density functions at two close intervals of time. If the response probability density function at an early close time is known a priori, its value at later close time can be evaluated. The roll dynamics of a ship subjected to impulse excitation, $W_P(t)$, and parametric random excitation, $W_0(t)$, may be described by the equation of motion:

$$\ddot{\phi} + 2\tilde{\zeta}\omega_n\dot{\phi} + a\dot{\phi}|\dot{\phi}| + \omega_n^2\phi - \delta\phi^3 + W_0(t)\phi = W_P(t). \tag{7.31}$$

$$\phi(0) = \phi_0,\ \dot{\phi}(0) = \dot{\phi}_0$$

where ϕ is the ship roll angle, ω_n is the natural frequency of the ship roll oscillation, $\tilde{\zeta}$ is the linear viscous damping factor, the third term represents nonlinear damping, and $\omega_n^2\phi-\delta\phi^3$ represents the restoring moment. Note that $W_0(t)$ represents the pitch angle which is assumed to be a random stationary process and the right hand side represents a Poisson random process and describes the moment acting on the ship due to sea waves. When $\omega_n^2\phi-\delta\phi^3 = 0$, the ship experiences capsizing. If we consider $W_0(t)$ as a normal white noise, the ship roll dynamics is captured by the solution of a single oscillator under parametric normal white noise acting simultaneously with an additive external Poisson white noise.

Equation (7.31) may be rewritten in terms of state variables as follows

$$\dot{\mathbf{Z}}(t) = \mathbf{D}\mathbf{Z}(t) + \mathbf{f}(Z,t) + \mathbf{L}W_P(t), \tag{7.32}$$

Where $\mathbf{Z}(t) = \begin{Bmatrix} Z_1(t) \\ Z_2(t) \end{Bmatrix} = \begin{Bmatrix} \phi(t) \\ \dot{\phi}(t) \end{Bmatrix}$; $\mathbf{D} = \begin{bmatrix} 0 & 1 \\ -\omega_n^2 & -2\tilde{\zeta}\omega_n \end{bmatrix}$;

$\mathbf{f}(Z,t) = \begin{Bmatrix} 0 \\ -aZ_2|Z_2| + \delta Z_1^3 - W_0(t)Z_1 \end{Bmatrix}$; and $\mathbf{L} = \begin{Bmatrix} 0 \\ 1 \end{Bmatrix}$.

The Chapman Kolmogorov equation may be written in the form:

$$p_{\mathbf{Z}}(\mathbf{Z}, t+\tau) = \int_{-\infty}^{\infty} p_{\mathbf{Z}}(\mathbf{Z}, t+\tau|\overline{\mathbf{Z}})p_{\mathbf{Z}}(\overline{\mathbf{Z}}, t)d\overline{Z}_1 d\overline{Z}_2 \tag{7.33}$$

The conditional *pdf* in equation (7.33) may be derived by considering the *pdf* of the response in τ of the following system

$$\begin{cases} \dot{\overline{\mathbf{Z}}} = \mathbf{D}\mathbf{Z}(\rho) + \mathbf{f}(Z,\rho) + \mathbf{L}W_P(t+\rho) \\ \overline{\mathbf{Z}} = \overline{\mathbf{z}} \end{cases} \tag{7.34}$$

where the initial conditions are the considered deterministic, $\overline{\mathbf{z}}^T = \{\, \overline{z}_1\ \overline{z}_2 \,\}$. Equations (7.34) may be rewritten in the state vector form

$$\dot{\overline{Z}}_1(\rho) = \overline{Z}_2(\rho)$$

$$\dot{\overline{Z}}_2(\rho) = -2\tilde{\zeta}\omega_n \overline{Z}_2(\rho) - \omega_n^2 \overline{Z}_1(\rho) - a\overline{Z}_2(\rho)|\overline{Z}_2(\rho)| + \delta \overline{Z}_1^3(\rho) - \overline{Z}_1(\rho)(t+\rho) + W_P(t+\rho) \tag{7.35}$$

$$\overline{Z}_1(0) = \overline{z}_1, \qquad \overline{Z}_2(0) = \overline{z}_2$$

For small τ, equation (7.35) yields the following statistics

$$E\left[\overline{Z}_1(\tau)\right] = \overline{z}_1 + \overline{z}_2\tau, \qquad \sigma^2_{\overline{Z}_1}(\tau) = 0 \tag{7.36}$$

$$p_{\overline{Z}_1}(z_1, \overline{z}_1, \overline{z}_2; \tau) = \delta(z_1 - y_1(\overline{z}_1, \overline{z}_2)) \tag{7.37}$$

Furthermore, there are two possible situations over the time interval $(t, t+\tau)$. The first does not contain spikes in the presence of the normal white noise, and this happens in mean $1 - \lambda(t)\tau$ times. The second includes one spike, which occurs simultaneously with the normal white noise. The second case occurs $\lambda(t)\tau$ times. In the first case we set $W_P(t) = 0$, and thus one can write

$$E\left[\overline{Z}_2(\tau)\right] = \overline{z}_2 - \left(2\tilde{\zeta}\omega_n\overline{z}_2 + \omega_n^2\overline{z}_1 + a\overline{z}_2|\overline{z}_2| + \delta\overline{z}_1^3\right)\tau = y_1(\overline{z}_1, \overline{z}_2) \tag{7.38a}$$

$$\sigma^2_{\overline{Z}_2}(\tau) = E\left[\overline{Z}_2^2(\tau)\right] - \left(E\left[\overline{Z}_2(\tau)\right]\right)^2 = \overline{z}_2^2 q\tau \tag{7.38b}$$

Since $\overline{Z}_1$ is deterministic, the two processes $\overline{Z}_1$ and $\overline{Z}_2$ are independent. In the absence of spikes, the contribution for the whole conditional *pdf* in equation (7.33) is given in the form

$$p_Z(z_1, z_2, t+\tau|\overline{z}_1, \overline{z}_2, t)_{\text{no spikes}} = (1-\lambda\tau)\frac{\delta\left(z_1 - (\overline{z}_1 + \overline{z}_2\tau)\right)}{\sqrt{2\pi\overline{z}_1^2 q\tau}} \times \exp\left\{-\frac{\left[z_2 - y_1(\overline{z}_1, \overline{z}_2)\right]^2}{2\overline{z}_1^2 q\tau}\right\} \tag{7.39}$$

The contribution due to the presence of one spike as well as a normal white noise is given by the convolution integral of both:

$$p_Z(z_1, z_2, t+\tau|\overline{z}_1, \overline{z}_2, t)_{\text{one spikes}} = \lambda\tau\frac{\delta\left(z_1 - (\overline{z}_1 + \overline{z}_2\tau)\right)}{\sqrt{2\pi\overline{z}_1^2 q\tau}} \times \int_{-\infty}^{\infty} \exp\left\{-\frac{\left[\xi - y_1(\overline{z}_1, \overline{z}_2)\right]^2}{2\overline{z}_1^2 q\tau}\right\} p_R(z_2 - \xi)d\xi \tag{7.40}$$

The *cpdf* of the kernel of equation (7.33) is

$$p_Z(z_1, z_2, t+\tau|\overline{z}_1, \overline{z}_2, t) = \delta\left(z_1 - (\overline{z}_1 + \overline{z}_2\tau)\right)\left\{\frac{(1-\lambda\tau)}{\sqrt{2\pi\overline{z}_1^2 q\tau}}\times\right.$$

$$\exp\left\{-\frac{\left[z_2 - y_1(\overline{z}_1, \overline{z}_2)\right]^2}{2\overline{z}_1^2 q\tau}\right\} +$$

$$\left.\frac{\lambda\tau}{\sqrt{2\pi\overline{z}_1^2 q\tau}}\int_{-\infty}^{\infty}\exp\left\{-\frac{\left[\xi - y_1(\overline{z}_1, \overline{z}_2)\right]^2}{2\overline{z}_1^2 q\tau}\right\} p_R(z_2-\xi)d\xi\right\} \tag{7.41}$$

The ship roll response *pdf* is determined for $\zeta = 0.1, \omega_n = 1, a = 0.005$, and $\delta = 0.006$. The external Poisson white noise, $W_P(t)$, has a jump Gaussian distribution, $p_R(x)$, with zero mean and standard deviation $\sigma_R = 0.07$, and the mean rate arrival $\lambda = 1/3$. The parametric normal white noise, $W_0(t)$, is assumed to possess an intensity parameter $q = 0.05$. The initial condition $p_{\phi\dot{\phi}}(\phi, \dot{\phi}; 0)$ is assumed to be a bivariate normal distribution with vector mean $(0.05, 0.025)$ and covariance matrix $\Sigma = \begin{bmatrix} 0.133^2 & 0 \\ 0 & 0.133^2 \end{bmatrix}$. Integrals have been performed numerically on a grid of $\Delta\phi = \Delta\dot{\phi} = 0.01$ and time step $\tau = 0.1$ sec.

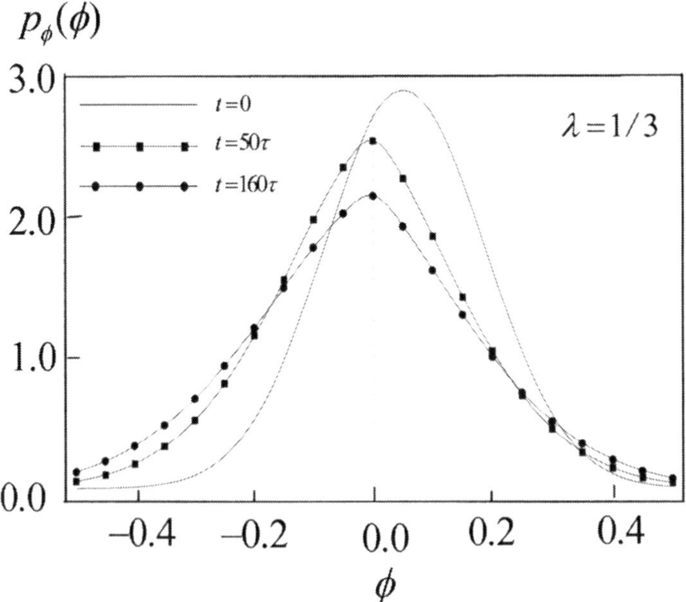

Fig. 7.14. Response *pdf* of ship roll angle for three different values of time instants, [191].

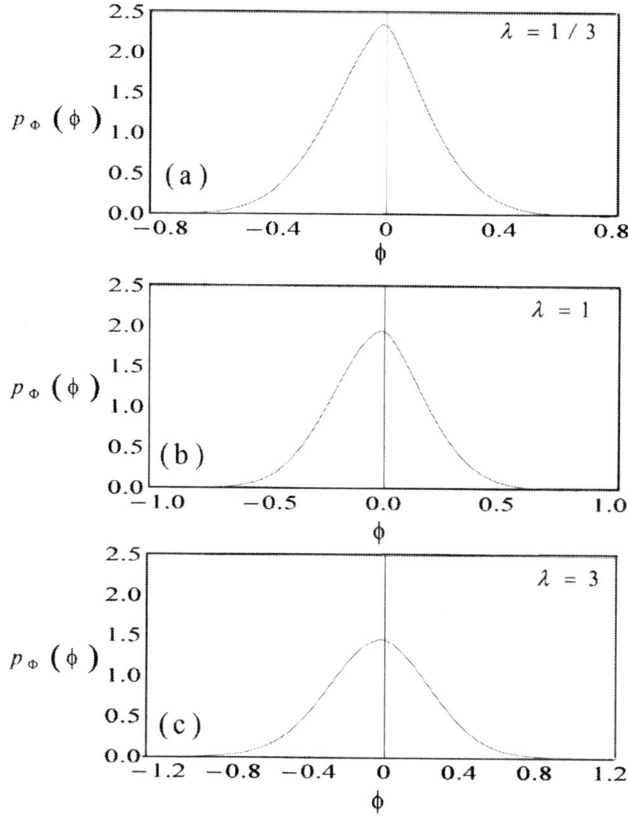

Fig. 7.15. Influence of the mean rate arrival parameter λ on the response *pdf* of ship rll angle, [191].

Fig. 7.14 shows the evolution of the response *pdf* at various time instants starting from the initial condition at $t = 0$. It is seen that as the time progresses, the peak value of the *pdf* is reduced and approaches the equilibrium position, $\phi = 0$. The response *pdf* is essentially non-Gaussian. In order to explore the influence of the mean rate arrival parameter λ on the response *pdf* at the extreme roll angle, the numerical integration was performed for three different values of $\lambda = 1/3$, 1, and 3. Fig. 7.15 shows three different plots of the response *pdf*. As the mean rate arrival increases, the peak of the response pdf is reduced and the *pdf* is spread over larger range of the roll angle. In other words, the probability of the ship response to reach the capsizing angle increases as the mean rate arrival increases.

7.6 Closing Remarks

The response statistics of random excitation of vibro–impact systems have been developed for few cases. The strong nonlinearity arises from impact loading introduces an additional complexity to an already difficult problem. Non–smooth coordinate transformations due to Zhuravlev and Ivanov for purely elastic and inelastic impacts have proven to be powerful tools to convert the vibro–impact governing equations of motion to equations without impact. This facilitates the analysis such that one can apply of the well established techniques in the theory of nonlinear random vibration. The method of path integral is convenient for systems subjected to random impulsive loading described by a Poisson arrival process. There are open avenues of potential research need to be considered for future research. These include the reliability and first passage problem of vibro–impact systems. Equally important is the stochastic stability analyses of these systems under different types of random excitation and constraint stiffness.

Chapter 8
Impact Dampers

8.1 Introduction

One of the most significant benefits of vibro–impact dynamics is the passive control of vibrating systems. The applications of impact masses as an absorbing source of undesirable vibration of structures, machines, and multi–storey buildings have been introduced over the last fifty years. This chapter provides an overview of the basic concept of impact dampers and their applications. Analogue computer, numerical simulations, and analytical techniques were employed to uncover their effectiveness in suppressing unwanted vibrations under deterministic and random excitations. Design considerations of impact dampers are assessed together with active and semi–active control techniques.

8.2 Basic Concept and Applications

Impact dampers, also referred as acceleration dampers, have been used to eliminate undesirable oscillations in mechanical systems ([591], [592], [913], [74]). In the absence of any source of energy dissipation, these systems act as vibro–impact isolators due energy transfer from the main system to the impact mass. The impact dampers can take the form of one mass impact damper (Fig. 8.1(a)), particle damper (Fig. 8.1(b)), and liquid sloshing damper (Fig. 8.1(c)). The liquid sloshing dampers are not discussed in this chapter and the reader may consult reference [436]. Impacts may occur as soon as the displacement of the main system exceeds the clearance between the mass damper and its clearance. Every collision produces some energy dissipation and exchange of momentum between colliding bodies. Energy dissipation is useful in attenuating the excessive vibration amplitudes of the main structure, but the important control mechanism is the exchange of momentum during collisions. For an adequate choice of clearance, the main structure and impact mass move in opposite directions before collision. The direction of motion of the smaller impact damper is reversed after collision, whereas

R.A. Ibrahim: Vibro-Impact Dynamics: Model., Map. & Appl., LNACM 43, pp. 217–232.
springerlink.com

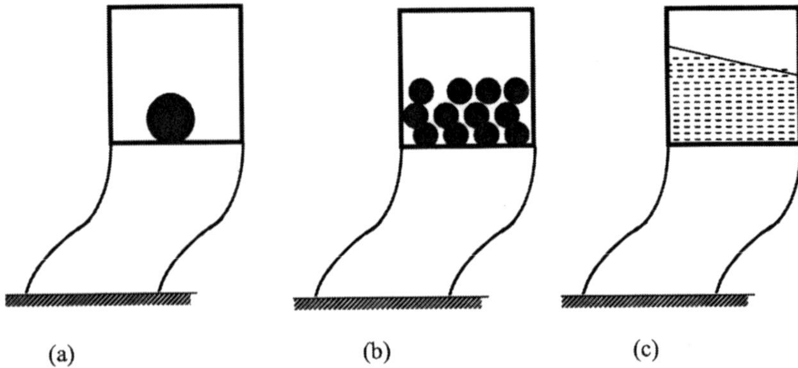

Fig. 8.1. Three different types of impact dampers: (a) Impact damper, (b) Particle damper, (c) Liquid sloshing damper.

the velocity of the main structure is only reduced due to its larger inertia. As a result of the reduced velocity, the main system attains smaller displacement amplitude than it would have without impacts.

The concept of suppressing vibration of mechanical systems using a mass moving between two walls of a tank is believed to be conceived by Lieber and Jensen [591]. Lieber and Jensen [591] introduced the impact damper to control the flutter of aeroelastic structures. They considered the case of two impacts per cycle and found that maximum flutter suppression occurs if the phase angle between the motion of the impact mass and the main system is 180°. Later, Grubin [379] assumed the existence of symmetric two impacts per cycle and determined the behavior of the main system by adding the effects of many impacts. It was found that the impact damper is most effective at resonance and large values of the coefficient of restitution result in more beneficial damping. Warburton [1061] allowed the impact at the right end of the container to occur at zero time and modified the excitation force to have a phase component. Arnold [30] represented the acting impact force by a Fourier series.

Noise generated from impact can be reduced using non–metallic layers (and/or inserts) on the surface of one of the impacting bodies ([155], [156], [92]). A rheological model of impact was proposed by Oledzki [749]. Oledzki et al [750] introduced an impact damper to suppress severe vibrations of long tubes in the structures of light aircraft. The numerical simulation of the rheological model showed a reduction in the resonance amplitude. This model was used in a mathematical model of impact damper and applied in the numerical simulation. They were found in good agreement with experimental results.

The response of a mass–spring damper exposed to repetitive impact was studied by Park [790]. The system was convenient to control equipment utilizing repetitive impact as driving force such as electric and pneumatic hammers, shakers, bin vibrators and pile drivers. Impact–dampers have been developed for industrial applications such as turbine blades, aircraft wings, and lighting poles along high–speed highways. For example, Tokumaru and Kotera [1006] introduced a concentrated–mass–continuum system with an impact–body moving in a groove of the mass and strikes it at opposite ends of the groove. They obtained periodic solutions under the base harmonic excitation. The analytical model resulted in a partial integro–differential equation. Other applications include impact dampers to the carrying cylinder of a web–fed printing press [948]. Yamamoto et al [1093] applied a set of turrets, used for providing punches and dies, to a damper for a C–frame turret punch press in an attempt to reduce the vibration and noise. The damping characteristics can be changed by adjusting the preload. Over a wide frequency range, the effect of the sound pressure level reduction was found to be the same for either one turret or a set of turrets. Furthermore, the complex structure of the turret damper and the double wall eliminates the first tuning vibration, and was found to reduce the sound pressure level in the actual punching by about $11dB$.

Moore et al [692] described high speed rotor–dynamic dampers for rocket engine turbo–pumps operating at cryogenic temperatures, such as those used in the space shuttle main engines. An impact damper was designed and tested to obtain effective damping in a rotor–bearing system. Their analytical results revealed a strong amplitude dependence on the impact damper performance. The damper performance was characterized by an equivalent viscous damping coefficient. The test results indicated that the impact damper is a viable means to suppress vibration in cryogenic rotor–bearing systems.

Impact dampers may be used to control vibrations of buildings and structures subjected to earthquake motion, mechanisms, and machine tool vibration [296]. For example, particle dampers have been utilized by Sato et al [915] to reduce the vibration in pantograph–support systems, and by Sims et al [944] to improve the chatter stability of a machining process. They have recently been used to control chaos in systems with limited power supply [222]. Fuse [352] utilized the impact dampers to eliminate the effects of resonance in mechanical systems by making the impact of the main system and the additional vibration system to mutually collide with each other at an opposite phase. Gibson [362] and Torvik and Gibson [1009] employed the impact dampers for space applications. The decay rate and the minimum effective amplitude were found to be the main two critical parameters that govern the damper design. This special class comprises a container filled with thousands of small granular particles which dissipate energy by friction and impact when the container vibrates. The resulting behavior is highly nonlinear but can provide very high levels of damping across a wide frequency range. Ogawa et al [747] introduced an impact mass damper as a damping device to suppress wind–induced vibration of a single pylon of a cable–stayed bridge.

Aiba and Murata [1] described an impact damper having a spring supported impactor without gap and pre–pressure. The transient response of the damper was obtained for various conditions based on experiments and numerical simulations. It was found that the damper without gap and pre–pressure practically has no residual vibration and the damping effects are not dependent on amplitude. A variable–attractive–force impact damper was introduced by Aiba et al [2] to eliminate chatter vibrations in the cutting process. The system was applied to eliminate chatter vibrations in the face milling process for very weak workpieces. The cutting tests showed that the system is very effective and versatile for stopping chatter vibrations.

Friend and Kinra ([346], [347]), Marhadi and Kinra [642], and Olson [751] considered particle impact damping for achieving high structural damping by the use of a particle–filled enclosure attached to the structure in a region of high displacements. The particles absorb kinetic energy of the structure and convert it into heat through inelastic collisions between the particles and the enclosure, and amongst the particles. Particle impact damping was measured for a cantilevered aluminum beam with the damping enclosure attached to its free end. It was found that the impact damping is highly nonlinear.

8.3 Analogue Computer and Numerical Simulations

Early research of impact dampers relied on analogue computer simulation. For example, Masri [650], Bhattacharyya and Chatterjee [111] and Peterka ([809], [810], [812]) used analogue computers to study the effectiveness of impact dampers in reducing vibration–amplitude levels under sinusoidal and random excitations. Peterka [812] demonstrated the symmetric periodic damper motion with two impacts of masses per period. He was able to explain the complex non–symmetric, beat, and multi–impact types of motion. Mansour and Teixeira Filho [638] conducted a parametric study of the modes and transitions of the auxiliary masses for impact dampers with Coulomb friction using analogue and digital simulations. It was shown that for shorter containers, Coulomb friction may introduce stability to the motion of the primary mass, improve the response for harmonic excitation and accelerate the attenuation of the amplitude of oscillations for step inputs.

Bapat and Sankar [91] studied the single unit impact damper under free and forced vibrations using numerical simulations. The simulations were used to predict the effects of mass ratio, coefficient of restitution and gap size on the free vibrations. In the study of forced motion, charts were generated to provide useful information such as optimum gaps and corresponding displacement amplitude reduction within the resonant frequency range. These results were later confirmed by Ema and Marui [290] who showed that the damping capability of impact dampers arises from collision between the free mass and the main mass. The optimum damping effect was achieved in combinations of the mass ratio (ratio of the mass damper to the main system mass) and a clearance. It was found that the use of a free mass of only 25% of the main mass and a clearance of 0.6–mm can improve the damping capability of the main vibratory system at least 10 times or more, even though the clearance and the free mass are not adjusted to amplitude of the main vibratory system.

The damping effects of an impact damper for the vibration neutralizer of pillar bodies were examined by Saito et al [909]. They considered a beam representing a pillar body together with an auxiliary mass coupled to the beam by means of a spring–dashpot system. The system includes a gap between the beam and the auxiliary mass, and they collide with each other at one point when the system vibrates. The impact force was modeled using Hertz' contact theory and the normal–mode approach was employed in the simulation. The numerical simulation results for free and random forced excitation were experimentally verified. The effects of mass ratio, natural frequency ratio, gap size and impact point for free and random responses were obtained by numerical simulation. Akl and Butt [5] demonstrated that impact dampers can increase the intrinsic damping of a lightly–damped flexible structure. The test structure consists of a slender flexible beam supported by a pin–type support at one end and supported by a linear helical flexible spring at another location. Sinusoidal excitation spanning the first three natural frequencies was applied in the horizontal plane. The orientation of the excitation and the test structure in the horizontal plane minimizes the effect of gravity on the behavior of the test structure. The results showed that the impact damper significantly increases the damping ratio of the test structure.

Duncan et al [279] presented computer simulations to examine the damping performance of a single particle vertical impact damper over a wide range of excitation frequencies and amplitudes. The influences of other parameters such as particle–to–structure mass ratio, lid clearance ratio, structural damping ratios and coefficient of restitution on the damping performance were also examined. Maximum damping at a fixed oscillation frequency was found to occur at an optimal lid height that increases with increasing mass ratio, increasing structural damping ratio, but decreases with coefficient of restitution. The corresponding maximum degree of damping was found to increase with increasing mass ratio and coefficient of restitution, but decreases with increasing structural damping ratio.

8.4 Analytical Results

Exact analytical solutions for the steady–state motion of undamped harmonic oscillator equipped with two–particle single–container impact damper were obtained by Masri ([648], [651]). The results were compared with those of the same damper with single–particle impact damper. Later, Masri ([652], [653], [654], [656]) obtained an exact solution for symmetric two impacts/particle/cycle motion of n–unit impact damper attached to a sinusoidally excited primary system. It was found that properly designed multiple–unit impact dampers are more efficient than equivalent single–unit impact dampers in regard to vibration reduction and noise level of operating units.

Masri and Caughey [659] analytically derived an exact solution and stable regions for symmetric two–impacts–per–cycle motion of impact dampers. The stability analysis defined the domains over which the modulus of all eigenvalues of certain matrix relating conditions after each of two conservative impacts is less than unity. Masri [657] conducted analytical and experimental forced excitation studies to examine the dynamic response of a system with a motion–limiting stop. The response under harmonic excitation was found to exhibit the jump resonance phenomena. Nigm and Shabana [730] studied the steady state vibrational motion of a multi–degree of freedom system equipped with an impact damper. The steady state of the damper was characterized by the existence of as many as three modes for a given excitation frequency.

Brown and North [144] and Brown [143] studied the free decay of impact–damped oscillators under a wide range of oscillator amplitudes. Three operating regions were reported:

- Low–amplitude range with less than one impact per cycle and very low damping.
- Useful middle amplitude range with a finite number of impacts per cycle.
- High–amplitude range with an infinite number of impacts per cycle and progressively decreasing damping.

Bapat et al [90] and Bapat and Bapat [88] predicted stability regions of two equi–spaced impacts/periodic motion of an impact–pair under a prescribed periodic displacement. Some techniques for estimating the clearance and impact forces were proposed by Lin and Bapat [594] based on a describing function and an optimization approach for a harmonically excited vibro–impact system exhibiting periodic oscillations. Bapat ([84], [86]) analyzed multi–stable impacts per period of an inclined impact damper with friction and collision on either one or both sides of the main mass with identical and non–identical coefficients of restitution. Bapat [87] developed the nonlinear equations governing multi–impact periodic motions of a single–degree–of–freedom oscillator under sinusoidal and bias force contacting rigid amplitude constraints on one or both sides. Exact closed form expressions were derived for one and two equi–spaced and non–equi–spaced impacts per cycle.

The problem of quenching self–excited vibrations by an impact damper composed of a ball impacting with its container walls was considered by Yoshitake and Sueoka [1114]. The Runge–Kutta–Gill method with variable time increment was applied to the numerical analysis of the self–excited system of Rayleigh's type. The response was found to exhibit periodic and chaotic motions after period doubling bifurcation and intermittency. It was indicated that for an optimum design of impact dampers there exists a certain relationship between the coefficient of restitution and clearance. An experiment was performed in order to quench vortex induced vibration by an impact damper.

Friction and impacts during oscillation lead to discontinuities of the velocity and internal forces in the time–domain and result in changes in the number of degrees of freedom. The analytical procedure for integrating such non–smooth motions involves the computation of the history dependent separation times and patching a sequence of solutions for successive smooth problems ([859], [860]). However, this procedure has its limitations even for a relatively low number of generalized coordinates because of the required computation. Regularization techniques were used with finite element to avoid the exact computation of all discontinuities by smoothing. However, there is a significant uncertainty in the choice of the regularization parameters needed for a sufficiently correct description of oscillations. Stationary solutions of two forced mass–spring oscillators were used to calibrate the regularization parameters by comparing analytical results with regularized ones. This allows one to compute the self–excitation of a continuous system and validate the phenomena with known experimental data. Koizumi et al [520] studied the dynamic characteristics of a steady–state repetitive impact motion. The impulse intensity was adjusted to be proportional to the pressure force level required to support the follower.

A relationship between the coefficient of restitution and impact damping ratio was developed by Cheng and Wang [165] and Cheng and Xu [166]. It was shown that the effective reduction of the response amplitude is nearly independent of the number of impacts, but primarily related to the type of collision. Furthermore, the results revealed that the clearance of an effective impact damper should be smaller than twice of the initial displacement of the main mass if the system is stimulated by an initial displacement only. Blazejczyk–Okolewska [119] and Peterka and Blazejczyk–Okolewska [825] considered a two–degree–of–freedom system whose parameters were properly selected to act as an impact damper. The regions of bifurcation diagrams and motion trajectories of different kinds of impact motion were estimated. The concept of targeted energy transfer was applied to seismic mitigation by Nucera et al [740]. It was demonstrated that a single–degree–of–freedom nonlinear energy absorber with non–smooth vibro–impact nonlinearities can passively absorb and locally dissipate a significant portion of the seismic energy of the primary structure.

Ekwaro Osire and Desen [286], Semercigil et al [919] and Shaw and Pierre [933] experimentally and analytically examined different types of impact vibration absorbers. Shaw and Pierre [933] considered absorbers using centrifugally induced restoring forces so that their non–impacting dynamics are tuned to a given order of rotation, whereas their large amplitude dynamics involve impacts with the primary flexible system. A class of symmetric impacting motions was analyzed and used to predict the effectiveness of the absorber when operating in its impacting mode. It was observed that two different types of grazing bifurcations take place as the rotor speed varies through resonance. Mikhlin and Reshetnikova [678] considered a nonlinear two–degrees–of–freedom system consisting of a linear oscillator with a relatively large mass and a vibro–impact oscillator with a relatively small mass. Their analysis showed that a stable localized vibration mode exists in a large region of the system parameters. Sung and Yu [972] used Poincaré mapping to detect bifurcation phenomena in studying the existence and stability of subharmonic motions in a vibro–impact two–degree–of–freedom damper. The evolution from the period–doubling sequence to chaotic motions was demonstrated. Ekwaro Osire et al [287] considered the dynamic characteristics of a bi–unit impact damper referred to as bi–unit impact vibration absorber. They experimentally analyzed the performance of the impact vibration absorber using digital image processing.

Ramachandran and Lesieutre ([880], [881], [882]) made an attempt to understand the dynamics of a particle impact damper as well as parameters governing its dynamic behavior. The base was harmonically excited in the vertical direction. A discrete event approach was used, wherein the variables at one 'event' (or impact) uniquely dictate the variables at the next 'event,' leading to a two–dimensional difference map. Periodic impact motions and 'irregular' motions were observed. The results revealed a peak for certain combinations of parameters. These parameters correspond to a region of parameter space where two–impact–per–cycle motions were observed over a wide range of base acceleration. The range of gap clearance over which two–impact–per–cycle solutions were found to increase as the coefficient of restitution increases.

8.5 Experimental Results

Numerous analytical and experimental studies of impact dampers have been reported in the literature. The main purpose of these studies was to validate the predicted results. Furthermore, experimental investigations have been conducted to explore the effectiveness of the impact dampers under different damper parameters. For example, Veluswami and Crossley [1028] and Veluswami et al [1029] used three different materials for coating the impacting plates in their impact dampers. It was found that the soft materials possess a lower coefficient of restitution and provide a smaller amount of damping at resonance. Sadek and Mills [904] and Sadek et al [905] examined the effects of gravity on impact dampers and found that the dampers are most effective under zero gravity, which occurs when the damper is excited in a direction perpendicular to gravity. Effective damping was also found when two symmetric and equal impacts take place per cycle and the effect of gravity resulted in impulses at the ends of the container. In the vicinity of resonance the steady state of two un–equi–spaced impacts prevail.

Cempel and Lotz [157] presented experimental results of vibration damping of shot–filled containers with some empty volume acting as the shot damper clearance. The energy dissipation by shot was found to depend on the internal and the external (container walls) impacts of shot particles associated with their internal and external friction. Hollkamp and Gordon [421] utilized particle damping in the form of metallic or ceramic particles inside structural cavities. As the cavity vibrates, energy is dissipated through particle collisions. Yokomichi et al ([1109], [1111], [1110], [1112]) and Saeki ([906], [907]) studied the performance of the shot impact damper applied to vibrating systems subjected to harmonic excitations. The periodic motion of the self–excited vibration accompanied by the damper mass was analytically and numerically determined. It was found that heavier damper mass provides greater damping effect on the fully built–up vibrations, and that a lighter mass is effective at the onset of the vibration. The optimum clearances under which the response amplitudes are reduced to a minimum for suppressing the vibrations were also determined. Yang [1096] and Yang et al [1097] developed design curves to predict the damping characteristics of particle impact dampers.

Li [587] conducted a series of experimental investigations to find out the effect of an impact damper on a multi–degree–of–freedom system. The effects of the size of the impact mass, clearance between the impact mass and the stops, excitation type and location were considered. The results in some instances did not correspond to those found for control of single–degree–of–freedom systems. In particular, increasing the size of impact mass was found not necessarily lead to an increase in damping for all modes.

Mao et al [640] employed a three–dimensional discrete element method for characterizing the performance of particle damping. It was found that the particle damping can achieve a very high value of specific damping capacity. The particle damping is a combination of the impact and friction damping. The damping was found to be highly nonlinear as the rate of energy dissipation depends on amplitude. Particularly, the damping effect results in a linear decay in amplitude over a finite period of time. It was concluded that the particle damping is a combination of these two damping mechanisms and the relative significance of these damping mechanisms depends on a particular arrangement of the damper. The problem was further studied by Xu et al [1090] who confirmed that the mechanisms of energy dissipation of particle damping are primarily related to friction and impact phenomena. They considered elastic beam and plate structures with drilled longitudinal holes filled with damping particles. Particular attention was given to the form of damping due to shear friction induced by strain gradient along the length of the structure. Experimental tests of beam and plate structures for different damping treatments revealed that the particle damping is remarkably strong for a broadband frequency range. Moreover, the shear friction was found to be the major contributing mechanism of damping especially at a high volumetric packing ratio. The numerical and experimental results suggested that the best damping effect can be achieved by using a design of multiple particle chambers involving an appropriate combination of the impact, friction and shear mechanisms.

Experimental investigations on a platform wedge damper were conducted to compare its effectiveness with that of the impact damper ([271], [272],

[273], [274], [275]). The self–tuning impact damper combines the tuned mass damper and the impact damper. It consists of a ball located within a cavity in the blade. The ball rolls back and forth on a spherical trough under centrifugal load (tuned mass damper) and can strike the walls of the cavity (impact damper). The ball rolling natural frequency is proportional to the rotor speed and can be designed to follow an engine–order line (integer multiple of rotor speed). The excitation was provided by three different sources: 1) an eddy–current engine–order excitation, 2) electromechanical shakers, and 3) magnetic bearing excitation. The eddy–current system consists of magnets located circumferentially around the rotor. As a blade passes a magnet, a force is imparted on the blade. The number of used magnets can be varied to change the desired engine order of the excitation. The other two methods apply force on the rotating shaft itself at frequencies independent of the rotor speed. Later, Duffy [270] analytically and experimentally investigated a self–tuning impact damper as a device to inhibit vibration and increase the fatigue life of rotating components in turbomachinery.

8.6 Random Excitation

The effectiveness of impact dampers under random excitation has been considered in few studies. For example, the response statistics of an impact damper to a white noise Gaussian excitation using numerical simulation and analogue computer were estimated by Masri and Ibrahim ([661], [662]), Masri and Stott [664] and Semercigil et al [918]. Masri and Ibrahim [660] obtained an approximate analytical solution for the stationary response of a highly nonlinear auxiliary mass damper attached to a single–degree–of–freedom oscillator subjected to a white noise random excitation. It was found that the impact damper is substantially more effective than the conventional dynamic vibration neutralizer in controlling the response of stochastically excited primary systems.

Papalou and Masri ([786], [787]) presented experimental and analytical results describing the performance of granular material dampers with tungsten powder, as an impacting mass, under a wide–band random excitation. The influence of the auxiliary mass ratio, container dimensions and excitation intensity were investigated using a small building model under base excitation. An approximate analytical solution based on the concept of an equivalent single–unit impact damper was developed. Comparison between experimental and analytical results revealed accurate estimates of the *rms* response of a primary system under stationary random excitation.

Nayeri et al [721] developed and evaluated practical design strategies for maximizing the damping efficiency of multi–unit particle dampers under random excitation. Both stationary and nonstationary excitations were considered. They performed high–fidelity simulation with a variable number of multi–unit dampers ranging from 1 to 100. The magnitude of the

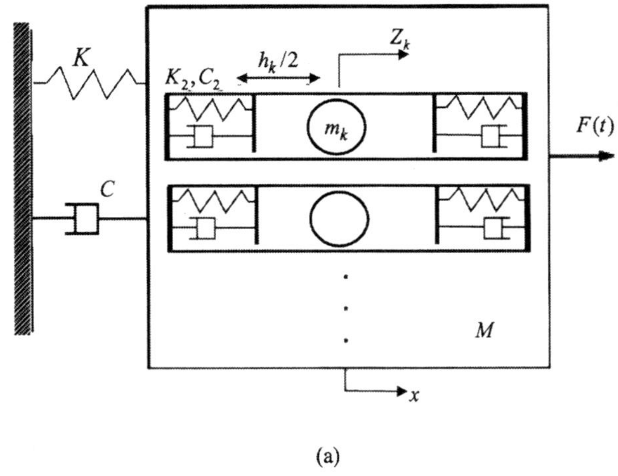

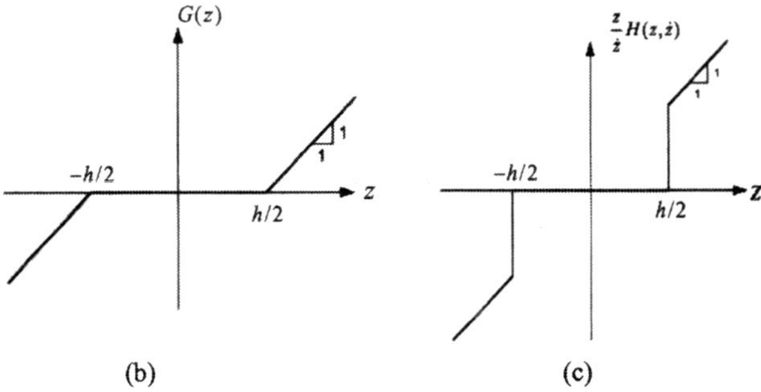

Fig. 8.2. Multiple unit impact damper and associated characteristic functions: (a) schematic diagram of the multiple unit impact damper, (b) nonlinear function $G(z_k)$ and (c) nonlinear function $H(z_k, \dot{z}_k)$, [721].

"dead–space" nonlinearity was considered as a random variable with a prescribed probability distribution. Computational results were calibrated with experimental measurements for a single–unit/single–particle, single–unit/multi–particle, and multiple–unit/multi–particle dampers. It was shown that wide latitude exists in the trade–off between high vibration attenuation over a narrow range of damper gap size versus slightly reduced attenuation over a much broader range. The study considered the multiple impact damper shown in Fig. 8.2. The primary system has mass M with a nonlinear auxiliary multi–unit impact damper. Each unit of the damper consists of mass m_k, which is coupled to the main mass by a piecewise linear dashpot c_2 and spring k_2 with a dead space of clearance h_2. The analytical model according to Nayeri et al [721] is described by the equations of motion

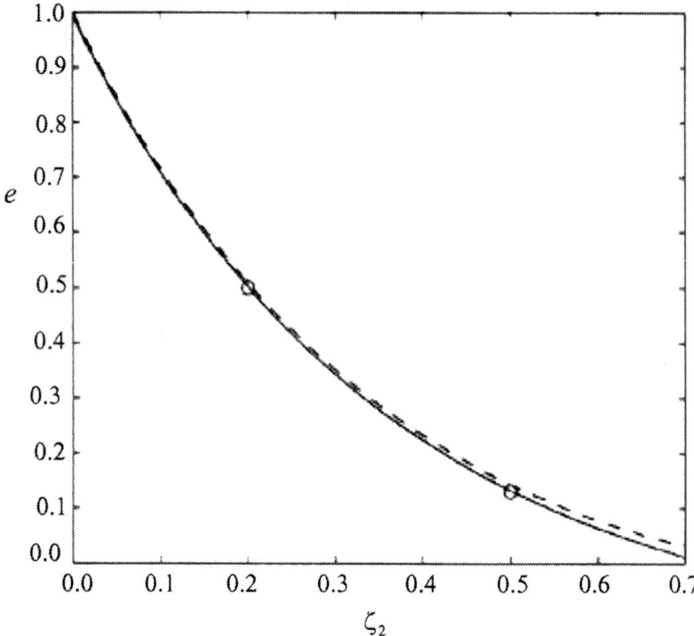

Fig. 8.3. Dependency of the coefficient of restitution on the damping parameter ζ_2: —— $\omega_2/\omega_n = 5$, - - - $\omega_2/\omega_n = 20$, $\circ$ experimental measurements, [721].

$$\ddot{x} + 2\zeta\omega_n\dot{x} + \omega_n^2 x$$

$$-\sum_{k=1}^{N}\left\{\mu_k\left[\omega_2^2 G(z_k) + +2\zeta_2\omega_2 H\left(z_k, \dot{z}_k\right) + \mu_s g sgn\left(\dot{z}_k\right)\right]\right\} = \frac{f(t)}{M} \quad (8.1)$$

$$\ddot{z}_k + 2\zeta_2\omega_2 H\left(z_k, \dot{z}_k\right) + \omega_2^2 G(z_k) + \mu_s g sgn\left(\dot{z}_k\right) - \ddot{x} = 0, \quad (8.2)$$

$$k = 1, 2, ..., N,$$

where x is the displacement of mass M, z_k is the relative displacement of the $k-$th particle with respect to the primary system, and $f(t)$ is the external excitation force. ω_n and ζ are the natural frequency and damping factor of the primary system, respectively. ω_2 and ζ_2 are the natural frequency and damping factor of the impact damper stops, respectively. $\mu_k = m_k/M$ is the mass ratio of the $k-$th particle, and μ_s is the friction coefficient between particles and primary mass. g is the gravitational acceleration, $G(z_k)$ and $H(z_k, \dot{z}_k)$ are nonlinear functions shown in Figs. 8.2(b) and 8.2(c), respectively. Note that the damping factor ζ_2 and the function $H(z_k, \dot{z}_k)$ provide means for simulating inelastic impacts, ranging from purely plastic to the purely elastic ones. Accordingly, the value of the coefficient of restitution, e, can be adjusted by selecting

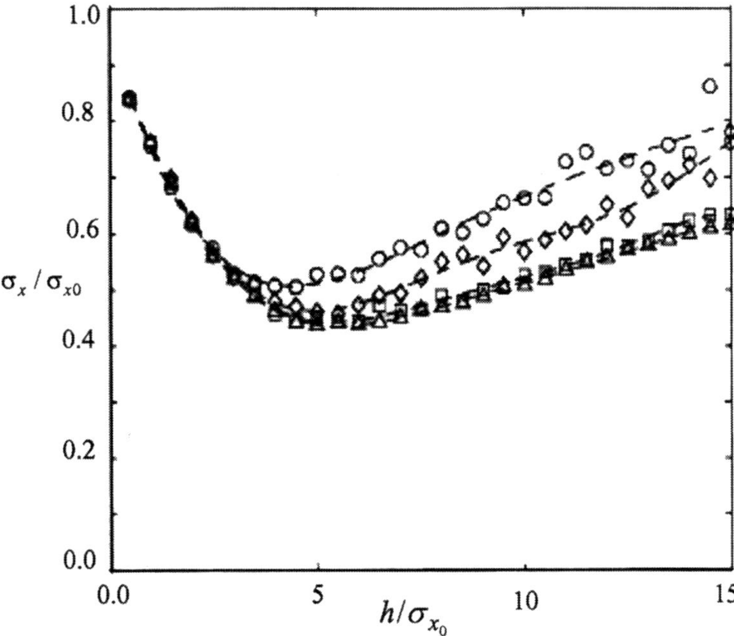

Fig. 8.4. Dependence of the response *rms* displacement of the main system for $e = 0.75$, $\mu = 0.1$, and $\zeta = 0.01$: ○ one particle, ◇ 2 particles, ⊡ 10 particles, Δ 100 particles, [721].

the appropriate value of ζ_2 as shown in Fig. 8.3. Under stationary Gaussian random excitation, the response statistics were estimated numerically and Fig. 8.4 shows the dependence of *rms* ratio of the response displacement, σ_x/σ_{x0}, on the clearance ratio h/σ_{x0}, where σ_{x0} is the *rms* of the primary mass displacement in the absence of the multi–unit impact dampers. Fig. 8.4 was obtained for four different number of particle units. For all cases, the *rms* level of the response exhibits a definite minimum for certain clearance ratios in the neighborhood of $h/\sigma_{x0} \approx 5.0$. It was reported that as the number of unit particles increases, the sensitivity of vibration attenuation to changes in h decreases.

8.7 Design Considerations

Particle vibration damping combines impact suppression and friction damping ([776], [777], [778], [779]). The particle damper shown in Fig. 8.1(b) has been extensively studied by Araki et al ([24], [25], [26], [27], [28], [29]) to determine the characteristics of the damper with granular materials with the purpose of reducing the vibration of a single degree–of–freedom system. They also determined the characteristics of the impact damper for reducing the vibration of the system under horizontal excitation. The main design

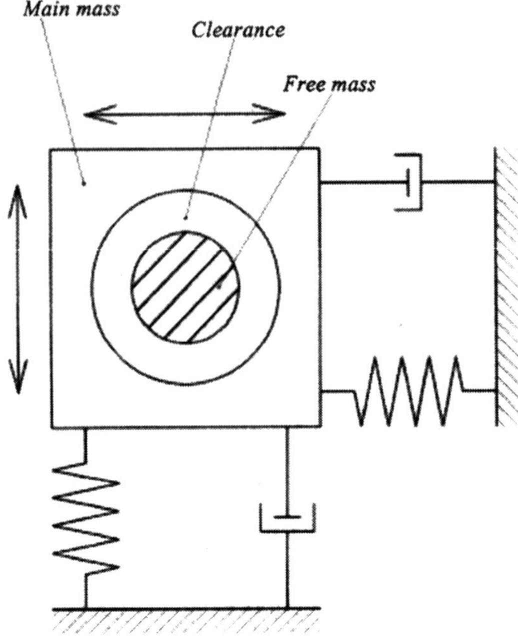

Fig. 8.5. Impact damper used in boring tools, [293].

parameters of these dampers were discussed by Estabrook and Plunckett [301] and Dokainish and Elmaraghy [263]. The influence of mass ratio, particle size, particle/slot clearance, excitation levels and direction of excitation was studied in several references (see, e.g., [861], [863], [775], [788]). Impact motions were found to be very sensitive to small fluctuations in the clearance between masses and the stiffness and loading of the oscillator near its linear resonant frequency. It was concluded that the rigid mass acts as an effective damper at or just above this frequency condition. Furthermore, a plastic bean bag filled with lead shot exhibited much greater damping effectiveness and softer impacts than a single lead slug of equal mass [339]. Popplewell and Liao [862] introduced approximate techniques to simplify the design of an impact damper intended to reduce resonant displacements.

Impact dampers capable to control machine tool chatter and to improve the damping capability of boring tools were developed in the literature ([903], [902], [290], [291], [292], [293]). These dampers consist of a free mass and a clearance as shown in Fig. 8.5. It was demonstrated that the damping capability of boring tools is considerably improved using impact dampers. All types of impact dampers used in the experiment can considerably suppress the vibration of boring tools in the vertical direction (principal force direction), but hardly suppress it in the horizontal direction (thrust force direction) where the amplitude is extremely small.

Thomas et al [991] found that the impact damper boring bar is very effective and provides a stable metal removal rates. A theoretical analysis for predicting the effectiveness of the impact damper with a spring supported impacting mass was presented. The results of this analysis were used to select the optimum mass ratio and gap setting for specific characteristics of the vibrating systems to which the damper is fitted. A systematic approach based on a univariate search optimization method was used to determine the best design parameters for suppressing self–excited vibrations [32]. Optimum parameters for complete quenching of such vibrations were obtained using quasi–static as well as dynamic variations of the bifurcation parameter for both supercritical and subcritical Hopf bifurcation.

Liu et al [598] studied the influence of parameters governing the design of a bean bag damper. The basic parameters include the size of beans, the mass ratio of the bean bag to the structure to which it is attached, the clearance distance and the position of the bag. It was found that reducing the size of beans would increase the exchange of momentum in the system due to the increase in the effective contact areas. Within the range of values of mass ratio studied, the damping performance of the damper was found to improve with higher values of mass ratio. There was an optimum clearance for any specific damper whereby the maximum attenuation could be achieved. It was shown that an appropriately configured bean bag damper was capable of reducing the amplitude of vibration by 80% to 90%.

8.8 Semi–Active and Active Control of Impact Dampers

Active and semi–active controls of impact dampers are usually introduced to improve their performance. Karyeaclis and Caughey ([494], [495]) examined the stability of a semi–active impact damper. All solutions of the system were shown to be bounded when the input is bounded. Emphasis was given to the case of two impacts/cycle periodic solutions. Sensitivity to clearance was experimentally studied by Papalou and Masri [788]. They showed that particle dampers, even with a small mass ratio, can be very effective in attenuating the vibrations of lightly damped structures.

A dynamic damper with a preview action was proposed by Tanaka and Kikushima [979] with the purpose of verifying the control effect of the dynamic damper for the transient vibration caused by an impact force. The influence of the maximum impact force upon the control effect for suppressing the first wave of the transient vibration was examined. Later, Tanaka and Kikushima [980] proposed a new semi–active damper driven by the motion of a released damper mass from an initial displacement for suppressing impact vibration. Using the principle of impact vibration control, the optimal design conditions of the semi–active damper were derived.

Li and Darby [588] considered an impact damper in the form of a freely moving mass constrained by stops and located on a dynamic structural system. At the point of impact, large values of acceleration were imparted to the structure. In order to reduce the associated high accelerations, it was proposed to incorporate a buffer region between the mass and the stop. The performance of the buffered impact damper was compared with that of a conventional rigid impact damper. It was found that the buffered impact damper not only significantly reduces the accelerations, contact force and the associated noise generated by a collision but also enhances the level of vibration control.

Chatterjee et al ([160], [161]) and Chatterjee and Mallik [159] studied the performance of different types of nonlinear oscillators including a hard Duffing oscillator impact damper under harmonic excitation. Both elastic and inelastic collisions were considered. The use of an impact damper as an on–off migration controller for a guided transition from the resonance to the non–resonance branch of the solution was proposed. Bifurcations of self–excited oscillators with an impact damper were determined. The persistent bifurcation structure of codimension–one which is independent of the exact functional form of the self–excitation mechanism was determined. Yasuda et al [1100] and Kamiya et al [488] introduced an elastic impact damper made of an elastic impacting body and a container. Possibility of vibration suppression of resonances for several modes of the beam was numerically studied. They also experimentally measured the influence of the system parameters on its performance.

Hundal [432] analyzed an impact absorber consisting of a linear spring in parallel with a hydraulic damper with variable area orifice. The orifice area was made to vary in two stages in order to overcome the deteriorating effect of fluid compressibility. Collette [185] and Collette et al [186] studied the control of excessive transient vibrations of a light and flexible secondary system with a tuned absorber and impact damper combination. The particular modified tuned absorber concept was mounted on a single degree–of–freedom primary structure.

8.9 Closing Remarks

Although vibro–impact is detrimental and undesirable for the safe operation of mechanical systems, it has one bright and attractive advantage in suppressing severe vibration of structural and mechanical systems. This has been achieved by the advent of impact dampers. The design of these dampers has taken different schemes and approaches such as mass dampers, particle dampers, bean bag dampers, and liquid sloshing dampers. An optimum design of impact dampers is based on appropriate selection of the coefficient of restitution and clearance. The damping performance of bean bag dampers was found to improve if the designer increases the mass ratio of the bean bag to the structure mass. This chapter provided a brief overview; however, the subject matter deserves an independent research monograph.

References

1. Aiba, T., Murata, R.: Research on reducing of nonlinear behavior of an impact damper in transient vibration. J. Japan Society for Precision Engineering 60(8), 1107–1111 (1994) (in Japanese)
2. Aiba, T., Murata, R., Henmi, N., Nakamura, Y.: An investigation on variable–attractive–force impact damper and application for controlling cutting vibration in milling process. J. Japan Society of Precision Engineering 61(1), 75–79 (1995) (in Japanese)
3. Aidanpää, J.O., Gupta, R.B.: Periodic and Chaotic behavior of a threshold–limited two–degree–of–freedom system. J. Sound & Vib. 165(2), 305–327 (1993)
4. Aidanpää, J.O., Shen, H.H., Gupta, R.B.: Stability and bifurcations of a stationary state for an impact oscillator. Chaos 4(4), 621–630 (1994)
5. Akl, F.A., Butt, A.S.: Application of Impact Dampers in Vibration Control of Flexible Structures. NASA Johnson Space Center, Washington, DC, p. 15 (1995)
6. Alabuzhev, P., Gritchin, A., Kim, L., Migirenko, G., Chon, V., Stepanov, P.: Vibration Protecting and Measuring Systems with Quasi–Zero Stiffness. Hemisphere Publishing Co., New York (1989)
7. Alabuzhev, P., Gritchin, A., Stepanov, P., Khon, V.F.: Studies of vibro–protecting systems with stiffness correction. Physico–Tech Probl Exploitation of Mineral Resources 3, 136–149 (1977)
8. Alexander, R.M., Noah, S.T., Franck, C.G.: Parametric identification of a vibratory system with a clearance. Proc. ASME Design Eng Div, Modal Analysis, Modeling, Diagnostics, and Control – Analytical and Experimental 38, 47–57 (1991)
9. Alighanbari, H.: Aeroelastic response of an airfoil–aileron combination with free–play in aileron hinge. J. Aircraft 39(4), 711–713 (2002)
10. Alzate, R., di Bernardo, M., Montanaro, U., Santini, S.: Experimental and numerical verification of bifurcations and chaos in cam–follower impacting systems. Nonlin. Dyn. 50(3), 409–429 (2007)
11. Amelkin, V.V., Kalitin, B.S.: Ideal 'two–impact' model of the clock with an analytical strongly isochronous oscillator of free oscillations. Appl. Math. Mech (PMM) 62(1), 92–99 (1998) (in Russian)
12. Andreaus, U., Casini, P.: Dynamics of SDOF oscillators with hysteretic motion–limiting stop. Nonlin. Dyn. 22(2), 155–174 (2000)

13. Andrianov, I.V., Awrejcewicz, J.: Asymptotic behavior of a system with damping and high power–form nonlinearity. J. Sound & Vib. 267, 1169–1174 (2003)
14. Antoine, J.F., Visa, C., Sauvey, C., Abba, G.: Approximate analytical model for Hertzian elliptical contact problems. ASME J. Tribology 128, 660–664 (2006)
15. Antunes, J., Axisa, F., Vento, M.A.: Experiments on vibro–impact dynamics under fluidelastic instability. In: Proc. ASME Pressure Vessels and Piping Division, Flow–Induced Vibration, New York, NY, PVP, vol. 189, pp. 127–138 (1990)
16. Antunes, J., Axisa, F., Vento, M.A.: Experiments on tube/support interaction with feedback–controlled instability. ASME J. Pressure Vessel Technology 114(1), 23–32 (1992a)
17. Antunes, J., De Langre, E., Vento, M.A., Axisa, F.: Theoretical model for the vibro–impact motions of tube bundles under fluidelastic instability. In: Proc. ASME Pressure Vessels and Piping Division, Cross–Flow Induced Vibration of Cylinder Arrays, 1992, Anaheim, CA, PVP, vol. 242, pp. 135–150 (1992b)
18. Antunes, J., Axisa, F., Beaufils, B., Guilbaud, D.: Coulomb friction modeling in numerical simulations of vibration and wear work rate of multispan tube bundles. J. Fluids & Struct. 4, 287–304 (1992c)
19. Antunes, J., Paulino, M., Piteau, P.: Remote identification of impact forces on loosely supported tubes: Part 2 – Complex vibro–impact motions. J. Sound & Vib. 215(5), 1043–1064 (1998)
20. Antunes, J., Paulino, M., Izquierdo, P.: Blind identification of impact forces from multiple remote vibratory measurements. Int. J. Nonlin. Sci. & Num. Simul. 2(1), 1–20 (2001)
21. Aoki, S., Watanabe, T.: Forced vibration analysis of cantilever beam with unsymmetrical stop. Nippon Kikai Gakkai Ronbunshu, C Hen/Trans JSME, Part C 61(588), 3190–3195 (1995) (in Japanese)
22. Aoki, S., Watanabe, T.: Forced vibration of continuous system with unsymmetrical collision characteristics. Nonlin. Dyn. 17(2), 141–157 (1998)
23. Aoki, S., Watanabe, T.: Practical response analysis of a mass–spring impact system with hysteresis damping. Nuclear Eng. & Design 234, 1–9 (2004)
24. Araki, Y., Yokomichi, I., Inoue, J.: Impact damper with granular materials (2nd report, both sides impact in a vertical oscillating system). Bulletin JSME 28(241), 1466–1472 (1985a)
25. Araki, Y., Yuhki, Y., Yokomichi, I., Jinnouchi, Y.: Impact damper with granular materials (3rd Report, Indicial Response). Bulletin JSME 28(240), 1211–1217 (1985b)
26. Araki, Y., Yokomichi, I., Jinnouchi, Y.: Impact damper with granular materials (4th report, frequency response in a horizontal system). Bulletin JSME 29(258), 4334–4338 (1986)
27. Araki, Y., Jinnouchi, Y., Inoue, J.: Impact damper with granular materials. ASME Pressure Vessels and Piping Div. 133, 879–893 (1988)
28. Araki, Y., Jinnouchi, Y., Inoue, J., Yokomichi, I.: Indicial response of impact damper with granular material. Seismic. Eng. 182, 73–79 (1989)
29. Araki, Y., Jinnouchi, Y., Yokomichi, I., Inoue, J.: Impact damper with granular materials for multibody system. In: Proc. ASME Pressure Vessels and Piping, DOE Facilities Programs, Systems Interaction, and Active/Inactive Damping, vol. 229, pp. 139–145 (1992)

30. Arnold, R.N.: Response of an impact vibration absorber to forced vibration. Actes IX Congress on Appl. Mech., Brussels 7, 407–418 (1957)
31. Arnold, V.I.: Geometrical Methods in the Theory of Differential Equations. Springer, Berlin (1983)
32. Asfar, K.R., Akour, S.N.: Optimization analysis of impact viscous damper for controlling self–excited vibrations. J. Vib. & Cont. 11(1), 103–120 (2005)
33. Astashev, V.K.: On the dynamics of an oscillator impacting against a stop. Machine Studies (2) (1971)
34. Astashev, V.K., Babitsky, V.I.: Ultrasonic cutting as a nonlinear (vibro–impact) process. Ultrasonics 36(1-5), 89–96 (1998)
35. Astashev, V.K., Babitsky, V.I.: Ultrasonic Processes and Machines: Dynamics, Control and Applications. Springer, Berlin (2007)
36. Astashev, V.K., Babitsky, V.I., Kolovsky, M.: Dynamics and Control Machines. Springer, Berlin (2000)
37. Astashev, V.K., Krupenin, V.L.: Experimental research of oscillations of strings, interacting with dot stops. Doklady Akademii Nauk 379(3), 329–334 (2001a)
38. Astashev, V.K., Krupenin, V.L.: Experimental investigation of vibrations of strings interacting with point obstacles. Doklady Physics 46(7), 522–525 (2001b)
39. Astashev, V.K., Krupenin, V.L., Tresvyatskii, A.N.: Experimental investigation of synchronization of impacts in distributed systems with parallel impact pairs. Physics – Doklady 41(11), 532 (1996)
40. Au–Yang, M.K.: Joint and cross acceptance for cross–flow–induced vibration, Part I: Theoretical and finite element formulations. ASME J. Pressure Vessel Techn. 122, 349–354 (2000a)
41. Au–Yang, M.K.: Joint and cross acceptance for cross–flow–induced vibration, Part II: Charts and applications. ASME J. Pressure Vessel Techn. 122, 355–361 (2000b)
42. Avranov, K.V.: Analysis of bifurcations of the vibro–impact system on the basis of amplitude surfaces method. Prikl Mekh. 38(9), 138–144 (2002)
43. Awrejcewicz, J., Tomczak, K.: Stabilization of the vibro–impact systems. In: Proc. 5th Int Symp on Methods and Models in Automation and Robotics, Part 2, Miedzyzdroje, Poland, vol. 2, pp. 515–520 (1998)
44. Axisa, F., Antunes, J., Villard, B.: Overview of numerical methods for predicting flow–induced vibration and wear of heat–exchanger tubes. In: Proc. ASME Conf. Pressure Vessels and Piping, Chicago, IL, PVP, vol. 104, pp. 147–159 (1986)
45. Axisa, F., Antunes, J.: Vibration response of loosely supported tubes to fluidelastic instability. In: CEA Centre d'Etudes Nucleaires de Saclay, Gif–sur–Yvette (France) Report: Int. Conf. engineering–aero–hydroelasticity, Prague (Czechoslovakia), December 5–8, CEA–CONF–10249; CONF–891244, 1989, p. 17 (1989)
46. Axisa, F., Dessaux, A., Gilbert, R.J.: Experimental study of tube/support impact forces in multi–span PWR steam generator tubes. In: ASME Symp. on Flow–Induced Vibr., New York, NY, vol. 3, pp. 139–148 (1984)
47. Axisa, F., Izquierdo, P.: Experiments on vibro–impact dynamics of loosely supported tubes under harmonic excitation. In: ASME Symp. on Flow–Induced Vibration and Noise, Anaheim, CA, PVP, vol. 242(2), pp. 281–299 (1992)

48. Azar, R.C., Crossley, F.R.E.: Digital simulation of impact phenomenon in spur gear systems. ASME J. Eng. Indust. 99, 792–798 (1977)
49. Azeez, M.A.F., Vakakis, A.F.: Proper orthogonal decomposition (POD) of a class of vibro–impact oscillations. J. Sound & Vib. 240(5), 859–889 (2001)
50. Azeez, M.A.F., Vakakis, A.F., Manevich, L.I.: Exact solutions of the problem of the vibro–impact oscillations of a discrete system with two degrees of freedom. J. Appl. Math. Mech. 63, 527–530 (1999)
51. Azejczyk, B., Kapitaniak, T., Wojewoda, J., Barron, R.: Experimental–observation of intermittent chaos in a mechanical system with impacts. J. Sound & Vib. 178, 272–275 (1994)
52. Babitsky, V.I.: Vibro–impact motions of pendulum with inertial suspension in vibrating container. Analysis and Synthesis of Automatic Machines, Nauka, Moscow, 20–30 (1966)
53. Babitsky, V.I.: Theory of Vibro–Impact Systems: Approximate Methods, Nauka, Moscow. Springer, Berlin (1978) (Revised English translation 1998)
54. Babitsky, V.I.: Autoresonant mechatronic systems. Mechatronics 5, 483–495 (1995)
55. Babitsky, V.I.: Hand–held percussion machine as discrete nonlinear converter. J. Sound & Vib. 214(1), 165–178 (1998)
56. Babitsky, V.I., MZh, K.: Response of a linear system with barriers to random excitation. Mechanics of Solids 3, 147–161 (1967) (in Russian)
57. Babitsky, V.I., Kolovsky, M.: Investigation of resonances in vibro–impact systems. Mechanics of Solids (Mekhanika Tverdogo Tela 4, 88–91 (1976) (in Russian)
58. Babitsky, V.I., Kovaleva, A.S., Krupenin, V.L.: Analysis of quasi–conservative vibro–impact systems by the method of averaging. In: Proc. USSR Academy of Sciences, MTT (Mechanics of Solids), vol. 1, pp. 41–50 (1982) (in Russian)
59. Babitsky, V.I., Krupenin, V.L.: Analyzing models of vibration–impact systems. Mechanics of Solids (Mekhanika Tverdogo Tela) 12(6), 19–26 (1977)
60. Babitsky, V.I., Krupenin, V.L.: Vibrations in Strongly Nonlinear Systems, Nauka, Moscow (1985) (in Russian)
61. Babitsky, V.I., Krupenin, V.L.: Vibration of Strongly Nonlinear Discontinuous Systems. Springer, Berlin (2001)
62. Babitsky, V.I., Veprik, A.M.: Universal bumpered vibration isolator for severe environment. J. Sound & Vib. 218, 269–292 (1998)
63. Babitsky, V.I., Veprik, A.M.: Structure–borne vibro–impact resonances and periodic Green functions. In: Proc. ASME Design Eng. Tech. Conf., Pittsburgh, PA, vol. 6 (B), pp. 1179–1188 (2001)
64. Babitsky, V.I., Sokolov, I.J.: Autoresonant homeostat concept for engineering application of nonlinear vibration modes. Nonlin. Dyn. 50(3), 447–460 (2007)
65. Badertscher, J., Cunefare, K.A., Ferri, A.A.: Braking impact of normal dither signals. ASME J. Vib. & Acoust. 129, 17–23 (2007)
66. Bae, J.S., Lee, I.: Limit cycle oscillation of missile control fin with structural nonlinearity. J. Sound & Vib. 269(3–5), 669–687 (2004)
67. Bae, J.S., Inman, D.J., Lee, I.: Effects of structural nonlinearity on subsonic aeroelastic characteristics of an aircraft wing with control surface. J. Fluids & Struct. 19, 747–763 (2004b)
68. Bae, J.S., Kim, D.K., Shin, W.H., Lee, I., Kim, S.H.: Nonlinear aeroelastic analysis of deployable missile control fin. J. Spacecraft & Rockets 41(2), 264–271 (2004a)

69. Bae, J.S., Yang, S.M., Lee, I.: Linear and nonlinear aeroelastic analysis of fighter–type wing with control surface. J. Aircraft 39, 697–708 (2002)
70. Baksys, B., Puodziuniene, N.: Modeling of vibrational non–impact motion of mobile–based body. Int. J. Nonlin. Mech. 40, 861–873 (2005)
71. Baksys, B., Puodziuniene, N.: Modeling of vibrational impact motion of mobile–based body. Int. J. Nonlin. Mech. 42, 1092–1101 (2007)
72. Balachandran, B.: Dynamics of an elastic structure excited by harmonic impactor motions. J. Vib. & Cont. 9(3,4), 265–279 (2003)
73. Balandin, D.V., Bolotnik, N.N., Piley, W.D.: Limiting performance analysis of impact isolation systems for injury prevention. Shock & Vib. Dig. 33, 453–472 (2001)
74. Ballo, I.: Contribution to investigation of dynamical properties of nonlinear systems with acceleration damper. Polska Akademia Nauk – Instytut Podstawowych Problemow Techniki – Zaklad Badania Drgan 5, 86–92 (1963)
75. Banakh, L., Nikiforov, L.: Vibro–impact regimes and stability of system 'Rotor–Sealing Ring'. J. Sound & Vib. 308, 785–793 (2007)
76. Banerjee, S., Grebogi, C.: Border collision bifurcations in two–dimensional piecewise smooth maps. Phys. Rev. E 59, 4052–4061 (1999)
77. Banerjee, S., Grebogi, C.: Border collision bifurcations at the change of state–space Dimension. Chaos. 12, 1054–1069 (2002)
78. Banerjee, S., Karthik, M.S., Yuan, G.H., Yorke, J.A.: Bifurcations in one–dimensional piecewise smooth maps – theory and applications in switching circuits. IEEE Trans. Circuits Syst. I 47(3), 389–394 (2000)
79. Banerjee, S., Ma, Y., Agarwal, M.: Border collision bifurcations in a soft impact system. Phys. Lett. A 354, 281–287 (2006)
80. Banerjee, S., Ott, E., Yorke, J.A., Yuan, G.H.: Anomalous bifurcations in dc–dc converters: Borderline collisions in piecewise maps. In: Proc. 28th Annual IEEE Power Electronics Specialists' Conference, St Louis, Missouri, pp. 1337–1344 (1997)
81. Banerjee, S., Parui, S., Gupta, A.: Dynamical effects of missed switching in current–mode controlled dc–dc converters. IEEE Trans. Circuit Syst. II 51(12), 649–654 (2004)
82. Banerjee, S., Ranjan, P., Grebogi, C.: Bifurcations in two–dimensional piecewise smooth maps–theory and applications in switching circuits. IEEE Trans. Circuit Syst. I 47(5), 633–643 (2000)
83. Banerjee, S., Verghese, G.C. (eds.): Nonlinear Phenomena in Power Electronics: Attractors, Bifurcations, Chaos, and Nonlinear Control. IEEE Press, New York (2001)
84. Bapat, C.N.: The general motion of an inclined impact damper with friction. J. Sound & Vib. 184(3), 417–427 (1995a)
85. Bapat, C.N.: Duffing oscillator under periodic impulses. J. Sound & Vib. 179(4), 725–732 (1995b)
86. Bapat, C.N.: The general motion of an inclined impact damper with friction. J. Sound & Vib. 184(3), 417–427 (1995c)
87. Bapat, C.N.: Periodic motions of an impact oscillator. J. Sound & Vib. 209, 43–60 (1998)
88. Bapat, C.N., Bapat, C.: Impact–pair under periodic excitation. J. Sound & Vib. 120(1), 53–61 (1988)
89. Bapat, C.N., Popplewell, N.: Several similar vibro–impact systems. J. Sound & Vib. 113(1), 17–28 (1987)

90. Bapat, C.N., Popplewell, N., McLachlan, K.: Stable periodic motions of an impact pair. J. Sound & Vib. 87(1), 19–40 (1983)
91. Bapat, C.N., Sankar, S.: Single unit impact damper in free and forced vibration. J. Sound & Vib. 99(1), 85–94 (1985a)
92. Bapat, C.N., Sankar, S.: Multi–unit impact damper, re–examined. J. Sound & Vib. 103(4), 457–469 (1985b)
93. Barauskas, R.: Dynamic analysis of structures with unilateral constraints: numerical integration and reduction of structural equations. Int. J. Num. Meth. Eng. 37(2), 323–342 (1994)
94. Barkan, D.D., Shekhter, O.Y.: On the theory of forced oscillation of a vibrator with limiter. J. Technical Physics 25(13) (1955) (in Russian)
95. Batako, A.D., Babitsky, V.I., Halliwell, N.A.: A self–excited system for percussive–rotary drilling. J. Sound & Vib. 259(1), 97–118 (2003)
96. Batako, A.D., Babitsky, V.I., Halliwell, N.A.: Modelling of vibro–impact penetration of self–exciting percussive–rotary drill bit. J. Sound & Vib. 271, 209–225 (2004)
97. Batako, A.D.L., Lalor, M.J., Piiroinen, P.T.: Numerical bifurcation analysis of a friction–driven vibro–impact system. J. Sound & Vib. 308, 392–404 (2007)
98. Batako, A.D., Piiroinen, P.T.: Friction–driven vibro–impact systems for percussive–rotary drilling: A numerical study of the system dynamics. Proc. IMech E Part C: J Mechanical Engineering Science 222, 1925–1934 (2008)
99. Bayly, P.V., Virgin, L.N.: An experimental study of an impacting pendulum. J. Sound & Vib. 164, 364–374 (1993)
100. Beatty, R.F.: Differentiating rotor response due to radial rubbing. ASME J. Vib. & Acoust, Stress & Reliab in Design 107, 151–160 (1985)
101. Bedout, J.M., Francheck, M.A., Bajaj, A.K.: Robust control of chaotic vibrations for impacting heat exchanger tubes in crossflow. J. Sound & Vib. 227, 183–204 (1999)
102. Begley, C.J., Virgin, L.N.: On the OGY control of an impact–friction oscillator. J. Vib. & Cont. 7(6), 923–931 (2001)
103. Behringer, K., Kostic, L., Seifritz, W.: Observation of in–core instrument tube vibrations in a boiling water reactor by evaluating reactor noise data. Prog. Nucl. Energy 1(2–4), 183–185 (1977)
104. Belik, O., Bishop, R.E.D., Price, W.G.: Influence of bottom and flare slamming on structural responses. Trans. Royal Inst. Naval. Arch. (RINA) 130, 325–337 (1988)
105. Belik, O., Price, W.G.: Comparison of slamming theories in the time simulation of ship responses. Int. Shipbuilding Progress 29(335), 173–187 (1982)
106. Bellizzi, S., Bouc, R.: Spectral response of asymmetrical random oscillators. Probab. Eng. Mech. 11, 51–59 (1996)
107. Bently, D.: Forced subrotative speed dynamic action of rotating machinery. In: Petroleum Mechanical Engineering Conf., Dallas, Texas, ASME Paper No. 74–PET–16, p. 8 (1974)
108. Berzi, P., Beccu, R., Lundberg, B.: Identification of a percussive drill rod joint from its response to stress wave loading. Int. J. Impact Eng. 18(3), 281–290 (1996)
109. Bespalova, L.V.: On the theory of the vibro–impact system. Izv. AN SSSR, OTN 5, 3–14 (1957)

110. Bespalova, L.V., Yul, N., Feigin, M.I.: Dynamical systems with impact interactions and the theory of nonlinear oscillations. Eng. J. Mech. Solids 1, 151–159 (1966) (in Russian)
111. Bhattacharyya, A., Chatterjee, A.B.: Analogue simulation of impact dampers. Annals CIRP 21(1), 117–118 (1972)
112. Bhutani, N., Kulkarni, S., Bapat, C.N.: Vibro–impacts of a Duffing oscillator under sinusoidal force. J. Sound & Vib. 210(3), 407–411 (1998)
113. Bishop, S.R.: Impact oscillators. Phil. Trans. Royal Soc. London A 347, 347–351 (1994)
114. Bishop, S.R., Thomson, M.G., Foale, S.: Prediction of period–1 impacts in a driven beam. Proc. Royal. Society of London A 452, 2579–2592 (1996)
115. Bishop, R., Wagg, D.J., Xu, D.: Use of control to maintain period–1 motions during wind–up or wind–down operations of an impacting driven beam. Chaos, Solitons and Fractals 9, 261–269 (1998)
116. Bishop, S.R., Xu, D.: The use of control to eliminate subharmonic and chaotic impacting motions of a driven beam. J. Sound & Vib. 205(2), 223–234 (1997)
117. Black, H.F.: Interaction of a whirling rotor with a vibrating stator across a clearance annulus. J. Mech. Eng. Sci. 10, 1–12 (1968)
118. Blazejczyk–Okolewske, B.: Study of the impact oscillator with elastic coupling of masses. Chaos, Solitons and Fractals 11(15), 2487–2492 (2000)
119. Blazejczyk–Okolewske, B.: Analysis of an impact damper of vibrations. Chaos, Solitons and Fractals 12(11), 1983–1988 (2001)
120. Blazejczyk–Okolewske, B., Kapitaniak, T.: Dynamics of impact oscillator with dry friction. Chaos, Solitons and Fractals 7, 1455–1459 (1996)
121. Blazejczyk–Okolewske, B., Kapitaniak, T.: Co–existing attractors of impact oscillator. Chaos, Solitons and Fractals 9(8), 1439–1443 (1998)
122. Blazejczyk–Okolewske, B., Brindly, J., Kapitaniak, T.: Practical riddling in mechanical systems. Chaos, Solitons and Fractals 11, 2511–2514 (2000)
123. Blazejczyk–Okolewske, B., Kapitaniak, T., Wojewoda, J., Barron, R.: Experimental observation of intermittent chaos in a mechanical system with impacts. J. Sound & Vib. 178, 272–275 (1994)
124. Blazejczyk Okolewska, B., Peterka, F.: An investigation of the dynamic system with impacts. Chaos, Solitons and Fractals 9(8), 1321–1338 (1998)
125. Blinov, A.P.: Motion of a gyroscopic pendulum with an ideal unilateral constraint with respect to the angle of nutation. Mechanics of Solids (Mekhanika Tverdogo Tela) 20(6), 53–56 (1985)
126. Bocquet, L.: The physics of stone skipping. Amer. J. Phys. 71(2), 150–155 (2003)
127. Bolotnik, N.N., Melikyan, A.A.: Time–optimal control of a vibro–impact system with one degree of freedom. In: Proc. Int. Conf. Control of Oscillations and Chaos, St. Petersburg, Russia, vol. 2, pp. 249–252 (2000)
128. Bonheure, D., Fabry, C.: Periodic motions in impact oscillators with perfectly elastic bounces. Nonlinearity 15(4), 1281–1297 (2002)
129. Bouc, R., Defilippi, M.: Spectral response of a beam–stop system under random excitation. In: Proc. IUTAM Symp Advances in Nonlinear Stochastic Mechanics, Trondheim, Norway, July 3–7, 1995, pp. 69–78. Kluwer, Dordrecht (1996)
130. Bouc, R., Defilippi, M.: Multimodal nonlinear spectral response of a beam with impact under random input. Prob. Eng. Mech. 12(3), 163–179 (1997)

131. Boucher, K., Taylor, C.: Tube support effectiveness and wear damage assessment in the U–bend region of nuclear steam generators. In: Proc. ASME Pressure Vessel & Piping Conf., Montreal, Canada, ASME PVP, vol. 318, pp. 285–296 (1996)
132. Brach, R.M.: Mechanical Impact Dynamics. In: Rigid Body Collisions. Wiley, New York (1991)
133. Bratus, A., Dimentberg, M.F., Iourtchenko, D.: Optimal bounded response control for a second–order system under a white–noise excitation. J. Vib. & Cont. 6(5), 741–755 (2000)
134. Brindeu, L.: Stability of the periodic motions of the vibro–impact systems. Chaos, Solitons and Fractals 11, 2493–2503 (2000)
135. Brogliato, B.: Non–smooth Impact Mechanics: Dynamics and Control. Lecture Notes in Control and Information Sciences, vol. 220. Springer, Berlin (1996a)
136. Brogliato, B.: On dynamics and feedback control of non–smooth impacting mechanical systems. In: Proc. Modelling and Control of Mechanisms and Robots, Bertinoro, Italy, pp. 185–226 (1996b)
137. Brogliato, B.: Impacts in Mechanical Systems – Analysis and Modelling. Lecture Notes in Physics, vol. 551. Springer, Heidelberg (2000)
138. Brogliato, B.: Some perspectives on the analysis and control of complementarity systems. IEEE Transactions on Automatic Control 48, 918–935 (2003)
139. Brogliato, B., Acary, V.: Numerical Methods for non–smooth Dynamical Systems: Applications in Mechanics and Electronics. Lecture Notes in Applied and Computational Mechanics, vol. 35. Springer, Berlin (2008)
140. Brogliato, B., Ten Dam, A.A., Paoli, L., Genot, F., Abadie, M.: Numerical simulation of finite dimensional multibody non–smooth mechanical systems. ASME Appl. Mech. Rev. 55(2), 107–115 (2002)
141. Brogliato, B., Daniilidis, A., Lemarechal, C., Acary, V.: On the equivalence between complementarity systems, projected systems and differential inclusions. Systems & Control Lett. 55(1), 45–51 (2006)
142. Brogliato, B., Zavala–Rio, A.: On the control of complementary–slackness juggling mechanical systems. IEEE Trans. Automatic Control 45(2), 235–246 (2000)
143. Brown, G.V.: Survey of impact damper performance. In (Army Aviation Systems Command, St. Louis, MO.) Sponsor: National Aeronautics and Space Administration, Washington, DC, p. 10 (1988)
144. Brown, G.V., North, C.M.: Impact damped harmonic oscillator in free decay. NASA Tech. Memo, p. 22 (September 1987)
145. Budd, C.J.: Grazing in impact oscillations. In: Brannet, B., Hijorth, P. (eds.) Real and Complex Dynamical Systems, pp. 47–64. Kluwer Academic, Dordrecht (1995)
146. Budd, C.J., Dux, F.: Intermittency in impact oscillators close to resonance. Nonlinearity 7, 1191–1224 (1994a)
147. Budd, C.J., Dux, F.: Chattering and related behavior in impact oscillators. Phil. Trans. Royal Soc. A 347, 365–389 (1994b)
148. Budd, C.J., Cliffe, K.A., Dux, F.: The effect of frequency and clearance variations on one–degree–of–freedom impact oscillators. J. Sound & Vib. 184(3), 475–502 (1995)
149. Budd, C.J., Lee, A.G.: Double impact orbits of periodically forced impact oscillators. Proc. Royal Society of London A 452, 2719–2750 (1996)

150. Budd, C.J., Piiroinen, P.T.: Corner bifurcations in non–smoothly forced impact oscillators. Physica D 220, 127–145 (2006)
151. Cammaert, A.B., Tsinker, G.P.: Impact of large ice floes and icebergs on marine structures. In: Proc. Int. Conf. Port and Ocean Engineering under Arctic Conditions–POAC 1981, Quebec, vol. 2, pp. 653–662 (1981)
152. Carson, R.M., Johnson, K.L.: Surface Corrugations Spontaneously Generated in a Rolling Con–tact Disc Machine. Wear 17, 59–72 (1971)
153. Casas, F., Chin, W., Grebogi, C., Ott, E.: Universal grazing bifurcations in impact oscillators. Phys. Rev. E 53(1), 134–139 (1996)
154. Castravete, S.C., Beloiu, D.M., Bulkrani, G., Ibrahim, R.A.: Mechanism of brake squeal and contact stiffness measurement. Final report for General Motors, Wayne State University, Detroit, MI, January 15 (2002)
155. Cempel, C.Z.: The multi–unit impact damper: Equivalent continuous force approach. J. Sound & Vib. 34(2), 199–209 (1974)
156. Cempel, C.: Receptance model of the multi–unit vibration impact neutralizer (MUVIN). J. Sound & Vib. 40(2), 249–266 (1975)
157. Cempel, C., Lotz, G.: Efficiency of vibrational energy dissipation by moving shot. J. Struct. Eng. 119(9), 2642–2652 (1993)
158. Chatterjee, A.: On realism of complementarity conditions in rigid body collisions. Nonlin. Dyn. 20, 159–168 (1999)
159. Chatterjee, S., Mallik, A.K.: Bifurcations and chaos in autonomous self–excited oscillators with impact damping. J. Sound & Vib. 191, 539–562 (1996)
160. Chatterjee, S., Mallik, A.K., Ghosh, A.: On impact dampers for nonlinear vibrating systems. J. Sound & Vib. 187, 403–420 (1995)
161. Chatterjee, S., Mallik, A.K., Ghosh, A.: Impact dampers for controlling self–excited oscillation. J. Sound & Vib. 193(5), 1003–1014 (1996)
162. Chattopadhyay, S., Saxena, R.: Dynamic response of a loosely restrained structure. In: ASME Winter Annual Meeting, Atlanta, GA, Paper ASMSA4, December 1–6, pp. 1–6 (1991)
163. Chen, R., Wan, C., Xue, S., Tang, H.: Impact response of an unrestrained modified Timoshenko beam. Chinese J. Theor. Appl. Mech. 38(2), 262–269 (2006) (in Chinese) Lixue Xuebao
164. Chen, S.S.: Flow– induced instability of an elastic tube. ASME Paper 71–Vibr–39, September 8–10, p. 9 (1971)
165. Cheng, C.C., Wang, J.Y.: Free vibration analysis of a resilient impact damper. Int. J. Mech. Sci. 45(4), 589–604 (2003)
166. Cheng, J., Xu, H.: Inner mass impact damper for attenuating structure vibration. Int. J. Solids & Struct. 43(17), 5355–5369 (2006)
167. Cheng, J., Xu, H.: Periodic motions, bifurcation, and hysteresis of the vibro–impact system. Mech. Based Design of Struct. & Mach. 35(2), 179–203 (2007)
168. Childs, D.W.: Rub–induced parametric excitation in rotors. ASME J. Mech. Design 101(4), 640–644 (1979)
169. Childs, D.W.: Fractional–frequency rotor motion due to nonsymmetric clearance effects. ASME J. Energy and Power 104, 533–541 (1982)
170. Chikatani, Y., Suehiro, A.: Reduction of idling rattle noise in trucks. SAE Paper No. 911044, Traverse City, MI, 49–56 (1991)
171. Chillingworth, D.R.J.: Discontinuity geometry for an impact oscillator. Dyn. Systems 17, 380–420 (2002)
172. Chin, W., Ott, E., Nusse, H.E., Grebogi, C.: Grazing bifurcations in impact oscillators. Phys. Rev. E 50, 4427–4444 (1994)

173. Chin, W., Ott, E., Nusse, H.E., Grebogi, C.: Universal behavior of impact oscillators near grazing incidence. Phys. Lett. A 201, 197–204 (1995)
174. Chirikov, B.V.: A universal instability of many–dimensional oscillator systems. Phys. Rev. 52, 263–379 (1979)
175. Choi, Y.S., Noah, S.T.: Nonlinear steady–state response of a rotor–support system. ASME J. Vib., Acoust, Stress, and Rel. Des. 109, 255–261 (1987)
176. Choy, F.K., Padovan, J.: Nonlinear Transient analysis of rotor–casing rub events. J. Sound & Vib. 113, 529–545 (1987)
177. Choy, F.K., Padovan, J., Manos, M.: Component effects on rotating equipment dynamics. In: Proc. ASME Pressure Vessels and Piping, Seismic Engineering, Pittsburgh, PA, PVP, vol. 144, pp. 151–160 (1988)
178. Choy, F.K., Padovan, J., Batur, C.: Rub interactions of flexible casing rotor systems. ASME J. Eng. Gas Turb. & Power 111(4), 652–658 (1989a)
179. Choy, F.K., Padovan, J., Li, W.H.: Seismic induced nonlinear rotor–bearing–casing interaction of rotating nuclear components. ASME J. Vib, Acoust, Stress & Reliab Design 111(1), 11–16 (1989b)
180. Chu, F., Lu, W.: Experimental observation of nonlinear vibrations in a rub–impact rotor system. J. Sound & Vib. 283(3–5), 621–643 (2005)
181. Chu, F., Zhang, Z.: Periodic, quasi–periodic and chaotic vibrations of a rub–impact rotor system supported on oil film bearings. Int. J. Eng. Sci. 35(10), 963–973 (1997)
182. Chu, F., Zhang, Z.: Bifurcation and chaos in a rub–impact Jeffcott rotor system. J. Sound & Vib. 210(1), 1–18 (1998)
183. Clanet, C., Hersen, F., Bocquet, L.: Secrets of successful stone–skipping. Nature 427, 29 (2004)
184. Clark, B.K., Rosa Jr., E., Hall, A.D., Shepherd, T.R.: Dynamics of an electronic impact oscillator. Phys. Lett. A 318(6), 514–521 (2003)
185. Collette, F.S.: A combined tuned absorber and pendulum impact damper under random excitation. J. Sound & Vib. 216, 199–213 (1998)
186. Collette, F., Huynh, D., Semercigil, S.E.: Further results with tuned absorber–impact damper combination. In: Proc. Int. Modal Analysis Conf (IMAC), San Antonio, TX, February 7–10, 2000, vol. 1, pp. 404–410 (2000)
187. Comparin, R.J., Singh, R.: nonlinear frequency response characteristics of an impact pair. J. Sound & Vib. 134, 259–290 (1989)
188. Comparin, R.J., Singh, R.: An analytical study of automotive neutral gear rattle. ASME J. Mech. Design 112, 237–245 (1990)
189. Cone, K.M., Zadoks, R.I.: A numerical study of an impact oscillator with the addition of dry friction. J. Sound & Vib. 188(5), 659–683 (1995)
190. Cooker, M.J., Peregrine, D.H.: Model for breaking wave impact pressures. In: Proc. 22nd Int. Conf. Coastal Eng., Delft, vol. 2, pp. 1473–1486. ASCE Publications, The Netherlands (1991)
191. Cottone, G., Di Paola, M., Ibrahim, R., Pirrotta, A., Santoro, R.: Ship roll motion under stochastic agencies handled by Path Integral Solution. In: Proc. Vibro–Impact Dynamics Symposium 2008, Troy, Michigan, USA, October 2–3. Springer, Berlin (2008) (in press)
192. Cox, P.A., Bowles, E.B., Bass, R.L.: Evaluation of liquid dynamic loads in slack LNG cargo tanks, Southwest Research Institute, San Antonio, Texas, Tech Report SSC–297 (1980)
193. Craig, K.J., Kingselev, T.C.: Design optimization of containers for sloshing and impact. Structural and Multidisciplinary Optimization 33(1), 71–87 (2007)

194. Croteau, P., Rojansky, M., Gerwick, B.C.: Summer ice floe impact against caisson–type exploratory and production platforms. ASME J. Energy Res. Techn. 106(2), 169–175 (1984)
195. Cusumano, J.P.: Spatial coherence in the chaotic dynamics of multi–degree–of–freedom elastic impact oscillators. In: Mechanics Computing in 1990's and Beyond, Columbus, Ohio, pp. 776–780 (1991)
196. Cusumano, J.P., Bai, B.Y.: Period–infinity periodic motions, chaos, and spatial coherence in a 10 degree of freedom impact oscillator. Chaos, Solitons & Fractals 3(5), 515–535 (1993)
197. Cusumano, J.P., Sharkady, M.T., Kimble, B.W.: Experimental measurements of dimensionality and spatial coherence in the dynamics of a flexible–beam impact oscillator. Phil. Trans. Royal. Soc., Series A (Physical Sciences and Engineering) 347(1683), 421–438 (1994)
198. Czolczynski, K.: On the existence of a stable periodic motion of two impacting oscillators. Chaos, Solitons and Fractals 15(2), 371–379 (2003)
199. Czolczynski, K.: On the existence of a stable periodic solution of an impacting oscillator with damping. Chaos, Solitons & Fractals 19(5), 1291–1311 (2004)
200. Czolczynski, K., Kapitaniak, T.: Influence of the mass and stiffness ratio on a periodic motion of two impacting oscillators. Chaos, Solitons & Fractals 17(1), 1–10 (2003)
201. Czolczynski, K., Kapitaniak, T.: On the existence of a stable periodic solution of two impacting oscillators with damping. Int. J. Bifur & Chaos in Appl. Sci. & Eng. 14(11), 3931–3947 (2004)
202. Dalrymple, T.O.: Numerical solutions to vibro–impact via an initial value problem formulation. J. Sound & Vib. 132(1), 19–32 (1989)
203. Dankowicz, H., Piiroinen, P.T.: Exploiting discontinuities for stabilization of recurrent motions. Dyn. Syst. 17, 317–342 (2002)
204. Dankowicz, H., Piiroinen, P.T., Nordmark, A.B.: Low–velocity impacts of quasi–periodic oscillations. Chaos, Solitons & Fractals 14, 241–255 (2002)
205. Dankowicz, H., Jerrelind, J.: Control of near–grazing dynamics in impact oscillators. Proc. Royal Society of London A 461(2063), 3380–4465 (2005)
206. Dankowicz, H., Svahn, F.: On the stabilizability of near–grazing dynamics in impact oscillators. Int. J. Robust & Nonlin. Control 17, 1405–1429 (2007)
207. Dankowicz, H., Zhao, X.: Local analysis of codimension–one and codimension–two grazing bifurcations in impact microactuators. Physica D 202, 238–257 (2005)
208. Davies, H.G.: Random vibration of beam impacting stops. J. Sound & Vib. 68, 479–487 (1980)
209. Davies, H.G., Rogers, R.J.: The vibration of structures elastically constrained at discrete points. J. Sound & Vib. 63(3), 437–447 (1979)
210. Davies, M.A., Balachandran, B.: Impact dynamics in the milling of thin–walled structures. Nonlin. Dyn. 22(4), 375–392 (2000)
211. De Aroujo, M., Antunes, J., Piteau, P.: Remote identification of impact forces on loosely supported tubes. I. Basic theory and experiments. J. Sound & Vib. 215(5), 1015–1041 (1998)
212. De Kraker, A., Van de Wouw, N., Van den Bosch, H.L.A., Van Campen, D.H.: Identification of nonlinear phenomena in a stochastically excited beam system with impact. In: Proc. 23rd Int Conf. Noise and Vibration Eng. (ISMA), Leuven, Belgium, September 16–18, pp. 569–576 (1998)

213. De Langre, E., Doveil, F., Porcher, G., Axisa, F.: Chaotic and periodic motion of a nonlinear oscillator in relation with flow–induced vibrations of loosely supported tubes. In: Proc. ASME Conf. Pressure Vessels and Piping, PVP, New York, NY, vol. 189, pp. 119–125 (1990)
214. De Langre, E., Lebreton, G.: Experimental and numerical analysis of chaotic motion in vibration with impact. In: Proc. ASME Pressure Vessels and Piping Division, PVP, Flow–Induced Vibration, Montreal, Canada, vol. 328, pp. 317–325 (1996)
215. Demeio, L., Lenci, S.: Asymptotic analysis of chattering oscillations for an impacting inverted pendulum. Quart J. Inst. Mech. & Appl. Math. 59(3), 419–434 (2006)
216. De Oliveira Rosa, M., Pereira, J.C., Grellet, M., Alwan, A.: A contribution to simulating a three–dimensional larynx model using the finite element method. J. Acoust. Soc. Amer. 114(5), 2893–2905 (2003)
217. Deshpande, S., Mehta, S., Jazar, G.N.: Jump avoidance conditions for piecewise linear vibration isolator. In: Proc. ASME 20th Biennial Conf. Mechanical Vibration and Noise, Long Beach, CA, USA, vol. 1, pp. 1955–1962 (2005a)
218. Deshpande, S., Mehta, S., Jazar, G.N.: Sensitivity of jump avoidance condition of a piecewise linear vibration isolator to dynamical parameters. In: Proc. ASME 20th Biennial Conf. Mechanical Vibration and Noise, Long Beach, CA, vol. 118(2), pp. 1108–1113 (2005b)
219. De Souza, S.L.T., Batista, A.M., Caldas, I.L., Viana, R.L., Kapitaniak, T.: Noise–induced basin hopping in a vibro–impact system. Chaos, Solitons & Fractals 32, 758–767 (2007)
220. De Souza, S.L.T., Caldas, I.L.: Basins of attraction and transient chaos in a gear–rattling model. J. Vib. & Cont. 7, 849–862 (2001)
221. De Souza, S.L.T., Caldas, I.L.: Controlling chaotic orbits in mechanical systems with impacts. Chaos, Solitons & Fractals 19, 171–178 (2004)
222. De Souza, S.L.T., Caldas, I.L., Viana, R.L., Balthazar, J.M., Brasil, R.M.: Impact dampers for controlling chaos in systems with limited power supply. J. Sound & Vib. 279, 955–967 (2005)
223. De Souza, S.L.T., Caldas, I.L., Viana, R.L.: Damping control law for a chaotic impact oscillator. Chaos, Solitons & Fractals 32(2), 745–750 (2007)
224. De Souza, X., Silvio, L.T., Caldas, I.L.: Calculation of Lyapunov exponents in systems with impacts. Chaos, Solitons & Fractals 19(3), 569–579 (2004)
225. De Weger, J., Van de Water, W., Molenaar, J.: Grazing impact oscillations. Phys. Rev. E 62(2), 2030–2041 (2000)
226. De Weger, J., Binks, D., Molenaar, J., Van de Water, W.: Generic behavior of grazing impact oscillators. Phys. Rev. Lett. 76(21), 3951–3954 (1996)
227. De Weger, J., Binks, D., Van de Water, W., Molenaar, J.: Universal behavior of oscillators that undergo low velocity impacts. In: Proc. Int. Conf. Control of Oscillations and Chaos, St. Petersburg, Russia, vol. 1, pp. 166–167 (1997)
228. Di Bernardo, M.: Normal forms of border collisions in high dimensional non–smooth maps. In: Proc. IEEE Int. Symp. Circuits & Systems, Bangkok, Thailand, vol. 3, pp. 76–79 (2003)
229. Di Bernardo, M., Budd, C.J., Champneys, A.R.: Grazing, skipping and sliding: analysis of the non–smooth dynamics of the DC/DC buck converter. Nonlinearity 11(4), 859–890 (1998)
230. Di Bernardo, M., Budd, C.J., Champneys, A.R.: Corner–collision implies border–collision bifurcation. Physica D 154, 171–194 (2001a)

231. Di Bernardo, M., Budd, C.J., Champneys, A.R.: Grazing and border–collision in piecewise smooth, systems: a unified analytical framework. Phys. Rev. Lett. 86, 2553–2556 (2001b)
232. Di Bernardo, M., Budd, C.J., Champneys, A.R.: Grazing bifurcations in n–dimensional piecewise–smooth dynamical systems. Physica D 160, 222–254 (2001c)
233. Di Bernardo, M., Budd, C.J., Champneys, A.R., Kowalczyk, P., Nordmark, A.B., Olivar, G., Piiroinen, P.T.: Bifurcations in Non–smooth Dynamical Systems (2006), `http://www.enm.bris.ac.uk/staff/ptp/Temp/PWSreview.pdf`
234. Di Bernardo, M., Budd, C.J., Champneys, A.R.: Normal form maps for grazing bifurcation n–dimensional piecewise–smooth dynamical systems. Physica D 160, 222–254 (2000)
235. Di Bernardo, M., Feigin, M.I., Hogan, S.J., Homer, M.E.: Local analysis of C–bifurcation in n–dimensional piecewise–smooth dynamical systems. Chaos, Solitons & Fractals 10(11), 1881–1908 (1999)
236. Di Bernardo, M., Fossas, E., Olivar, G., Vasca, F.: Secondary bifurcations and high periodic orbits in voltage controlled buck converter. Int. J. Bifurc & Chaos 7, 2755–2771 (1997)
237. Di Bernardo, M., Garofalo, F., Glielmo, L., Vasca, F.: Switchings, bifurcations and chaos in DC/DC converters. IEEE Trans. Circuits and Systems, Part I 45, 133–141 (1998)
238. Di Bernardo, M., Garofalo, F., Ianelli, L., Vasca, F.: Bifurcations in piecewise–smooth feedback systems. Int. J. Control 75(16), 1243–1259 (2002a)
239. Di Bernardo, M., Johansson, K.H., Vasca, F.: Self–oscillations in relay feedback systems: symmetry and bifurcations. Int. J. Bifurc & Chaos 11, 1121–1140 (2001d)
240. Di Bernardo, M., Kowalczyk, P., Nordmark, A.: Bifurcations of dynamical systems with sliding: derivation of normal–form mappings. Physica D 170(3–4), 175–205 (2002b)
241. Di Bernardo, M., Nordmark, A., Olivar, G.: Discontinuity–induced bifurcations of equilibria in piecewise–smooth and impacting dynamical systems. Physica D 237(1), 119–136 (2008)
242. Dimentberg, M.F.: Statistical Dynamics of Nonlinear and Time–Varying Systems. Research Studies Press/ John Wiley, Somerset/ England (1988)
243. Dimentberg, M.F.: Pseudolinear vibro–impact systems: Non–white random excitation. Nonlin. Dyn. 9(4), 327–332 (1995)
244. Dimentberg, M.F.: Random vibrations of an iscochronous SDOF bilinear system. Nonlin. Dyn. 11, 401–405 (1996a)
245. Dimentberg, M.F.: Pseudolinear vibro–impact systems: non–white random excitation. Nonlin. Dyn. 9(4), 327–332 (1996b)
246. Dimentberg, M.F., Gaidai, O., Naess, A.: random vibrations with inelastic impacts. In: Proc. Vibro–Impact Dynamics of Ocean Systems and Related Problems Symposium, Troy, Michigan, USA, October 2–3. Springer, Berlin (2008) (in press)
247. Dimentberg, M.F., Haenisch, H.G.: Pseudo–linear vibro–impact system with a secondary structure: Response to a white–noise excitation. ASME J. Appl. Mech. 65, 772–774 (1998)

248. Dimentberg, M.F., Hou, Z., Noori, M.: Spectral density of a nonlinear SDOF system's response to a white–noise random excitation: A unique case of an exact solution. In: Proc. ASME Design Eng. Div., DE, Stochastic Dynamics and Reliability of Nonlinear Ocean Systems, Chicago, IL, vol. 77, pp. 35–38 (1994)
249. Dimentberg, M.F., Iourtchenko, D.V.: Towards incorporating impact losses into random vibration analysis: A model problem. Prob. Eng. Mech. 14, 323–328 (1999)
250. Dimentberg, M.F., Iourtchenko, D.V.: Energy balance for random vibrations of piecewise–conservative systems. J. Sound & Vib. 248(5), 913–923 (2001)
251. Dimentberg, M.F., Iourtchenko, D.V.: Random vibrations with impacts: A review. Nonlin. Dyn. 36, 229–254 (2004)
252. Dimentberg, M.F., Iourtchenko, D.V., Van Ewijk, O.: Subharmonic response of a quasi–isochronous vibro–impact system to a randomly disordered periodic excitation. Nonlin. Dyn. 17, 173–186 (1998)
253. Dimentberg, M.F., Menyailov, A.: Certain stochastic problems of vibro–impact systems. Mech. Solids 4 (1976)
254. Dimentberg, M.F., Menyailov, A.: Response of a single–mass vibro–impact system to white–noise random excitation. Zeit Ang. Math. Mech. (ZAMM) 59, 709–716 (1979)
255. Ding, W.C., Xie, J.H., Sun, Q.G.: Interaction of Hopf and period doubling bifurcations of a vibro–impact system. J. Sound & Vib. 275(1–2), 27–45 (2004)
256. Ding, W., Xie, J.: Dynamical analysis of a two–parameter family for a vibro–impact system in resonance cases. J. Sound & Vib. 287, 101–115 (2005)
257. Ding, W., Xie, J.: Torus T2 and its routes to chaos of a vibro–impact system. Phys. Lett. A 349, 324–330 (2006)
258. Di Paola, M., Pirrotta, A.: Nonlinear systems under impulsive parametric input. Int. J. Nonlin. Mech. 34, 843–851 (1999)
259. Di Paola, M., Santoro, R.: Nonlinear systems under Poisson white noise handled by path integral solution. J. Vib. & Control 14(1–2), 35–49 (2008a)
260. Di Paola, M., Santoro, R.: Path integral solution for nonlinear system enforced by Poisson White Noise. Probab. Eng. Mech. 23(2-3), 164–169 (2008b)
261. Do, Y.: A mechanism for dangerous border collision bifurcations. Chaos, Solitons & Fractals 32(2), 352–362 (2007)
262. Dobry, M.W., Brzezinski, J.: Vibro–isolation of pneumatic hammer. Revue Francaise de Mecanique 3, 439–444 (1993) (in French)
263. Dokainish, M.A., Elmaraghy, H.: Optimum design parameters for impact dampers. In: Proc. ASME Design Engineering & Technical Conf., vol. 61, pp. 1–7 (1973)
264. Dongping, J., Haiyan, H.: Piecewise analysis of oblique vibro–impacting systems. Acta Mechanica Sinica (English Series) 19(6), 579–584 (2003)
265. Du, Z., Zhang, W.: Melnikov method for homoclinic bifurcation in nonlinear impact oscillators. Comput. & Math. with Appl. 50(3-4), 445–458 (2005)
266. Du, Z., Li, Y., Zhang, W.: Type I periodic motions for nonlinear impact oscillators. Nonlin. Analysis, Theory, Methods & Appl. 67(5), 1344–1358 (2007)
267. Dubowsky, S., Freudenstein, F.: Dynamic analysis of mechanical systems with clearances, Part 1: Formulation of dynamic model. ASME J. Eng. Indust B 93, 305–309 (1971a)

268. Dubowsky, S., Freudenstein, F.: Dynamic analysis of mechanical systems with clearances, Part 2: Dynamic response. ASME J. Eng. Indust B 93, 310–316 (1971b)
269. Duckwald, C.S.: Impact damping for turbine buckets. General Engineering Laboratory, General Electric, Report No. R55GL108 (1955)
270. Duffy, K.P.: Self–tuning impact damper for rotating blades. J. Acoust. Soc. Amer 117(5), 2690 (2005)
271. Duffy, K.P., Brown, G.V., Mehmed, O.: Impact damping of rotating cantilever plates. In: 3rd National Turbine Engine High Cycle Fatigue Conf., San Antonio, TX, February 2–5 (1998)
272. Duffy, K.P., Bagley, R.L., Mehmed, O.: On a self–tuning impact vibration damper for rotating turbomachinery. NASA TM–2000–210215 (2000); also AIAA Paper 2000–3100, AIAA Joint Propulsion Conf
273. Duffy, K.P., Bagley, R.L., Mehmed, O.: A self–tuning impact damper for rotating blades. NASA Tech Briefs TSP LEW–168333, 1–15 (2001a)
274. Duffy, K.P., Mehmed, O., Johnson, D.: Self–tuning impact dampers for fan and turbine blades. In: 6th National Turbine Engine High Cycle Fatigue Conf., Wright–Patterson AFB, Ohio, March 5–8 (2001b)
275. Duffy, K.P., Mehmed, O.: Self–tuning impact dampers for turbine blades. In: Proc. 7th National Turbine Engine High Cycle Fatigue Conf., Palm Beach Gardens, FL, May 14–17 (2002)
276. Dufour, R., Der Hagopian, J., Pompei, M., Garnier, C.: Shock and sine response of rigid structures on nonlinear mounts. In: ASME Design Engineering Division, Structural Vibration and Acoustics, Miami, FL, USA, vol. 34, pp. 171–176 (1991)
277. Dumont, Y.: Some remarks on a vibro–impact scheme. Numerical Algorithms 33, 227–240 (2003)
278. Dumont, Y., Lubuma, J.M.S.: Non–standard finite–difference methods for vibro–impact problems. Proc. Royal Soc London, Series A (Mathematical, Physical and Engineering Sciences) 461(2058), 1927–1950 (2005)
279. Duncan, M.R., Wassgren, C.R., Krousgrill, C.M.: The damping performance of a single particle impact damper. J. Sound & Vib. 286(1–2), 123–144 (2005)
280. Duthinh, D.: The head–on impact of an iceberg on a vertical, gravity–based structure. In: Proc. Specialty Conf. Computer Methods in Offshore Engineering, Halifax, N.S., pp. 397–412 (1984)
281. Dutta, P.S., Routroy, B., Banerjee, S., Alam, S.S.: On the existence of low–period orbits in n–dimensional piecewise linear discontinuous maps. Nonlin. Dyn. 53, 369–380 (2008)
282. Earles, S.W.E., Seneviratne, L.D.: Design guidelines for predicting contact loss in revolute joints of planar linkages mechanisms. Proc. Inst. Mech. Eng. 204, 9–18 (1990)
283. Edelstein, W.S., Chen, S.S.: Flow–induced instability of an elastic tube with a variable support. Nuclear Eng. & Design 84(1), 1–11 (1985)
284. Ehrich, F.: The dynamic stability of rotor/stator radial rubs in rotating machinery. ASME J. Eng. Indust 91, 1025–1028 (1969)
285. Ehrich, F.: High–order subharmonic response of high speed rotors in bearing clearance. ASME J. Vib., Acoust, Stress & Reliab Design 110, 9–16 (1988)
286. Ekwaro Osire, S., Desen, I.C.: Experimental study on an impact vibration absorber. J. Vib. & Cont. 7, 475–493 (2001)

287. Ekwaro Osire, S., Nieto, E., Gungor, F., Gumus, E., Ertas, A.: Performance of a Bi–Unit Damper using Digital Image Processing. In: Proc. Vibro–Impact Dynamics Symposium 2008, Troy, Michigan, USA, October 2–3. Springer, Berlin (2008) (in press)
288. El Aroudi, A., Debbat, M., Martinez–Salamero, L.: Poincaré maps modeling and local orbital stability analysis of discontinuous piecewise affine periodically driven systems. Nonlin. Dyn. 50(3), 431–445 (2007)
289. El–Sayad, M.A., Hanna, S.N., Ibrahim, R.A.: Parametric excitation of nonlinear elastic systems involving hydrodynamic sloshing impact. Nonlin. Dyn. 18(1), 25–50 (1999)
290. Ema, S., Marui, E.: Fundamental study on impact dampers. Int. J. Mach. Tools & Manuf. 34(3), 407–421 (1994)
291. Ema, S., Marui, E.: Damping characteristics of an impact damper and its application. Int. J. Mach. Tools & Manuf. 36(3), 293–306 (1996)
292. Ema, S., Marui, E.: Suppression of chatter vibration in drilling. ASME J. Manuf. Sci. & Eng. 120(1), 200–2–2 (1998)
293. Ema, S., Marui, E.: Suppression of chatter vibration of boring tools using impact dampers. Int. J. Mach. Tools & Manuf. 40(8), 1141–1156 (2000)
294. Emaci, E., Nayfeh, T.A., Vakakis, A.F.: Numerical and experimental study of nonlinear localization in a flexible structure with vibro–impacts. Zeit Ang. Math. Mech. (ZAMM) 77(7), 527–541 (1997)
295. Emad, J., Vakakis, A.F., Miller, N.: Experimental nonlinear localization in a periodically forced repetitive system of coupled magneto–elastic beams. Physica D 137, 192–201 (2000)
296. Engleder, T., Vielsack, P., Spiess, H.: Damping by impacts, an application of non–smooth dynamics. In: Proc. Conf. Nonlin. Oscillations in Mechanical Systems, St. Petersburg, Russia, pp. 134–144 (1998)
297. Engleder, T., Vielsack, P., Schweizrhof, K.: FE–regularization of non–smooth vibrations due to friction and impacts. Comput. Mech. 28(2), 162–168 (2002)
298. Erington, A.C.: Transverse impact on beams and plates. ASME J. Appl. Mech. 20(4), 461–468 (1953)
299. Ervin, E.K., Wickert, J.A.: Experiments on a beam–rigid body structure repetitively impacting a rod. Nonlin. Dyn. 50(3), 701–716 (2007a)
300. Ervin, E.K., Wickert, J.A.: Repetitive impact response of a beam structure subjected to harmonic base excitation. J. Sound & Vib. 307(1–2), 2–19 (2007b)
301. Estabrook, L.H., Plunckett, R.: Design parameters for impact dampers. General Engineering Laboratory, General Electric, Report No. R55GL250 (1955)
302. Evans, R.J., Parmerter, R.R.: Ice forces due to impact on a sloping structure. In: Proc. ASCE Conf. Arctic 1985, San Francisco, CA, pp. 220–229 (1985)
303. Everson, R.M.: Chaotic dynamics of a bouncing ball. Physica D 19, 355–383 (1986)
304. Ezovskikh, V.E.: Method of non–smooth transformations as applied to vibration–impact systems with moving limiters. Mech. Solids 20(5), 50–54 (1985)
305. Falsone, G., Muscolino, G.: Response of a beam–stop system under random excitations by an equivalent nonlinearization approach. In: Int. Conf. structures under shock and impact, Cambridge, vol. 8(6), pp. 661–623 (2000)
306. Faltinsen, O.M., Chezhian, M.: A generalized Wagner method for three–dimensional slamming. J. Ship Res. 49(4), 279–287 (2005)

307. Fang, W., Wickert, J.A.: Response of a periodically driven impact oscillator. J. Sound & Vib. 170(3), 397–409 (1994)
308. Fathi, A., Popplewell, N.: Improved approximations for a beam impacting a stop. J. Sound & Vib. 170(3), 365–375 (1994)
309. Fathi, A., Young, D.R., Popplewell, N.: Computer aided design of vibro–impact stops. In: 10th IMACS World Congress on System Simulation and Scientific Computation, Montreal, Quebec, Canada, vol. 3, pp. 177–179 (1982)
310. Feely, O., Chua, L.O.: Nonlinear dynamics of a class of analog–to–digital converters. Int. J. Bifurc. Chaos 22(2), 325–340 (1992)
311. Fegelman, K.J.L., Grosh, K.: Acoustic radiation by a rattling plate: Theoretical and experimental analysis. In: Proc. National Conf. Noise Control Engineering, Ypsilanti, MI, pp. 163–168 (1998)
312. Feigin, M.I.: Resonance behavior of a dynamical system with collisions. Appl. Math. Mech. (PMM) 30, 942–946 (1966)
313. Feigin, M.I.: Doubling of the oscillation period with C–bifurcation in piecewise–continuous system. Appl. Math. Mech. (PMM) 34, 822–829 (1970)
314. Feigin, M.I.: On the generation of sets of subharmonic modes in a piecewise–continuous system. Appl. Math. Mech. (PMM) 38, 810–818 (1974)
315. Feigin, M.I.: On the behavior of dynamic systems in the vicinity of existence of periodic motions. Appl. Math. Mech. (PMM) 41, 628–636 (1977)
316. Feigin, M.I.: On the structure of C–bifurcation boundaries of piecewise continuous systems. J. Appl. Math. Mech. (PMM) 42(5), 885–895 (1978)
317. Feigin, M.I.: Forced Oscillations in Systems with Discontinuous Nonlinearities, Nauka, Moscow (1994) (in Russian)
318. Feigin, M.I.: The increasing complex structure of the bifurcation tree of a piecewise–smooth system. J. Appl. Math. Mech. (PMM) 59(6), 853–863 (1995)
319. Feng, Q., He, H.: Modeling of the mean Poincaré map on a class of random impact oscillators. European J Mechanics, A/Solids 22(2), 267–281 (2003)
320. Feng, Q., Pfeiffer, F.: Stochastic model of a rattling system. J. Sound & Vib. 215(3), 439–453 (1998)
321. Fermi, E.: On the origin of the cosmic radiation. Phys. Revs. 75, 1169–1174 (1949)
322. Fidlin, A.: Nonlinear oscillations in Mechanical Engineering. Springer, Heidelberg (2005a)
323. Fidlin, A.: On the strongly nonlinear behavior of an oscillator in a clearance. In: van Campen, D.H., Lazurko, M.D., van der Oever, W.P.J.M. (eds.) Proc. ENOC 2005, Eindhoven, Netherlands, Technical University of Eindhoven, August 7-12, pp. 389–398 (2005b)
324. Field Jr., R.V., Epp, D.S.: Development and calibration of a stochastic dynamics model for the design of a MEMS inertial switch. Sensors & Actuators A 134(1), 109–118 (2007)
325. Filippov, A.F.: Differential Equations with Discontinuous Right-hand Sides. Kluwer Academic Publishers, Dordrecht (1988)
326. Fisher, N., Olesen, M., Rogers, R., Ko, P.: Simulation of a tube–to–support dynamic interaction in heat exchange equipment. ASME J. Pressure Vessel Techn. 111, 378–384 (1989)
327. Fisher, N., Pettigrew, M., Roger, R.: Fretting wear damage prediction in the inlet region of nuclear steam generators. In: Mech. Eng. Int. Cnf. Flow–Induced Vibrations Brighton, U. K., pp. 149–158 (1991)

328. Fisher, N., Tromp, J., Smith, B.: Measurement of dynamic interaction between a vibrating fuel element and its support. In: Proc. ASME Pressure Vessel and Piping Conf., Montreal, Canada, ASME PVP, vol. 318, pp. 271–283 (1996)
329. Flanagan, J.L.: Speech Analysis, Synthesis and Perception. Springer, New York (1972)
330. Flanagan, J.L., Landgraf, L.L.: Self oscillating source for vocal tract synthesizers. IEEE Trans. Audio & Electro–acoustics AU 16, 57–64 (1968)
331. Flores, P., Ambrosio, J., Claro, J.C.P., Lankarani, H.M.: Spatial revolute joints with clearances for dynamic analysis of multi–body systems. Proc. Inst. Mech. Eng., Part K (J. Multi–Body Dynamics) 220(K4), 257–271 (2006)
332. Foale, S.: Analytical determination of bifurcations in an impact oscillator. Phil. Trans. Royal Soc. London 347, 353–364 (1994)
333. Foale, S., Bishop, S.R.: Dynamical complexities of forced impacting systems. Phil. Trans. Royal Soc. London, Series A 338, 547–556 (1992)
334. Foale, S., Bishop, S.R.: Bifurcations in impact oscillators. Nonlin. Dyn. 6, 285–299 (1994a)
335. Foale, S., Bishop, S.R.: Bifurcations of impact oscillators: Theory and experiment. In: Thompson, J.M.T., Bishop, S.R. (eds.) Proc. Nonlinearity and Chaos in Engineering Dynamics. Wiley, Chichester (1994b)
336. Fogli, M., Bressolette, P.: Spectral response of a stochastic oscillator under impacts. Meccanica 32(1), 1–12 (1997)
337. Fogli, M., Bressolette, P., Bernard, P.: The dynamics of a stochastic oscillator with impacts. European J. Mech. A/Solids 15(2), 213–241 (1996)
338. Folkmans, S.L., Ferney, B.D., Bingham, J.G., Dutson, J.D.: Friction and impact damping in a truss using pinned joints. In: Guran, A., Pfeiffer, F., Popp, K. (eds.) Dynamics with Friction. World Scientific, Singapore (1996)
339. Fowler, B.L., Flint, E.M., Olson, S.E.: Effectiveness and predictability of particle damping. In: Proc. Int. Society for Optical Engineering, Newport Beach, CA, vol. 3989, pp. 356–367 (2000)
340. Fredriksson, M.H.: Grazing bifurcations in multibody systems. Nonlin. Anal. Theory, Methods & Appl. 30, 4475–4483 (1997)
341. Fredriksson, M.H., Nordmark, A.B.: Bifurcations caused by grazing incidence in many degrees of freedom impact oscillators. Proc. Royal Society of London A 453, 1261–1276 (1997)
342. Fredriksson, M.H., Nordmark, A.B.: On normal form calculations in impact oscillators. Proc. Royal Society of London A 456, 315–330 (2000)
343. Fredriksson, M.H., Borglund, D., Nordmark, A.B.: Experiments on the onset of impacting motion using a pipe conveying fluid. Nonlin. Dyn. 19, 261–271 (1999)
344. Freire, E., Ponce, E., Rodrigo, F., Torres, F.: Bifurcation sets of continuous piecewise linear systems with two zones. Int. J. Bifurc. & Chaos 8(11), 2073–2097 (1998)
345. Fricker, A.J.: Numerical analysis of the fluidelastic vibration of a steam generator tube with loose supports. In: Proc. ASME Int. Symp. on Flow–Induced Vibration, Noise Flow Induced–Vibration Heat Transfer Equipment, Chicago, IL, November 27–December 2, pp. 105–120 (1988)
346. Friend, R.D., Kinra, V.K.: Measurement and Analysis of Particle Impact Damping. In: Proc. SPIE Conf. Passive Damping and Isolation, San Jose, CA, January 23–29, 1999, pp. 20–31 (1999)

347. Friend, R.D., Kinra, V.K.: Particle Impact Damping. J. Sound & Vib. 233(1), 93–118 (2000)
348. Frosch, H., Buttner, H.: Two coupled impact oscillators. Zeit Physik B (Condensed Matter) 58(4), 323–328 (1985)
349. Fuglem, M.K., Duthinh, D., Lever, J.H., Jordan, I.: Probabilistic determination of iceberg collision design loads for floating production vessels. In: Proc. IUTAM–IAHR Symp. on Ice–Structure Interaction, St. John's Newfoundland, pp. 459–482 (1991)
350. Fuglem, M., Jordaan, I.J., Crocker, G.: Iceberg–structure interaction probabilities for design. Canadian Journal of Civil Engineering 23(1), 231–241 (1996)
351. Fuglem, M., Muggeridge, K., Jordaan, I.J.: Design load calculations for iceberg impacts. Int. J. Offshore Polar Engineering 9(4), 298–306 (1999)
352. Fuse, T.: Prevention of resonances by impact damper. ASME Pressure Vessels and Piping Division, PVP-182, 57–66 (1989)
353. Gagnon, R., Jones, S.J., Frederking, R., Spencer, P.A., Masterson, D.M.: Large–scale hull loading of ice in Tuktoyaktuk harbour. In: 18th Int. Conf. Offshore Mechanics and Arctic Engineering, St. John's, Newfoundland, ASME Paper No. OMAE99/P&A–119 (1999) (CD–ROM)
354. Gagnon, R., Jones, S.J., Frederking, R., Spencer, P.A., Masterson, D.M.: Large–scale hull loading of sea ice, lake ice, and ice in Tuktoyaktuk harbour. ASME J. Offshore Mech. & Arctic Eng. 123, 159–169 (2001)
355. Gaidis, R.D., Raqulskis, K.M.: Component shape optimization in vibro–impact systems with distributed parameters. Vibration Engineering (English Translation of Vibrotekhnika) 4(2-4), 269–280 (1990)
356. Gajewsky, K., Radiszewski, B.: On the stability of impact systems. Bull Polish Acad Sci., Tech. Sci. 35, 183–188 (1987)
357. Garza, S., Ertas, A.: Experimental investigation of the dynamics and bifurcations of an impacting spherical pendulum. In: Proc. ASME 15th Biennial Conf. Mechanical Vibration and Noise, Boston, MA, DE, vol. 84(3), Pt A/1, pp. 181–189 (1995)
358. Gaudet, S.: Numerical simulation of circular disks entering the free surface of a fluid. Phys. Fluids 10(10), 2489–2499 (1998)
359. Gendelman, O.V.: Modeling of inelastic impacts with the help of smooth functions. Chaos, Solitons & Fractals 28, 522–526 (2006)
360. Gendelman, O.V., Meimukhin, D.: Response regimes of integrable damped strongly nonlinear oscillator under impact periodic forcing. Chaos, Solitons & Fractals 32(2), 405–414 (2007)
361. Gerner, I.I., Kim, L.I., Mokin, N.V.: Calculation of nonlinear corrector for vibration isolated suspension. In: Proc. Research for Railway Transportation, Nauchnii Institut Zheleznodoro–zhnogo Transporta (NIZhT), Novosibirsk No. 156, pp. 152–159 (1974)
362. Gibson, B.W.: Usefulness of Impact Dampers for Space Applications. Air Force Inst. of Tech., Wright–Patterson AFB, OH. School of Engineering, Report: AFIT/GA/AA/83M–2, 147 (March 1983)
363. Giusepponi, S., Marchesoni, F., Borromeo, M.: Randomness in the bouncing ball dynamics. Physica A 351(1), 142–158 (2005)
364. Glocker, C., Pfeiffer, F.: Multiple impacts with friction in rigid multibody systems. Nonlin. Dyn. 7(4), 471–497 (1995)

365. Glocker, C., Studer, C.: Formulation and preparation for numerical evaluation of linear complementarity systems in dynamics. Multibody system Dyn. 13, 463–477 (2005)
366. Goldman, P., Muszynska, A.: Dynamic effects in mechanical structures with gaps and impacting: order and chaos. ASME J. Vibr. & Acoust 116(4), 541–547 (1994a)
367. Goldman, P., Muszynska, A.: Chaotic behavior of rotor/stator systems with rubs. ASME J. Eng. Gas Turbines & Power 16(3), 692–701 (1994b)
368. Goldman, P., Muszynska, A.: Smoothing technique for rub or looseness-related rotor dynamic problems. In: Proc. ASME Biennial Conf. Mechanical Vibration and Noise, Boston, MA, Design Engineering Division DE, September 17–20, Vol. 84(3 Pt A/1), pp. 565–572 (1995)
369. Goldsmith, W.: Impact: Theory and Physical Behavior of Colliding Solids. Edward Arnold, London (1960)
370. Gontier, C.: Multiple degree of freedom impact oscillator (Electricite de France, Clamart. Direction des Etudes et Recherches.) Report: EDF–96–NB–00171, 63 (1996)
371. Gontier, C., Toulemonde, C.: Approach to the periodic and chaotic behavior of the impact oscillator by a continuation method. European J. Mech. A/Solids 16, 141–163 (1997)
372. Goverdovskiy, V.N., Gyzatullin, B.S., Petrov, V.A.: Method of vibration isolation for the man–operator of transport–technological machine. Patent 2115570, Russia (1998)
373. Grace, I.F., Ibrahim, R.A.: Modeling and analysis of ship roll oscillations interacting with stationary icebergs. IME J. Mech. Eng. Sci. (in print, 2008)
374. Gray, G.G., Johnson, K.L.: The Dynamic Response of Elastic Bodies in Rolling in Rolling Contact to Random Roughness of Their Surfaces. J. Sound & Vib. 22, 297–322 (1972)
375. Greco, M., Landrini, M., Faltinsen, O.M.: Impact Flows and Loads on Ship–Deck Structures. Inseam Italian Shop Model Basin, Rome, Italy, 13 (March 2003)
376. Greco, M., Landrini, M., Faltinsen, O.M.: Impact flows and loads on ship–deck structures. J. Fluids & Struct 18(suppl.), 251–275 (2004)
377. Greenwood, J.A., Williamson, J.B.P.: Contact of Nominally Flat Surfaces. Proc. Royal Society London 295, 300–319 (1966)
378. Gregory, R.W., Païdoussis, M.P.: Unstable oscillation of tubular cantilevers conveying fluid, I: Theory; II: Experiments. Proc. Royal Society London 293, 512–527, 528–542 (1966)
379. Grubin, C.: On the theory of the acceleration damper. ASME J. Appl. Mech. 23, 373–378 (1956)
380. Guckenheimer, J., Holmes, P.: Nonlinear Oscillations, Dynamic Systems, and Bifurcation of Vector Fields. Applied Mathematical Sciences, 2nd edn., vol. 42. Springer, New York (1986)
381. Gunter, H.: A mechanical model of vocal–fold collision with high spatial and temporal resolution. J. Acoust. Soc. Amer. 113(2), 994–1000 (2003)
382. Gutierrez, E., Arrowsmith, D.K.: Control of a double impacting mechanical oscillator using displacement feedback. Int. J. Bifurc & Chaos 14(9), 3095–3113 (2004)
383. Gutierrez, E., Arrowsmith, D.K.: The equivalence between feedback and dissipation in impact oscillators. Int. J. Bifurc & Chaos 17(1), 255–269 (2007)

384. Haller, G., Stepan, G.: Micro–chaos in digital control. J. Nonlin. Sci. 6, 415–448 (1996)
385. Halse, C., Homer, M., Di Bernardo, M.: C–bifurcations and period–adding in one–dimensional piecewise–smooth maps. Chaos, Solitons & Fractals 18(5), 953–976 (2003)
386. Halse, C.K., Wilson, R.E., Di Bernardo, M., Homer, M.E.: Coexisting solutions and bifurcations in mechanical oscillators with backlash. J. Sound & Vib. 305(4–5), 854–885 (2007)
387. Han, D., Gao, Z., Wang, H.W., Zhang, H.Q.: Analysis of the computational method and the helicopter blade characteristics with coupled rigid and elastic motions. Hangkong Dongli Xuebao/J. Aerospace Power 21(1), 36–40 (2006) (in Chinese)
388. Han, I., Gilmore, B.J.: Multi–body impact motion with friction analysis, simulation, and experimental validation. ASME J. Mech. Design 115, 412–422 (1993)
389. Han, Q., Zhang, Z., Wen, B.: Periodic motions of a dual–disc rotor system with rub–impact at fixed limiter. Proc. IMech E Part C: J. Mechanical Engineering Science 222, 1935–1946 (2008)
390. Han, W., Jin, D.P., Hu, H.Y.: Dynamics of an oblique–impact vibrating system of two degrees of freedom. J. Sound & Vib. 217, 795–822 (2004)
391. Han, W., Hu, H.Y., Jin, D.P.: Experimental study on dynamics of an oblique–impact vibrating system of two degrees of freedom. Nonlin. Dyn. 50(3), 551–573 (2007)
392. Hassan, M.: Simulation of fluidelastic vibrations of heat exchanger tubes with loose supports. In: Proc. ASME Pressure Vessels and Piping Division Conf., PVP2006/ICPVT–11, Vancouver, BC, Canada, p. 11 (2006)
393. Hassan, M.A., Hayder, M.: Modeling of fluidelastic vibrations of heat exchanger tubes with loose supports. Nucl. Eng. &Design 238, 2507–2520 (2008)
394. Hassan, M.A., Rogers, R.: Friction modeling of preloaded tube contact dynamics. Nuclear Eng. & Design 235(22), 2349–2357 (2005)
395. Hassan, M.A., Weaver, D.S., Dokainish, M.A.: A simulation of the turbulence response of heat exchanger tubes in lattice–bar supports. J. Fluids & Struct. 16(8), 1145–1176 (2002)
396. Hassan, M.A., Weaver, D.S., Dokainish, M.A.: The effects of support geometry on the turbulence response of loosely supported heat exchanger tubes. J. Fluids & Struct. 18(5), 529–554 (2003)
397. Hassan, M.A., Weaver, D.S., Dokainish, M.A.: A new tube/support impact model for heat exchanger tubes. J. Fluids & Struct. 21(5–7), 561–577 (2005)
398. Hassouneh, M.A., Abed, E.H.: Lyapunov and LMI analysis and feedback control of border collision bifurcation. Nonlin. Dyn. 50(3), 373–386 (2007)
399. Hassouneh, M.A., Abed, E.H., Nusse, H.E.: Robust dangerous border–collision bifurcation in piecewise smooth systems. Phys. Rev. Lett. 92, 070201–4 (2004)
400. Heiman, M.S., Sherman, P.J., Bajaj, A.K.: On the dynamics and stability of an inclined impact pair. J. Sound & Vib. 114(3), 535–547 (1987)
401. Heiman, M.S., Sherman, P.J., Bajaj, A.K.: Periodic motions and bifurcations in dynamics of an inclined impact pair. J. Sound & Vib. 124(1), 55–78 (1988)
402. Hendriks, F.: Bounce and chaotic motion in impact print hammers. IBM J. Res. & Devel 27(3), 273–280 (1983)

403. Hennig, K., Grunwald, G.: Treatment of flow–induced pendulum oscillations [PWR application]. In: Proc. 3rd Int. Conf. Vibration in Nuclear Plant, Keswick, UK, May 11–14, 1982, vol. 2, pp. 1100–1116. British Nuclear Energy Society, London (1983)
404. Hénon, M.: On the numerical computation of Poincaré maps. Physica D 5(2-3), 412–414 (1982)
405. Herbert, R.G., McWhannell, D.C.: Shape and frequency composition of pulses from an impact pair. ASME J. Eng. Indust. 99, 513–518 (1977)
406. Hertz, H.: Über die berü hrung fester elastischer Körper. J. für die reine und angew Math 92, 156–171 (1881)
407. Hertz, H.: Gesammelte Werke, Schriften Vermischten Inhalts, J. A. Barth, Leipzig, Germany, vol. 1 (1895)
408. Herzel, H., Knudsen, C.: Bifurcations and chaos in voice signals. ASME Appl. Mech. Rev. 46(7), 399–413 (1993)
409. Herzel, H., Knudsen, C.: Bifurcations in a vocal fold model. Nonlin. Dyn. 7(1), 53–64 (1995)
410. Herzel, H., Wendler, J.: Evidence of chaos in phonatory samples. In: Proc. EUROSPEECH 1991, 2nd European Conf. on Speech Communication and Technology, Genova, Italy, pp. 263–266 (1991)
411. Hess, D.P., Soom, A.: Normal Vibrations and Friction under Harmonic Loads, Part I: Hertzian Contact. ASME Journal Tribology 113, 80–86 (1991a)
412. Hess, D.P., Soom, A.: Normal Vibrations and Friction under Harmonic Loads, Part II: Rough Planar Contact. ASME J. Tribology 113, 87–92 (1991b)
413. Hess, D.P., Wagh, N.J.: Chaotic vibrations and friction at mechanical joints. In: Proc. ASME WAM, Anaheim, CA, Design Engineering Division, Friction–Induced Vibration, Chatter, Squeal, and Chaos, DE, vol. 49, pp. 149–156 (1992)
414. Hess, D.P., Wagh, N.J.: Chaotic normal vibrations and friction at mechanical joints with nonlinear elastic properties. ASME J. Vib. & Acoust. 116(4), 474–479 (1994)
415. Hess, M.M., Verdolini, K., Bierhals, W., Mansmann, U., Gross, M.: Enolaryngeal contact pressure. J. Voice 12, 50–67 (1998)
416. Hill, J.M., Jennings, M.J., To, D.V., Williams, K.A.: Dynamics of an elastic ball bouncing on an oscillating plane. Appl. Math. Model 24(10), 715–732 (2000)
417. Hindmarsh, M.B., Jefferies, D.J.: On the motions of the offset impact oscillator. J. Physics A 17, 1791–1803 (1984)
418. Hinrichs, N., Oestreich, M., Popp, K.: Dynamics of oscillators with impact and friction. Chaos, Solitons & Fractals 8(4), 535–558 (1997)
419. Hinrichs, N., Oestreich, M., Popp, K.: Experiments, modelling and analysis of friction and impact oscillators. Zeit Ang Math Mech (ZAMM) 79(1), S95–S96 (1999)
420. Hogan, S.J., Higham, L., Griffin, T.C.L.: Dynamics of a piecewise linear map with a gap. Proc. Royal Soc. London A 463, 49–65 (2007)
421. Hollkamp, J.J., Gordon, R.W.: Experiments with particle damping. In: Proc. SPIE – The Int Society for Optical Engineering. Smart Structures and Materials. Passive Damping and Isolation, San Diego, CA, March 2–3, 1998, vol. 3327, pp. 2–12 (1998)
422. Holmberg, E.B., Hillman, R.E., Perkell, J.S.: Glottal air–flow and transglottal air–pressure measurements for male and female speakers in soft, normal, and loud voice. J. Acoust. Soc. Amer. 84, 511–529 (1988)

423. Holmes, P.J.: The dynamics of repeated impacts with a sinusoidally vibrating table. J. Sound & Vib. 84(2), 173–189 (1982)
424. Hongler, M.O.: Chaos in vibro-transportation. In: Proc. a Workshop on Dynamics and Stochastic Processes, Theory and Applications, October 24–29, 1988, pp. 142–164. Springer, Berlin (1990)
425. Horáček, J., Švec, J.G.: Aeroelastic model of vocal–fold–shaped vibrating element for studying the phonation threshold. J. Fluids & Struct. 16, 931–955 (2002)
426. Horáček, J., Šidlof, P., Švec, J.G.: Numerical simulation of self–oscillations of human vocal folds with Hertz model of impact forces. J. Fluids & Struct. 20, 853–869 (2005)
427. Hu, H.Y.: Detection of grazing orbits and incident bifurcations of a forced continuous, piecewise–linear oscillator. J. Sound & Vib. 187, 485–493 (1995)
428. Hu, H., Xiong, Z.G.: Oscillations in an potential. J. Sound & Vib. 259, 977–980 (2003)
429. Huang, Z.L., Gao, L.X., Zhu, W.Q.: Random response of SDOF vibro–impact system to wide–band excitation. Zhejiang Daxue Xuebao (Gongxue Ban)/ J. Zhejiang University 37(1), 94–97 (2003) (in Chinese)
430. Huang, Z.L., Liu, Z.H., Zhu, W.Q.: Stationary response of multi–degree–of–freedom vibro–impact systems under white noise excitations. J. Sound & Vib. 275, 223–240 (2004)
431. HuiMing, Y.: Stability analysis of a class of three–degree–of freedom vibro–impact system. J. Mech. Strength 27(1), 6–11 (2005)
432. Hundal, M.S.: Impact absorber with two–stage, variable area orifice hydraulic damper. J. Sound & Vib. 50(2), 195–202 (1977)
433. Hunt, J.P., Sarid, D.: Kinetics of lossy grazing impact oscillators. Appl. Phys. Lett. 72(23), 2969–2971 (1998)
434. Hunt, K.H., Crossley, F.R.E.: Coefficient of restitution interpreted as damping in vibro–impact. ASME J. Appl. Mech. 97, 440–445 (1975)
435. Ibrahim, R.A.: Recent results in random vibrations of nonlinear mechanical systems. ASME J. Mech. Design & J. Vibr. & Acoust. 117, 222–233 (1995); special issue
436. Ibrahim, R.A.: Liquid Sloshing Dynamics: Theory and Applications. Cambridge University Press, Cambridge (2005)
437. Ibrahim, R.A.: Recent advances in nonlinear passive vibration isolators. J. Sound & Vib. 314, 371–452 (2008)
438. Ibrahim, R.A., Chalhoub, N.G., Falzarano, J.: Interaction of ships and ocean structures with ice loads and stochastic ocean waves. ASME Appl. Mech. Rev. 60, 1–44 (2007)
439. Ibrahim, R.A., El–Sayad, M.A.: Simultaneous parametric and internal resonances in systems involving strong nonlinearities. J. Sound & Vib. 225(5), 857–885 (1999)
440. Igarashi, T., Aimoto, T.: Studies on impact sound. Bull JSME 28, 1247–1254 (1985)
441. Igarashi, T., Aimoto, T.: Studies on impact sound. Nippon Kikai Gakkai Ronbunshu, C Hen Trans. JSME, Part C 55(520), 2992–2998 (1989) (in Japanese) (5th Report, effects of boundary condition of a plate on sound)
442. Imamura, H.: A new analytical approach to derive global function form of periodic solutions for impact oscillator. In: Proc. 47th Midwest Symp. Circuits and Systems (IEEE Cat. No.04CH37540), vol. 2, pt 2, pp. II-621–II-624 (2004)

443. Imamura, H.: Exact derivation of global form of all periodic solutions for zero stiffness impact oscillator using initial value correction periodic method (case of viscous damping free system). Nihon Kikai Gakkai Ronbunshu, C Hen/ Trans JSME, Part C 72(3), 706–713 (2006) (in Japanese)
444. Imamura, H., Suzuki, K.: General form of steady response and periodic solution for an impact oscillator having no sticking. JSME Int. J. Series C: Mech Systems, Machine Elem. & Manuf. 46(4), 1456–1463 (2003)
445. Ing, J., Pavlovskaia, E., Wiercigroch, M.: Dynamics of a nearly symmetrical piecewise linear oscillator close to grazing incidence: Modeling and experimental verification. Nonlin. Dyn. 46, 225–238 (2006)
446. Ing, J., Pavlovskaia, E., Wiercigroch, M., Banerjee, S.: Experimental study of impact oscillator with one–sided elastic constraint. Phil. Trans. Royal Soc. A 366(1866), 679–704 (2008)
447. Ioos, G.: Bifurcations of Maps and Applications. Mathematics Studies, vol. 36. North–Holland, Amsterdam (1979)
448. Iourtchenko, D.V., Dimentberg, M.F.: Energy balance for random vibration of piecewise–conservative systems. J. Sound & Vib. 248, 913–923 (2001)
449. Iourtchenko, D.V., Song, L.L.: Numerical investigation of a response probability density function of stochastic vibro–impact systems with inelastic impacts. Int. J. Nonlin. Mech. 41, 447–455 (2006)
450. Irie, T., Fukaya, K.: On the stationary impact vibration of a mechanical system with two–degree–of–freedom. Bull JSME 15(81), 299–306 (1972)
451. Ishizaka, K., Flanagan, J.L.: Synthesis of voiced sounds from a two–mass model of the vocal cords. Bell System Tech. J. 51, 1233–1268 (1972)
452. Isomaki, H.M., Von Boehm, J., Raty, R.: Devil's attractors and chaos of a driven impact oscillator. Phys. Lett. A 107A(8), 343–346 (1985a)
453. Isomaki, H.M., Von Boehm, J., Raty, R.: Chaotic oscillations of a particle with impacts. In: Proc. XIX Annual Conf. of the Finnish Physical Society, Oulu, Finland, 8–9 February, 4(4) (1985b)
454. Ivanov, A.P.: Stabilization of an impact oscillator near grazing incidence owing to resonance. J. Sound & Vib. 162, 562–565 (1993)
455. Ivanov, A.P.: Impact oscillations: Linear theory of stability and bifurcations. J. Sound & Vib. 178(3), 361–378 (1994)
456. Ivanov, A.P.: Dynamics of Systems with Mechanical Impacts. Int. Education Program, Moscow (1997) (in Russian)
457. Ivanov, A.P., Pereverzev, V.I.: Analysis of a two–mass vibro–impact system with an elastic bond. Mech. Solids 36(4), 32–39 (2001)
458. Iwankiewicz, R., Nielsen, S.R.K.: Dynamic response of nonlinear systems to Poisson distributed random impulses. J. Sound & Vibr. 156(3), 407–423 (1992a)
459. Iwankiewicz, R., Nielsen, S.R.K.: Dynamic response of hysteretic systems to Poisson distributed pulse trains. Probabilistic Engineering Mechanics 7, 135–148 (1992b)
460. Jackson, W.C., Poe Jr., C.C.: The use of impact force as a scale parameter for the impact response of composite laminates. NASA Tech. Memorandum 104189, US Army Aviation Systems Command (AVSCOM) Tech. Rept. 92–B–001 (January 1992)
461. Jacquart, G., Gay, N.: Computation of impact–friction interaction between a vibrating tube and its loose supports. EDF–Electricite de France, Clamart, France, 94NB00040, p. 19 (1993)

462. Jain, P., Banerjee, S.: Border–collision bifurcations in one–dimensional discontinuous maps. Int. J. Bifurc & Chaos 13(11), 3341–3351 (2003)
463. Jang, B.P., Kowbel, W., Jang, B.: Impact behavior and impact fatigue testing of polymer components. Compos. Sci. Technol. 44, 107–118 (1992)
464. Janin, O., Lamarque, C.H.: Comparison of several numerical methods for mechanical systems with impacts. Int. J. Num. Methods Eng. 51, 1101–1132 (2001)
465. Janin, O., Lamarque, C.H.: Stability of singular periodic motions in a vibro–impact oscillator. Nonlin. Dyn. 28, 231–241 (2002)
466. Jazar, G.N., Narmani, A., Golnaraghi, M.F., Swanson, D.A.: Practical frequency and time optimal design of passive linear vibration isolation mounts. J. Vehicle System Dyn. 39, 437–466 (2003)
467. Jazar, G.N., Aagaah, R.M., Mahinfalah, M., Nazari, G.: Comparison of exact and approximate frequency response of a piecewise linear vibration isolator. In: Proc. ASME 20th Bien Conf. Mechanical Vibration and Noise, Long Beach, CA, vol. 1A, pp. 107–115 (2005)
468. Jendrzejczyk, J.A., Chen, S.S.: Experiments on tubes conveying fluid. Thin–Walled Structures 3, 109–134 (1985)
469. Jerrelind, J., Dankowicz, H.: A Global Control Strategy for Efficient Control of a Braille Impact Hammer. ASME J. Vib. & Acoust. 128, 184–189 (2006)
470. Jiang, J.J., Diaz, C.E., Hanson, D.G.: Finite element modeling of vocal fold vibration in normal phonation and hyper functional dysphonia: Implications for the pathogenesis of vocal nodules. The Annals of otology, rhinology & laryngology (Suppl. 107), 603–610 (1998)
471. Jiang, J.J., Titz, I.R.: Measurement of vocal fold intraglottal pressure and impact stress. J. Voice 8, 132–144 (1994)
472. Jiang, J.J., Zhang, Y.: Chaotic vibration induced by turbulent noise in a two–mass model of vocal folds. J. Acoust. Soc. Amer. 112(5), 2127–2133 (2002)
473. Jin, J.D.: Stability and chaotic motions of a restrained pipe conveying fluid. J. Sound & Vib. 208, 427–439 (1997)
474. Jin, J.D., Zou, G.S.: Bifurcations and chaotic motions in the autonomous system of a restrained pipe conveying fluid. J. Sound & Vib. 260, 783–805 (2003)
475. Jing, H.S., Sheu, K.C.: Exact stationary solutions of the random response of a single–degree–of–freedom vibro–impact system. J. Sound & Vib. 141, 363–373 (1990)
476. Jing, H.S., Young, M.: Random response of a single–degree–of–freedom vibro–impact system with clearance. Earthquake Eng. & Struct. Dyn. 19, 789–798 (1990)
477. Jing, H.S., Young, M.: Impact interaction between two vibration systems under random excitation. Earthquake Eng. & Struct. Dyn. 20, 667–681 (1991)
478. Johansson, M., Rantzer, A.: Computation of piecewise quadratic Lyapunov functions for hybrid systems. IEEE Trans. Autom. Cont. 43(4), 555–559 (1998)
479. Johnson, A.A.: Impact Fatigue – an emerging field of study. Eng. Integrity 15, 14–20 (2004)
480. Johnson, D.C.: Synchronous whirl of a vertical shaft having clearance in one bearing. J. Mech. Eng. Sci. 4, 85–93 (1962)

481. Johnson, K.L.: Contact Mechanics. Cambridge University Press, Cambridge (1985)
482. Jones, D.I.G., Muszynska, A.: Harmonic response of a damped two–degree–of–freedom system with gaps. Nonlin. Vibr. Probl. 19, 153–178 (1979)
483. Jones, N.: Structural Impact. Cambridge University Press, New York (1997)
484. Jordaan, I.J.: Probabilistic analysis of environmental data for design of fixed and mobile arctic offshore structures. In: Reliability and Risk Analysis in Civil Engineering, Proceedings of the 5th International Conference on Applications of Statistics and Probability in Soil and Structural Engineering (ICASP5), Vancouver, vol. 2, pp. 1130–1137 (1987)
485. Jordan, R.W., Whiston, G.S.: Remote impact analysis by use of propagated acceleration signals, II: Comparison between theory and experiments. J. Sound & Vib. 97, 53–63 (1984)
486. Kahraman, A., Singh, R.: Nonlinear dynamics of a geared rotor–bearing system with multiple clearances. J. Sound & Vib. 144, 469–506 (1991)
487. Kalagnanam, J.R.: Controlling chaos: the example of an impact oscillator. ASME J. Dyn. Systems, Measur. & Control 116(3), 557–564 (1994)
488. Kamiya, K., Kouno, T., Kato, R., Yasuda, K.: Vibration suppression of a continuous system by an elastic impact damper. Nippon Kikai Gakkai Ronbunshu, C Hen/Trans JSME, Part C 70(11), 3046–3052 (2004) (in Japanese)
489. Kang, W., Wilcox, B., Dankowicz, H., Thota, P.: Bifurcation Analysis of a Micro-actuator Using a New Toolbox for Continuation of Hybrid System Trajectories. ASME J. Comput. & Nonlin. Dyn. (in print, 2008)
490. Kanso, E., Papadopoulos, P.: Pseudo–rigid ball impact on an oscillating rigid foundation. Int. J. Nonlin. Mech. 39(7), 1129–1145 (2004)
491. Kapitaniak, T., Czolczynski, K.: Impacting oscillators – the problem of visualization of basins of attraction. In: Proc. Int. Conf. Physics and Control, Saint Petersburg, Russia, vol. 2, pp. 657–662 (2003)
492. Karagiannis, K., Pfeiffer, F.: Theoretical and experimental investigations of gear–rattling. Nonlin. Dyn. 2, 367–387 (1991)
493. Karayannis, I., Vakakis, A.F., Georgiades, F.: Vibro–impact attachments as shock absorbers. Proc. IMech. E Part C: J. Mechanical Engineering Science 222, 1899–1908 (2008)
494. Karyeaclis, M.P., Caughey, T.K.: Stability of a semi–active impact damper. Part 1. Global behavior. ASME J. Appl. Mech. 56(4), 926–929 (1989a)
495. Karyeaclis, M.P., Caughey, T.K.: Stability of a semi–active impact damper, II: Periodic solutions. ASME J. Appl. Mech. 56(4), 930–940 (1989b)
496. Kascak, A.F., Tomko, J.J.: Effects of different rub models on simulated rotor dynamics. NASA TP 2220, AVSCOM Tr–83–C–8 (1983); also Proc. ASME Appl. Mech., Bioengineering, and Fluids Engineering Conf., Rotor Dynamical Instability, Houston, TX, USA, vol. 55, p. 35
497. Kascak, A.F., Tomko, J.J.: Effects of Different Rub Models on Simulated Rotor Dynamics. NASA, Washington, DC, Army Aviation Systems Command, St. Louis, MO. NASA–E–1801, 14 (Feburary 1984a)
498. Kascak, A.F., Tomko, J.J.: Effects of different rub models on simulated rotor dynamics. NASA Technical Paper, 11 pages (Feburary 1984b)
499. Kataoka, M., Ohno, S., Sugimoto, T.: A two–degree–of–freedom system including a clearance two–step hardening spring. JSME Int. J. Series II 34, 345–354 (1991)

500. Kawakami, M., Michimoto, J., Kobayashi, K.: Prediction of long term whipping vibration stress due to slamming of large full ships in rough seas. Int. Shipbuilding Progress 24, 83–110 (1977)
501. Kember, S.A., Babitsky, V.I.: Excitation of vibro–impact systems by periodic impulses. J. Sound & Vib. 227, 427–447 (1999)
502. Kheisin, D.I., Popov, Y.N. (eds.): Ice Navigation qualities of ships, Cold Regions Research and Engineering Laboratory (CRREL), Draft Translation 417, Hanover, New Hampshire (1973)
503. Kim, B., Pettigrew, M., Tromp, J.: Vibrations damping of heat exchanger tubes in liquid: Effects of support parameters. J. Fluids & Struct. 2, 593–614 (1988)
504. Kim, H., Edward, K.T.: Experimental and numerical analysis correlation of hail ice impacting composite structures. AIAA Journal 38(7), 1278–1288 (2000)
505. Kim, S.H., Lee, I.: Aeroelastic analysis of a flexible airfoil with a free–play nonlinearity. J. Sound & Vib. 193(4), 823–846 (1996)
506. Kim, T.C., Singh, R.: Dynamic interactions between loaded and unloaded gear pairs under rattle conditions. Soc. Automotive Eng. Trans. 110(6), 1934–1943 (2001)
507. Kim, T.C., Rook, T.E., Singh, R.: Effect of smoothening functions on the frequency response of an oscillator with clearance nonlinearity. J. Sound & Vib. 263, 665–678 (2003)
508. Kim, T.C., Rook, T.E., Singh, R.: Super– and sub–harmonic response calculations for a torsional system with clearance nonlinearity using the harmonic balance method. J. Sound & Vib. 281, 965–993 (2005)
509. Kim, Y.B., Noah, S.T.: Stability and bifurcation analysis of oscillators with piecewise–linear characteristics: A general approach. ASME J. Appl. Mech. 58(2), 545–553 (1990a)
510. Kim, Y.B., Noah, S.T.: Bifurcation analysis for a modified Jeffcott rotor with bearing clearances. Nonlin. Dyn. 1(3), 221–241 (1990b)
511. Knudsen, T.R., Feldberg, R., True, H.: Bifurcations and Chaos in a model of a rolling railway wheel-set. Phil. Trans. Royal Soc. London A 338, 455–469 (1992)
512. Knudsen, J., Massih, A.R.: Analysis of loosely supported beam under harmonic excitation. In: ASME Proc. Pressure Vessels & Piping Conf., Flow–Induced Vibration, Boston, MA. PVP, vol. 389, pp. 265–272 (1999)
513. Knudsen, J., Massih, A.R.: Vibro–impact dynamics of a periodically forced beam. ASME J. Pressure Vessel Technology 122, 210–221 (2000)
514. Knudsen, J., Massih, A.R.: Dynamic stability of weakly damped oscillators with elastic impacts and wear. J. Sound & Vib. 263(1), 175–204 (2003)
515. Knudsen, J., Massih, A.R., Johansson, L.: Calculation of vibro–impact dynamics of loosely supported rods. In: ASME Proc. Fluid–Structure Interaction, Aeroelasticity, Flow–Induced Vibration & Noise, Symp., Dallas, Texas, AD, vol. 53–1, pp. 229–237 (1997)
516. Ko, P.L., Basista, H.: Correlation of support impact force and fretting–wear for a heat exchanger tube. In: Proc. ASME Pressure Vessels and Piping Division, San Antonio, TX. PVP, vol. 82, pp. 221–233 (1984)
517. Kob, M.: Physical Modeling of the Singing Voice. Logos–Verlag, Berlin (2002)
518. Kobrinsky, A.E.: Dynamics of Mechanisms with Elastic Connections and Impact Systems, Iliffe, London (1969)

519. Kobrinsky, A.E., Kobrinsky, A.A.: Vibro–impact Systems. Nauka, Moscow (1973) (in Russian)
520. Koizumi, K., Sasaki, T., Sasaki, M., Okabe, S., Yokoyama, Y.: Repetitive impact motion in mass–spring system with impact follower by static pressure force suspension controlled – follower as time lag system of first order. Seimitsu Kogaku Kaishi/ J. Japan Soc. Precision Eng. 63(8), 1096–1100 (1997) (in Japanese)
521. Kollar, L.E., Stepan, G., Turi, J.: Dynamics of piecewise linear discontinuous maps. Int. J. Bifurc. Chaos 14, 2341–2351 (2004)
522. Korri, P., Varsta, P.: On the ice trial of a 14500 DWT tanker on the Gulf of Bothnia. NSTM–79, Soc Naval Architects in Finland (LARADI), Helsinki (1979)
523. Kotera, T., Peterka, F.: Laws of impact motion of mechanical systems with one degree of freedom – Part IV. analytical solution of the 2/n–impact motion and its stability. Acta. Technica. ČSAV (Ceskoslovensk Akademie Ved) 26(6), 747–758 (1981)
524. Kotera, T., Peterka, F.: Laws of impact motion of mechanical systems with one degree of freedom, Part VI. Analytical and analogue solution of the multi–impact motion and its stability. Acta Technica ČSAV (Ceskoslovensk Akademie Ved) 29(3), 255–279 (1984)
525. Kousaka, T., Kido, T., Ueta, T., Kawakami, H., Abe, M.: Analysis of border–collision bifurcation in a simple circuit. In: Proceedings – IEEE Int. Symp. on Circuits & Systems, Geneva, Switzerland, vol. 2, pp. II–481–II–484 (2000)
526. Kovaleva, A.S.: Investigation of vibro–impact resonances in systems with random parametric perturbation. Soviet Machine Science (English Transl Mashinovedenie) 2, 14–18 (1983)
527. Kovaleva, A.S.: Optimal Control of Mechanical Oscillations. Springer, Berlin (1999a) (Translation of Control of Vibrating and Vibro–Impact Systems, Nauka, Moscow, 1990)
528. Kovaleva, A.: Response of secondary structures in stochastic systems with impacts. In: Pfeiffer, F., Glockner, C. (eds.) Proc. IUTAM Symp on Unilateral Multibody Contacts, pp. 117–126. Kluwer, Dordrecht (1999b)
529. Kovaleva, A.: Response analysis of structures with motion limiters subjected to random loading. J. Structural Control 7, 191–202 (2000)
530. Kovaleva, A.S.: Stochastic dynamics of flexible systems with motion limiters. Nonlin. Dyn. 36, 313–327 (2004)
531. Kowalczyk, P., Di Bernardo, M., Champneys, A.R., Hogan, S.J., Homer, M., Kuznetsov, Y.A., Nordmark, A.P., Piiroinen, P.T.: Two–parameter non–smooth bifurcations of limit cycle: classification and open problems. Int. J. Bifur & Chaos 16(3), 601–629 (2006)
532. Köylüoğlu, H.U., Nielsen, S.R.K., Iwankiewicz, R.: Response and reliability of Poisson–driven systems by path integration. ASCE Journal of Engineering Mechanics 121(1), 117–130 (1995)
533. Kozol, J.E., Brach, R.M.: Two–dimensional vibratory impact and chaos. J. Sound & Vib. 148(2), 319–327 (1991)
534. Kozlov, V.V., Treshev, D.V.: Billiards: A Genetic Introduction on the Dynamics of Systems with Impacts, Ame.r Soc. Math., Providence, Rhod Island (1991)
535. Krechetnikov, R., Marsden, J.E.: On destabilizing effects of two fundamental non–conservative forces. Physica D 214, 25–32 (2006)

536. Krupenin, V.L.: On the theory of vibration–impact systems with distributed impact elements. Mech. Solids (Mekhanika Tverdogo Tela) 21(1), 51–60 (1986)
537. Krupenin, V.L.: Contribution of the theory of strongly nonlinear vibration guides. Sov. Mach. Sci. (Mashinovedenie) 1, 21–28 (1987)
538. Krupenin, V.L.: Transformation of vibrational modes of a string interacting with two straight extended obstacles. Sov. Phys. – Doklady 35(8), 743–745 (1990) (in Russian)
539. Krupenin, V.L.: On simulation of vibro–impact processes in systems with distributed impact elements. Mashinovedenie 3, 19–26 (1991) (in Russian)
540. Krupenin, V.L.: On calculation of resonance oscillations of flexible filament interacting with point limiter of stroke. Problemy Prochnosti i Nadezhnos'ti Mashin 2, 29–36 (1992) (in Russian)
541. Krupenin, V.L.: On vibrating impact processes transmission through mechanical filters. Problemy Prochnosti i Nadezhnos'ti Mashin 6, 22–28 (1993) (in Russian)
542. Krupenin, V.L.: On the analysis of vibration–impact processes in systems with a great deal of impact couples. In: Int. Congress on Bioceramics and the Human Body, vol. 2, pp. 97–105. Elsevier Science Publ. Ltd., Amsterdam (1994)
543. Krupenin, V.L.: A model for a highly nonlinear vibration–conductive medium with a distributed impact element. Physics–Doklady 40(8), 426–430 (1995a)
544. Krupenin, V.L.: Asymmetric vibrations in systems containing double obstacles. Physics–Doklady 40(3), 134–137 (1995b)
545. Krupenin, V.L.: Vibro–impact processes in systems with multiple impact pairs and distributed impact element. In: Babitsky, V.I. (ed.) Proc. Euromech Colloquium on Dynamics of Vibro–Impact Systems, Loughborough University, U.K, September 15–18, 1998, pp. 39–48. Springer, Berlin (1999)
546. Krupenin, V.L.: Periodic motions in an elastic systems family with boundary elements interacting through impacts. Problemy Mashinostraeniya i Nadezhnos'ti Mashin 3, 20–28 (2001) (in Russian)
547. Krupenin, V.L.: Description of dynamic effects accompanying vibrations of strings near T–beam stops. Doklady Physics 48(1), 46–50 (2003)
548. Krupenin, V.L.: Random collisions of an array structure with plane travel limiter. Problemy Mashinostraeniya i Nadezhnos'ti Mashin 4, 105–109 (2005a) (in Russian)
549. Krupenin, V.L.: On investigation of high nonlinear oscillations forms of vibroshock systems with distributed shock elements. Problemy Mashinostraeniya i Nadezhnos'ti Mashin 6, 31–38 (2005b) (in Russian)
550. Krupenin, V.L.: Vibrations of two–dimensional lattice structures in the presence of obstacles. Doklady Physics 51(1), 40–43 (2006)
551. Krupenin, V.L., Mel'nikova, G.V.: On simulation of vibro–impact processes in systems with distributed impact elements. Mashinovedenie 3, 19–26 (1991) (in Russian)
552. Kryzhevich, S.G.: Chaotic invariant sets of vibro–impact systems with one degree of freedom. Doklady Mathematics 74(2), 676–677 (2006)
553. Kryzhevich, S.G., Pliss, V.A.: Chaotic models of oscillation of a vibro–impact system. J. Appl. Math. Mech. 69, 13–26 (2005)
554. Kunert, A., Pfeiffer, F.: Stochastic modelling of chaotic rattle vibrations in gearboxes. Zeit. Ang. Math. Mech. (ZAMM) 70(4), 52–54 (1990a)

555. Kunert, A., Pfeiffer, F.: Stochastic model for rattling in gearboxes. In: Proc. Nonlin Dyn in Engineering Systems, pp. 173–180. Springer, Heidelberg (1990b)
556. Kunitsyn, A.L., Matveyev, M.V.: The Stability of a Class of Reversible Systems. J. Appl. Math. Mech. (PMM) 55(6), 780–788 (1991)
557. Kunitsyn, A.L., Muratov, A.S.: The stability of a class of quasi–autonomous periodic systems with internal resonance. J. Appl. Math. Mech. (PMM) 57(2), 247–255 (1993)
558. Kunze, M.: Non–smooth Dynamical Systems. Lecture Notes in Mathematics, vol. 1744. Springer, Heidelberg (2000)
559. Kurihara, C., Masuko, Y., Sakurai, A.: Sloshing impact pressure in roofed liquid tanks. In: Proc. ASME Pressure Vessel and Piping Conf, Fluid–Structure Vibration and Sloshing, New Orleans, LA, PVP, vol. 232, pp. 19–24 (1992)
560. Kuznetsov, Y.A.: Elements of Applied Bifurcation Theory. Springer, New York (1998)
561. Laggiard, E., Runkel, J., Stegemann, D.: One–dimensional bimodal model of vibration and impacting of instrument tubes in a boiling water reactor. Nuclear Sci. Eng. 115(1), 62–70 (1993)
562. Lamarque, C.H., Janin, O.: Modal analysis of mechanical systems with impact nonlinearities: Limitations to a modal superposition. J. Sound & Vib. 235(4), 567–609 (2000)
563. Lamba, H.: Regular, chaotic and unbounded behavior in the elastic impact oscillator. Physica D 82, 117–135 (1995)
564. Landa, P.S.: Vocal folds as a vibro–impact system. In: Babitsky, V.I. (ed.) Proc. Euromechanics Colloquium on Dynamics of Vibro–Impact Systems, Loughborough University, September 15–18, 1998, pp. 1–10. Springer, Heidelberg (1999)
565. Lansbury, A.N., Thompson, J.M.T., Stewart, H.B.: Basin corrosion in the twin–well Duffing oscillator: Two distinct bifurcation scenarios. Int. J. Bifurc. & Chaos 2, 505–532 (1992)
566. Laurenson, R.M., Trn, R.M.: Flutter analysis of missile control surfaces containing structural nonlinearities. AIAA Journal 18(10), 1245–1251 (1980)
567. Lauresen, T.A.: Computational Contact and Impact Mechanics: Fundamentals of Modeling Interfacial Phenomena in Nonlinear Finite Element Analysis. Springer, Berlin (2002)
568. Lazer, A.C., McKenna, P.J.: Periodic bouncing for a forced linear spring with obstacle. Differential– Integral Eq. 5, 165–172 (1992)
569. Lee, C.H., Byrne, K.O.: Impact statistics for a simple random rattling system. J. Sound & Vib. 119(3), 529–543 (1987)
570. Lee, C.M., Goverdovskiy, V.N.: Alternative vibration protecting systems for men–operators of transport machines: Modern level and prospects. J. Sound & Vib. 249, 635–647 (2002)
571. Lee, C.M., Goverdovskiy, V.N., Samoilenko, S.B.: Prediction of non–chaotic motion of the elastic system with small stiffness. J. Sound & Vib. 272, 643–655 (2004)
572. Lee, J.Y.: Motion behavior of impact oscillator. J. Marine Sci. & Tech. 13(2), 89–96 (2005)
573. Lee, J.Y., Nandi, A.K.: Signal processing of chaotic impacting series. IEE Colloquium (Digest) 393, 7/1–7/6 (1997)

574. Lee, J.Y., Nandi, A.K.: Blind deconvolution of impacting signals using higher–order statistics. Mech. Systems & Signal Proc. 12, 357–371 (1998)
575. Lee, J.Y., Yan, J.J.: Control of impact oscillator. Chaos, Solitons & Fractals 28(1), 136–142 (2006)
576. Lee, J.Y., Yan, J.J.: Position control of double–side impact oscillator. Mechanical Systems & Signal Processing 21(2), 1076–1083 (2007)
577. Leine, R.: Bifurcation in discontinuous mechanical systems of Filippov–type, Ph.D. Thesis, Teknische Universiteit Eindhoven, The Netherlands (2000)
578. Leine, R.I., Van Campen, D.H.: Bifurcation phenomena in non–smooth dynamical systems. European J. Mech. A/Solids 25, 595–616 (2006)
579. Leine, R.I., Glocker, C., Van Campen, D.H.: Nonlinear dynamics of the woodpecker toy. In: Proc. ASME Design Engineering Technical Conf., Pittsburgh, PA, vol. 6 C, pp. 2629–2637 (2001)
580. Leine, R.I., Nijmeijer, H.: Dynamics and Bifurcations of Non–smooth Mechanical Systems, 2nd edn. Lecture Notes in Applied and Computational Mechanics, vol. 18. Springer, Heidelberg (2006)
581. Leine, R.I., Van Campen, D.H., Glocker, C.: Nonlinear dynamics and modeling of various wooden toys with impact and friction. J. Vib. & Contr. 9(1,2), 25–78 (2003)
582. Lenci, S., Demeio, L., Petrini, M.: Response scenario and non–smooth features in the nonlinear dynamics of an impacting inverted pendulum. ASME J. Comput. & Nonlin. Dyn. 1, 56–64 (2006)
583. Lenci, S., Rega, G.: A procedure for reducing the chaotic response region in an impact mechanical system. Nonlin. Dyn. 15, 391–409 (1998)
584. Lenci, S., Rega, G.: Regular nonlinear dynamics and bifurcations of an impacting system under general periodic excitation. Nonlin. Dyn. 34, 249–268 (2003)
585. Lewis, A.D., Rogers, R.J.: Experimental and numerical study of forces during oblique impact. J. Sound & Vib. 125, 403–412 (1988)
586. Li, G.X., Païdoussis, M.P.: Impact phenomena of rotor–stator dynamical system. Nonlin. Dyn. 5, 53–70 (1994)
587. Li, K.: Experiments on the effect of an impact damper on a multiple–degree–of–freedom system. J. Vib. & Cont. 12(5), 445–464 (2006)
588. Li, K., Darby, A.P.: An experimental investigation into the use of a buffered impact damper. J. Sound & Vib. 291(3–5), 844–860 (2006)
589. Li, Q.X., Lu, Q.S.: Single rub–impacting periodic motions of a rigid constrained rotor system. Comm. Nonlin. Sci. & Numer. Simul. 5(4), 158–161 (2000)
590. Liberzon, D.: Switching in Systems and Control. Oxford University Press, Oxford (1999)
591. Lieber, P., Jensen, D.P.: An acceleration damper: Development and design, and some applications. Trans. ASME 67, 523–530 (1945)
592. Lieber, P., Tripp, F.: Experimental results on the acceleration damper. Rensselaer Polytechnic Institute Aeronautical Laboratory, Troy, New York, Report No. TR AE 5401 (1954)
593. Lifshits, P.S.: On the problem of forced vibration of a system impacting against a stop. Technical Physics, No. 6 (1952) (in Russian)
594. Lin, S.Q., Bapat, C.N.: Estimation of clearances and impact forces using vibro–impact response: sinusoidal excitation. J. Sound & Vib. 157(3), 485–513 (1992)

595. Lin, S.Q., Bapat, C.N.: Estimation of clearance and impact forces using vibro–impact response random excitation. J. Sound & Vib. 163(3), 407–421 (1993a)
596. Lin, S.Q., Bapat, C.N.: Extension of clearance and impact force estimation approaches to a beam–stop system. J. Sound & Vib. 163(3), 423–438 (1993b)
597. Lin, W., Qiao, N.: Nonlinear dynamics of a fluid–conveying curved pipe subjected to motion–limiting constraints and a harmonic excitation. J. Fluids & Struct. 24, 96–110 (2008)
598. Liu, A.Q., Wang, B., Choo, Y.S., Ong, K.S.: Effective design of bean bag as a vibro–impact damper. Shock & Vib. 7(6), 343–354 (2000)
599. Lo, C.C.: A cantilever beam chattering against a stop. J. Sound & Vib. 69(2), 245–255 (1980)
600. Long, Y., Nagaya, K., Niwa, H.: Vibration conveyance in spatial–curved tubes (Dynamic behaviors of a block in a vibrating spatial–curved tube). Nippon Kikai Gakkai Ronbunshu, C Hen/Trans JSME, Part C 59(558), 30–38 (1993)
601. Lu, L.Y., Lu, Z.H.: Periodicity of chaotic impact oscillators in Hausdorff phase spaces. J. Sound & Vib. 235(1), 105–116 (2000)
602. Lu, Q.S., Li, Q.X., Twizell, E.H.: The existence of periodic motions in rub–impact rotor systems. J. Sound & Vib. 264, 1127–1137 (2003)
603. Luck, J.M., Mehta, A.: Bouncing ball with a finite restitution: Chattering, locking, and chaos. Phys. Rev. E 48, 3988–3997 (1993)
604. Lugni, C., Brocchini, M., Faltinsen, O.M.: Wave impact loads: The role of the flip–through. Phys. Fluids 18(12), 122101–1–17 (2006)
605. Luo, A.C., Han, R.P.: Dynamics of a bouncing ball with a sinusoidally vibrating table revisited. Nonlin. Dyn. 10, 1–18 (1996)
606. Luo, A.C.J.: An unsymmetrical motion in a horizontal impact oscillator. ASME J. Vib. & Acoust 124(3), 420–426 (2002)
607. Luo, A.C.J.: Period–doubling induced chaotic motion in the LR model of a horizontal impact oscillator. Chaos, Solitons & Fractals 19(4), 823–839 (2004)
608. Luo, A.C.J.: Grazing and Chaos in a periodically forced, piecewise linear system. ASME J. Vib. & Acoust. 128, 28–34 (2006)
609. Luo, A.C.: A periodically forced, piecewise linear system. Part I: Local singularity and grazing bifurcation. Comm. Nonlin. Sci. & Numer. Simul. 12(3), 379–396 (2007a)
610. Luo, A.C.J.: A periodically forced, piecewise linear system, Part II: The fragmentation mechanism of strange attractors and grazing. Comm. Nonlin. Sci. & Numer. Simul. 12(6), 986–1004 (2007b)
611. Luo, G.W.: Hopf–flip bifurcations of vibratory systems with impacts. Nonlin. Anal: Real World Appl. 7, 1029–1041 (2006a)
612. Luo, G.W.: Dynamics of impacting forming machine. Int. J. Mech. Sci. 48(11), 1295–1313 (2006b)
613. Luo, G.W., Chu, Y.D., Zhang, Y.L., Zhang, J.G.: Double Neimark–Sacker bifurcation and torus bifurcation of a class of vibratory systems with symmetrical rigid stops. J. Sound & Vib. 298(1–2), 154–179 (2006f)
614. Luo, G.W., Gao, P., Yao, H.: Unusual routes from periodic motion to chaos for a vibro–impact system. Proc. SPIE – Int. Soc. Optical Eng. 4537, 453–456 (2001)
615. Luo, G.W., Lv, X.H.: Dynamics of a plastic–impact system with oscillatory and progressive motions. Int. J. Nonlin. Mech. 43, 100–110 (2008)

616. Luo, G.W., Xie, J.H.: Hopf bifurcation of a two–degree–of–freedom vibro–impact system. J. Sound & Vib. 213, 391–408 (1998)
617. Luo, G.W., Xie, J.H.: Bifurcations and Chaos in a system with impacts. Physica D 148, 183–200 (2001)
618. Luo, G.W., Xie, J.H.: Hopf bifurcations and chaos of a two–degree–of–freedom vibro–impact system in two strong resonance cases. Int. J. Nonlin. Mech. 37, 19–34 (2002)
619. Luo, G.W., Xie, J.H.: codimension–two bifurcation of periodic vibro–impact and chaos of a dual component system. Phys. Lett. A 313, 267–273 (2003)
620. Luo, G.W., Xie, J.H.: Stability of periodic motion, bifurcations and chaos of a two–degree–of–freedom vibratory system with symmetrical rigid stops. J. Sound & Vib. 273(3), 543–568 (2004)
621. Luo, G.W., Xie, J.H., Guo, S.H.L.: Periodic motions and global bifurcations of a two–degree–of–freedom system with plastic vibro–impact. J. Sound & Vib. 240, 837–858 (2001)
622. Luo, G.W., Xie, J.H., Zhu, A., Zhang, J.G.: Periodic motions and bifurcations of a vibro–impact system. Chaos, Solitons & Fractals 36(5), 1340–1347 (2008)
623. Luo, G.W., Yu, J.N., Yao, H.M., Chu, Y.D.: Periodic motions and bifurcations of a small vibro–impact pile driver. Gongcheng Lixue/Eng. Mech. 23(7), 105–129 (2006A)
624. Luo, G.W., Yu, J.N., Zhang, J.G.: Codimension two bifurcation and chaos of a vibro–impact forming machine associated with 1:2 resonance case. Acta Mechanica Sinica (English Series) 22(2), 185–198 (2006b)
625. Luo, G.W., Yu, J.N., Zhang, J.G.: Periodic impact motions and bifurcations of a dual component system. Nonlin Analys: real World Applic. 7, 813–828 (2006c)
626. Luo, G.W., Zhang, Y.L.: Analyses of impact motions of harmonically excited systems having rigid amplitude constraints. Int. J. Impact Eng. 34(11), 1883–1905 (2007)
627. Luo, G.W., Zhang, Y.L., Chu, Y.D., Zhang, J.G.: Codimension-two bifurcations of fixed points in a class of vibratory systems with symmetrical rigid stops. Nonlin. Analys: Real World Applic. 8, 1272–1292 (2007a)
628. Luo, G.W., Zhang, Y.L., Xie, J., Zhang, J.: Vibro–impact dynamics near a strong resonance point. Acta Mechanica Sinica/Lixue Xuebao 23(3), 329–341 (2007b)
629. Luo, G.W., Zhang, Y.L., Yu, J.N.: Dynamical behavior of vibro–impact machinery near a point of codimension–two bifurcation. J. sound & Vib. 292, 242–278 (2006d)
630. Luo, G.W., Zhang, Y.L., Xie, J.H., Zhang, J.G.: Periodic–impact motions and bifurcations of vibro–impact systems near 1:4 strong resonance point. Comm. Nonlin. Sci. & Numer. Simul. 13, 1002–1014 (2008)
631. Luo, G.W., Zhang, Y.L., Zhang, J.G.: Dynamical behavior of a class of vibratory systems with symmetrical rigid stops near the point of codimension-two bifurcation. J. Sound & Vib. 297, 17–36 (2006e)
632. Maezawa, S., Watanabe, T.: Steady impact vibrations in mechanical systems with broken–line collision characteristics. Nonlin. Vibr. Probl. 14, 473–500 (1973)
633. Maggio, G.M., Di Bernardo, M., Keneedy, M.P.: Non–smooth bifurcations in a piecewise–linear model of the colpitts oscillator. IEEE Trans. Circuits Syst. I 47(8), 1160–1177 (2000)

634. Mahinfala, M., Aagaah, M.R., Jazar, G.N., Mahmoudian, N.: Frequency response of vibration isolators with saturating spring elements. In: Proc. ASME 20th Biennial Conf. Mechanical Vibration and Noise, Long Beach, CA, pp. 1845–1854 (2005)
635. Makrides, G.A., Edelstein, W.S.: Finite element analysis of flow–induced instability of an elastic tube with a variable support. In: Proc. ASME Pressure Vessels and Piping Division, Sloshing, Pittsburgh, PA, PVP, vol. 145, pp. 41–46 (1988)
636. Manevich, L.I., Gendelman, O.V.: Oscillatory models of vibro–impact type for essentially nonlinear systems. Proc. IMech. E Part C: J. Mechanical Engineeringo Science 222, 2007–2043 (2008)
637. Mann, B.P., Carter, R.E., Hazra, S.S.: Experimental study of an impact oscillator with viscoelastic and Hertzian contact. Nonlin. Dyn. 50(3), 587–596 (2007)
638. Mansour, W.M., Teixeira Filho, D.R.: Impact dampers with Coulomb friction. J. Sound & Vib. 33(3), 247–265 (1974)
639. Mantha, S., Mongeau, L., Siegmund, T.: Dynamic digital image correlation of a dynamic physical model of the vocal folds. In: Proc. ASME Int. Mechanical Engineering Congress and Exposition, Orlando, FL, BED, vol. 57, pp. 77–78 (2005)
640. Mao, K., Wang, M.Y., Xu, Z., Chen, T.: Simulation and characterization of particle damping in transient vibrations. ASME J. Vibr. & Acoust. 126(2), 202–211 (2004)
641. Marghitu, D.B., Hurmuzlu, Y.: Nonlinear dynamics of an elastic rod with frictional impact. Nonlin. Dyn. 10, 187–201 (1996)
642. Marhadi, K.S., Kinra, V.K.: Particle impact damping: Effect of mass ratio, material and shape. J. Sound & Vib. 283(1), 433–448 (2005)
643. Marsden, C.C., Price, S.J.: The aeroelastic response of a wing section with a structural free-play nonlinearity: An experimental investigation. J. Fluids & Struct. 21, 257–276 (2005)
644. Marudachalam, K., Bursal, F.H.: Numerical study of an impact oscillator. In: Proc. ASME 15th Biennial Conf. Mech. Vibr. & Noise, Boston, MA, DE, vol. 84(3), Pt A/1, pp. 159–170 (1995)
645. Maruyama, S., Kato, T., Nagai, K.I., Yamagushi, T.: Experiments on chaotic vibrations of a cantilevered beam under vibro–impact. Nihon Kikai Gakkai Ronbunshu, C Hen/Trans. JSME, Part C 72(7), 2073–2079 (2006a) (in Japanese)
646. Maruyama, S., Nagai, K.I., Yamagushi, T., Kato, T.: Numerical analysis on chaotic vibrations of a cantilevered beam under vibro–impact. Nihon Kikai Gakkai Ronbunshu, C Hen/Trans. JSME, Part C 72(8), 2382–2389 (2006b) (in Japanese)
647. Mason, J., Homer, M., Wilson, R.E.: Mathematical models of gear rattle in rotor blower vacuum pumps. J. Sound & Vib. 308, 431–440 (2007)
648. Masri, S.F.: Analytical and Experimental Studies of Impact Dampers, Ph.D. Thesis, California Institute of Technology (1965)
649. Masri, S.F.: Effectiveness of two–particle impact dampers. J. Acoust Soc. America 41(6), 1553–1554 (1967a)
650. Masri, S.F.: Electric–analog studies of impact dampers. Experim. Mech. 7(2), 49–55 (1967b)

651. Masri, S.F.: Motion and stability of two–particle single–container impact dampers. ASME J. Appl. Mech. 34, 506–507 (1967c)
652. Masri, S.F.: Analytical and experimental studies of multiple–unit impact dampers. J. Acoust Soc. Amer. 45(5), 1111–1117 (1969)
653. Masri, S.F.: General motion of impact dampers. J. Acoust Soc. Amer. 47, 229–237 (1970a)
654. Masri, S.F.: Periodic excitation of multiple–unit impact dampers. J. Eng. Mech. Div. 96(5), 1195–1207 (1970b)
655. Masri, S.F.: Forced vibration of a class of nonlinear two–degree–of–freedom oscillators. Int. J. nonlinear Mech. 7, 663–674 (1972)
656. Masri, S.F.: Steady–state response of a multi–degree system with an impact damper. ASME J. Appl. Mech. 40(1), 127–132 (1973)
657. Masri, S.F.: Analytical and experimental studies of a dynamic system with a gap. ASME (Paper) No 77–DET–89, 26– 30 September, 7 (1977)
658. Masri, S.F.: Analytical and experimental studies of a dynamic system with a gap. ASME J. Mech. Design 100(3), 480–486 (1978)
659. Masri, S.F., Caughey, T.D.: On the stability of the impact damper. ASME J. Appl. Mech. 33, 586–592 (1966)
660. Masri, S.F., Ibrahim, A.M.: Stochastic excitation of a simple system with impact damper. Earthquake Eng. & Struct. Dyn. 1(4), 337–346 (1972)
661. Masri, S.F., Ibrahim, A.M.: Stochastic excitation of a simple system with impact damper. Earthquake Eng. & Struct. Dyn. 1, 337–346 (1973a)
662. Masri, S.F., Ibrahim, A.M.: Response of the impact damper to stationary random excitation. J. Acoust Soc. Amer. 53(1), 200–211 (1973b)
663. Masri, S.F., Mariamy, Y.A., Anderson, J.C.: Dynamic response of a beam with a geometric nonlinearity. ASME J. Appl. Mech. 48, 404–410 (1981)
664. Masri, S.F., Stott, S.J.: Random excitation of a nonlinear vibration neutralizer. ASME (Paper) No. 77–DET–94, September 26-30, p. 9 (1977)
665. Mathis, M.V., Fry, D.N., Robinson, J.C., Jones, J.E.: Neutron noise measurements to evaluate BWR–4 core modification to prevent instrument tube vibration. Trans. Am. Nucl. Soc. 23, 466 (1976)
666. Mehta, A., Luck, J.M.: Novel temporal behavior of a nonlinear dynamical system: The completely inelastic bouncing ball. Phys. Rev. Lett. 65, 393–396 (1990)
667. Meijaard, J.P.: A mechanism for the onset of chaos in mechanical systems with motion–limiting stops. Chaos, Solitons & Fractals 7(10), 1649–1658 (1996)
668. Mello, T.M., Tufillaro, N.B.: Strange attractors of a bouncing ball. Amer. J. Phys. 55(4), 316–320 (1987)
669. Mende, W., Herzel, H., Wernke, K.: Bifurcations and Chaos in newborn infant cries. Phys. Lett. A 145, 418–424 (1990)
670. Menini, L., Tornambe, A.: Control of (otherwise) uncontrollable linear mechanical systems through non–smooth impacts. Systems & Control Lett. 49(4), 311–322 (2003)
671. Metrikyn, V.S.: On the theory of vibro–impact devices with randomly varying parameters. Izvestiya Vysshikk Uchebnykh Zavedenii, Radiofizika 13, 4 (1970) (in Russian)
672. Mickens, R.E.: Oscillations in an $x^{4/3}$ potential. J. Sound & Vib. 246, 375–378 (2001)

673. Mickens, R.E.: Analysis of nonlinear oscillators having non–polynomial elastic terms. J. Sound & Vib. 255, 789–792 (2002)
674. Mickens, R.E.: Iteration method solutions for conservative and limit–cycle $x^{1/3}$ force oscillators. J. Sound & Vib. 292(3-5), 964–968 (2006)
675. Mickens, R.E.: Harmonic balance and iteration calculations of periodic solutions to $y + y^{-1} = 0$. J. Sound & Vib. 306(3–5), 968–972 (2007)
676. Mikhlin, Y.V.: Direct and inverse problem encountered in vibro–impact oscillations of a discrete system. J. Sound & Vib. 216, 227–250 (1998)
677. Mikhlin, Y.V., Reshetnikova, S.N.: Dynamical interaction of an elastic system essentially nonlinear absorber. J. Sound & Vib. 283(1–2), 91–120 (2005)
678. Mikhlin, Y.V., Reshetnikova, S.N.: Dynamical interaction of an elastic system and a vibro–impact absorber. Math. Probl. Eng., Article ID 37980 (2006)
679. Mikhlin, Y.V., Vakakis, A.F., Salenger, G.: Direct and inverse problems encountered in vibro–impact oscillations of a discrete system. J. Sound & Vib. 216(2), 227–250 (1998)
680. Milgram, J.H.: The motion of a fluid in a cylindrical container with a free surface following vertical impact. J. Fluid Mech. 37(3), 435–448 (1969)
681. Minowa, C.: Surface–sloshing behaviors of liquid storage tanks. In: Proc. ASME Pressure Vessels and Piping Conf., Sloshing and Fluid Structure Vibration, Honolulu, HI, PVP, vol. 157, pp. 165–171 (1989)
682. Minowa, C., Ogawa, N., Harada, I., Ma, D.C.: Sloshing roof impact tests of a rectangular tank. In: Proc. ASME Pressure Vess Piping Conf., Sloshing, Fluid–Structure Interaction and Struct Response Due to Shock and Impact Loads, Minneapolis, MN, PVP, vol. 272, pp. 13–21 (1994)
683. Mita, M., Arai, M., Tensaka, S., Kobayashi, D., Fujita, H.: A micromachined impact microactuator driven by electrostatic force. J. Microelectromechanical Systems 12, 37–41 (2003)
684. Modarres–Sadeghi, Y., Semler, C., Wadham–Gagnon, M., Païdoussis, M.P.: Dynamics of cantilevered pipes conveying fluid, Part 3: Three–dimensional dynamics in the presence of an end–mass. J. Fluids & Struct. 23, 589–603 (2007)
685. Moghisi, M., Squire, P.T.: Experimental investigation of the initial force of impact on a sphere striking a liquid surface. J. Fluid Mech. 108, 133–146 (1981)
686. Molenaar, J., De Weger, J.G., Van de Water, W.: Mapping of grazing–impact oscillators. Nonlinearity 14, 301–321 (2001)
687. Molin, B., Cointe, R., Fontaine, E.: On energy arguments applied to slamming force. In: Proc. 11th Int Workshop on Water Waves and Floating Bodies, Hamburg, Germany (1996)
688. Moon, F.C., Holmes, W., Khoury, P.: Symbol dynamic maps of spatial–temporal chaotic vibrations in a string of impact oscillations. Chaos 1(1), 65–68 (1991)
689. Moon, F.C., Li, G.X.: Experimental study of chaotic vibration in a pin–jointed space truss structure. AIAA Journal 28, 915–921 (1990)
690. Moon, F.C., Shaw, S.W.: Chaotic vibrations of a beam with nonlinear boundary conditions. Int. J. Nonlin. Mech. 18(6), 465–477 (1983)
691. Moore, D.B., Shaw, S.W.: The experimental response of an impacting pendulum system. Int. J. Nonlin. Mech. 25(1), 1–16 (1990)

692. Moore, J.J., Palazzolo, A.B., Gadangi, R., Nale, T.A., Klusman, S.A., Brown, G.V., Kascak, A.F.: Forced response analysis and application of impact dampers to rotor-dynamic vibration suppression in a cryogenic environment. ASME J. Vib. & Acoust. 117(3A), 300–310 (1995)
693. Moorthy, R.I.K., Kakodkar, A., Srirangarajan, H.R., Suryanarayan, S.: An assessment of the Newmark method for solving chaotic vibrations of impacting oscillators. Comp. & Struct. 49(4), 597–603 (1993)
694. Moreau, J.J.: Frictionless unilateral constraints and inelastic shocks. Comptes Rendus des Seances de l'Academie des Sciences, Serie II (Mecanique–Physique, Chimie, Sciences de la Terre, Sciences de l'Universe) 296(19), 1473–1476 (1983) (in French)
695. Morita, T., Yoshida, R., Okamoto, Y., Kurosawa, M.K., Higuchi, T.: A smooth impact rotation motor using a multi–layered torsional piezoelectric actuator. IEEE Trans. Ultrasonic Ferroelectrics & Frequency Control 46(6), 1439–1445 (1999)
696. Moshchuk, N.K., Ibrahim, R.A., Khasminskii, R.Z., Chow, P.L.: Ship Capsizing in Random Sea Waves and the Mathematical Pendulum. In: Naess, A., Krenk, S. (eds.) Proc. IUTAM Symp on Advances in Nonlinear Stochastic Mechanics, Trondheim, Norway, July 3–7, 1995, pp. 299–309. Kluwer, Dordrecht (1995)
697. Mosekilde, E., Zhusubalyev, Z.: Bifurcations and Chaos in Piecewise–Smooth Dynamical Systems. World Scientific, Singapore (2003)
698. Mott, J.E., Robinson, J.C., Fry, D.N., Brackin, M.P.: Detection of impacts of instrument tubes against channel boxes in BWR–4s using neutron noise analysis. Trans. Am. Nucl. Soc. 23, 465 (1976)
699. Mureithi, N., Itô, T., Nakamura, T.: Identification of fluidelastic instability under conditions of turbulence and nonlinear tube supports. In: Proc. ASME Pressure Vessel and Piping Conf., Montreal, Canada, Flow–Induced Vibration, PVP, vol. 318, pp. 19–24 (1996)
700. Murphy, K.D., Morrison, T.M.: Grazing instabilities and post–bifurcation behavior in an impacting string. J. Acoust Soc. Amer. 111(2), 884–892 (2002)
701. Muszynska, A.: Rotor–to–stationary element rub–related vibration phenomena in rotating machinery. Shock & Vib. Dig. 21, 3–11 (1989)
702. Muszynska, A., Bently, D.E., Franklin, W.D., Hayashida, R.D., Kingsley, L.M.: Influence of Rubbing on Rotor Dynamics, Part 1. NASA, Washington, DC Report: NAS 1.26:183648–PT–1, 523 pages (1989a)
703. Muszynska, A., Bently, D.E., Franklin, W.D., Hayashida, R.D., Kingsley, L.M.: Influence of Rubbing on Rotor Dynamics, Part 2. NASA, Washington, DC Report: NAS 1.26:183649–PT–2, 202 pages (1989b)
704. Naess, A., Johnsen, J.M.: Response statistics of nonlinear, compliant offshore structures by the path integral solution method. Probabilistic Engineering Mechanics 8(2), 91–106 (1993)
705. Naess, A., Moe, V.: Efficient Path Integration Methods for Nonlinear Dynamic Systems. Probabilistic Engineering Mechanics 15(2), 221–231 (2000)
706. Nagahiro, S., Hayakawa, Y.: Theoretical and numerical approach to magic angle of stone skipping. Phys. Rev. Lett. 94, 174501–(1–4) (2005)

707. Nakamura, T., Fujita, K.: Analysis of impact–vibration characteristics at the support of tube by modal method (Application of constraint mode method and the comparison with mode superposition method). Nippon Kikai Gakkai Ronbunshu, C Hen/Trans JSME, Part C 55(516), 1878–1883 (1989) (in Japanese)
708. Nakamura, T., Fujita, K.: Study on the characteristics of fluid–elastic vibration of a tube array. Nippon Kikai Gakkai Ronbunshu, C Hen/Trans JSME, Part C 56(527), 1680–1685 (1990) (2nd Report. Vibration during unstable conditions and impact analysis) (in Japanese)
709. Nakamura, T., Fujita, K., Kawanishi, K., Yamaguchi, N., Tsuge, A.: Study on the vibrational characteristics of a tube array caused by two–phase flow, 1: Random vibration. ASME J. Pressure Vessel Techn. 114(4), 472–478 (1992)
710. Namachchivaya, N.S., Park, J.H.: Stochastic dynamics of impact oscillators. ASME J. Appl. Mech. 72, 862–870 (2005)
711. Narimani, A., Golnaraghi, M.F.: Parameter optimizing for piecewise linear isolator. In: Proc. ASME Appl. Mech. Division 2004, Anaheim, California, USA, vol. 255, pp. 257–261 (2004)
712. Narimani, A., Golnaraghi, M.F., Jazar, G.N.: Frequency response of a piecewise linear vibration isolator. J. Vib. & Cont. 10, 1775–1794 (2004)
713. Nataraj, C., Nelson, H.D.: Periodic solutions in rotor dynamic systems with nonlinear supports: A general approach. In: Proc. ASME 11th Biennial Conf. Mechanical Vibration and Noise, Boston, MA, DE, vol. 3, pp. 51–58 (1987)
714. Natsiavas, S.: Periodic response and stability of oscillators with symmetric trilinear restoring force. J. Sound & Vib. 134(2), 315–331 (1989)
715. Natsiavas, S.: On the dynamics of oscillators with bi–linear damping and stiffness. Int. J. Nonlin. Mech. 25(5), 535–554 (1990a)
716. Natsiavas, S.: Stability and bifurcation analysis for oscillators with motion limiting constraints. J. Sound & Vib. 141(1), 97–102 (1990b)
717. Natsiavas, S.: Steady state oscillations and stability of nonlinear dynamic vibration absorbers. J. Sound & Vib. 156(2), 227–245 (1992)
718. Natsiavas, S.: Dynamics of multiple–degree–of–freedom oscillators with colliding components. J. Sound & Vib. 165(3), 439–453 (1993)
719. Natsiavas, S., Gonzalez, H.: Vibration of harmonically excited oscillators with asymmetric constraints. ASME J. Appl. Mech. 59, 284–290 (1992)
720. Nayak, P.R.: Contact Vibrations. J. Sound & Vib. 22, 297–322 (1972)
721. Nayeri, R.D., Masri, S.F., Caffrey, J.P.: Studies of the performance of multi–unit impact dampers under stochastic excitation. ASME J. Vib. & Acoust. 129(2), 239–251 (2007)
722. Nayfeh, A.H., Balachandran, B.: Applied Nonlinear Dynamics: Analytical, Computational, and Experimental Methods. Wiley Interscience, New York
723. Nelson, H.D., Meacham, W.L., Fleming, D.P., Kascak, A.F.: Nonlinear analysis of rotor–casing systems using component mode synthesis. J. Sound & Vib. 105, 267–294 (1974)
724. Neumann, N., Sattel, T.: Set–oriented numerical analysis of a vibro–impact drilling system with several contact interfaces. J. Sound & Vib. 308(3–5), 831–844 (2007)
725. Nevel, D.E.: Iceberg impact forces. In: Proceeding of the Int Association of Hydraulic Engineering and Research Symp. (IAHR) on Ice, Iowa City, vol. 3, pp. 345–365 (1986)

726. Nguyen, D.T., Noah, S.T., Kettleborough, C.F.: Impact behavior of an oscillator with limiting stops, Part I: A parametric study. J. Sound & Vib. 109(2), 293–307 (1986a)
727. Nguyen, D.T., Noah, S.T., Kettleborough, C.F.: Impact behavior of an oscillator with limiting stops. II: Dimensionless design parameters. J. Sound & Vib. 109(2), 309–325 (1986b)
728. Nguyan, V.D., Woo, K.C.: Nonlinear dynamic response of new electro–vibro–impact system. Rapid Communication, J. Sound & Vib. 310, 769–775 (2008)
729. Nguyan, V.D., Woo, K.C., Pavlovskaia, E.: Experimental study and mathematical modeling of a new vibro–impact moling device. Int. J. Nonlin. Mech. 43(6), 542–550 (2008)
730. Nigm, M.M., Shabana, A.A.: Effect of an impact damper on a multi–degree of freedom system. J. Sound & Vib. 89(4), 541–557 (1983)
731. Nordmark, A.B.: Non–periodic motion caused by grazing incidence in an impact oscillator. J. Sound & Vib. 145, 279–297 (1991)
732. Nordmark, A.B.: Grazing conditions and chaos in impacting systems, Ph.D. Thesis, Royal Institute of Technology, Stockholm, Sweden (1992a)
733. Nordmark, A.B.: Effects due to low velocity impact in mechanical oscillators. Int. J. Bifurc. & Chaos 2(3), 597–605 (1992b)
734. Nordmark, A.B.: Existence of periodic solutions in grazing bifurcations of impacting mechanical oscillators. J. Tech. Phys. 37(3-4), 531–534 (1996)
735. Nordmark, A.B.: Universal limit mapping in grazing bifurcations. Phys. Rev. E 55, 266–270 (1997)
736. Nordmark, A.B.: Existence of periodic orbits in grazing bifurcations of impacting mechanical oscillators. Nonlinearity 14, 1517–1542 (2001)
737. Nordmark, A.B.: Discontinuity mappings for vector fields with higher order continuity. Dynamic Systems 17(4), 359–376 (2002)
738. Nordmark, A., Dankowicz, H., Champneys, A.: non–smooth Bifurcations in Systems with Impact and Friction; I. Discontinuities in the Impact Law. Int. J. Nonlin. Mech (in press, 2008)
739. Nordmark, A.B., Kowalczyk, P.: A codimension–two scenario of sliding solutions in grazing–sliding bifurcations. Nonlinearity 19(1), 1–26 (2006)
740. Nucera, F., Vakakis, A.F., McFarland, D.M., Bergman, L.A., Kerschen, G.: Targeted energy transfers in vibro–impact oscillators for seismic mitigation. Nonlin. Dyn. 50(3), 651–677 (2007)
741. Nusse, H.E., Ott, E., Yorke, J.A.: Border–collision bifurcations: an explanation for observed bifurcation phenomena. Phys. Rev. E (Statistical Physics, Plasmas, Fluids, and Related Interdisciplinary Topics) 49(2), 1073–1076 (1994)
742. Nusse, H.E., Yorke, J.A.: Border–collision bifurcations including period two to period three for piecewise smooth systems. Physica D 57(1-2), 39–57 (1992)
743. Nusse, H.E., Yorke, J.A.: Border–collision bifurcations for piecewise smooth one–dimensional maps. Int. J. Bifurc. & Chaos 5, 189–207 (1995)
744. Ochi, M.K., Motter, L.E.: A method to estimate the slamming characteristics for ship design. Marine Techn. 8, 219–232 (1971)
745. Oestreich, M., Hinrichs, N., Popp, K.: Signal and model based analysis of systems with friction and impacts. In: Ferri, A.A., Flower, G.T., Sinha, S.C. (eds.) Proc. ASME Symp. on Elasto–Impact and Friction in Dynamic Systems, Atlanta, GA, DE, vol. 90, pp. 1–7 (1996)

746. Oestreich, M., Hinrichs, N., Popp, K., Budd, C.J.: Analysis and experimental investigation of an impact oscillator. In: ASME Proc. 16th Biennial Conf. Mechanical Vibrations and Noise, Sacramento, CA (1997)
747. Ogawa, K., Ide, T., Saitou, T.: Application of impact mass damper to a cable–stayed bridge pylon. J. Wind Eng. & Indust. Aerodyn. 72(1–3), 301–312 (1997)
748. Ohnishi, M., Inaba, N.: A singular bifurcation into instant chaos in a piecewise–linear circuit. IEEE Trans. Circuits & Systems I: Fundamental Theory & Applications 41(6), 433–442 (1994)
749. Oledzki, A.: A new kind of impact damper–from simulation to real design. Mechanism & Mach. Theory 16(3), 247–253 (1981)
750. Oledzki, A.A., Siwicki, I., Wisniewski, J.: Impact dampers in application for tube, rod and rope structures. Mechanism & Mach. Theory 34(2), 243–253 (1999)
751. Olson, S.E.: An analytical particle damping model. J. Sound & Vib. 264(5), 1155–1166 (2003)
752. Oppenheimer, C.H., Dubowsky, S.: A methodology for predicting impact–induced acoustic noise in machine systems. J. Sound & Vib. 266(5), 1025–1051 (2003)
753. Orzechowski, J., Murphree, M., Haggerty, P.: Use of nonlinear isolation mounts to reduce vehicle after shake. In: Proceedings – National Conf. Noise Control Engineering, Bellevue, Washington, USA, vol. 1, pp. 503–508 (1996)
754. Osakue, E., Rogers, R.J.: An experimental study of friction during planar elastic impact. In: Proc. ASME Pressure Vessels and Piping Division, Flow Induced Vibration, Atlanta, GA, vol. 420 II, pp. 11–20 (2001)
755. Ostasavichyus, V.V., Raqulskis, K.M., Gaidis, R.D.: Study of free decaying vibrations in vibro–impact systems with distributed parameters. Vib. Eng. (English Transl. Vibrotekhnika) 3(1), 139–147 (1989a)
756. Ostasavichyus, V.V., Raqulskis, K.M., Gaidis, R.D.: Study of vibro–impact systems with distributed parameters under harmonic excitation. Vib. Eng. (English Transl. Vibrotekhnika) 3(3), 425–437 (1989b)
757. Ostasavichyus, V.V., Povilaytis, S.Y., Raqulskis, K.M.: Dynamics of vibro–impact plate systems controlled by a magnetic field. Vib. Eng. (English Transl. Vibrotekhnika) 3(4), 531–538 (1989c)
758. Ott, E., Grebogi, C., Yorke, J.A.: Controlling chaos. Phys. Rev. Lett. 11, 1196–1199 (1990)
759. Padmanabhan, C., Barlow, R.C., Rook, T.E., Singh, R.: Computational issues associated with gear rattle analysis. ASME J. Mech. Design 117, 185–192 (1995)
760. Padmanabhan, C., Singh, R.: Dynamics of a piecewise nonlinear system subject to dual harmonic excitation using parametric continuation. J. Sound & Vib. 184(5), 767–799 (1995)
761. Padovan, J., Choy, F.K., Batur, C., Canilang, L.: Seismic induced impeller/blade rubs in rotating power plant components. In: Proc. ASME Pressure Vessels and Piping Division, Seismic Engineering: Recent Advances in Design, Analysis, Testing and Qualification Methods, San Diego, CA, PVP, vol. 127, pp. 21–27 (1987)
762. Paget, A.L.: Vibration in steam turbine buckets and damping by impacts. Engineering (London) 143, 305–307 (1937)

763. Païdoussis, M.P.: Flow–induced vibrations in nuclear reactors and heat exchangers: Practical experiences and state of knowledge. In: Naudascher, E., Rockwell, D. (eds.) Practical Experiences with Flow–Induced Vibrations, pp. 1–81. Springer, Berlin (1980)
764. Païdoussis, M.P.: Flow–induced instabilities of cylindrical structures. ASME Appl. Mech. Rev. 40, 163–175 (1987)
765. Païdoussis, M.P.: Fluid–Structure Interactions: Slender Structures and Axial Flow, vol. 1. Academic Press, London (1998)
766. Païdoussis, M.P.: The canonical problem of the fluid–conveying pipe and radiation of the knowledge gained to other dynamics problems across Appl. Mech. J. Sound & Vib. 310, 462–492 (2008)
767. Païdoussis, M.P., Cusumano, J.P., Copeland, G.S.: Low–dimensional chaos in a flexible tube conveying fluid. ASME J. Appl. Mech. 59(1), 196–205 (1992)
768. Païdoussis, M.P., Li, G.X.: Cross–flow–induced chaotic vibrations of hat–exchanger tubes impacting on loose supports. J. Sound & Vib. 152(2), 305–326 (1992)
769. Païdoussis, M.P., Li, G.X., Moon, F.C.: Chaotic oscillations of the autonomous system of a constrained pipe conveying fluid. J. Sound & Vib. 135(1), 1–19 (1989)
770. Païdoussis, M.P., Li, G.X., Rand, R.H.: Chaotic oscillations of a constrained pipe conveying fluid: Comparison between simulation, analysis, and experiment. ASME J. Appl. Mech. 58, 559–565 (1991)
771. Païdoussis, M.P., Moon, F.C.: Nonlinear and chaotic fluidelastic vibrations of a flexible pipe conveying fluid. J. Fluids & Struct. 3, 567–591 (1988)
772. Païdoussis, M.P., Semler, C.: Nonlinear and chaotic oscillations of a constrained pipe conveying fluid: A full nonlinear analysis. Nonlin. Dyn. 4, 655–670 (1993a)
773. Païdoussis, M.P., Semler, C.: Nonlinear dynamics of a fluid–conveying cantilevered pipe with an intermediate spring support. J. Fluids & Struct. 7, 269–298 (1993b)
774. Païdoussis, M.P., Semler, C., Wadham–Gagnon, M., Saaid, S.: Dynamics of cantilevered pipes conveying fluid, Part 2: Dynamics of the system with intermediate spring support. J. Fluids & Struct. 23, 569–587 (2007)
775. Pang, C., Popplewell, N., Semercigil, S.E.: An overview of a bean bag dampers effectiveness. J. Sound & Vib. 133(2), 359–363 (1989)
776. Panossian, H.V.: Structural damping/acoustic attenuation optimization via Nopd. JANNAF Propulsion Meeting (1990)
777. Panossian, H.V.: An overview of Nopd: A passive damping technique. Shock & Vib. 1(6), 4–10 (1991a)
778. Panossian, H.V.: Nonobstructive particle damping (Nopd) performance under compaction forces. Mach. Dyn. Element. Vib. 36, 17–20 (1991b)
779. Panossian, H.V.: Structural damping enhancement via non–obstructive particle damping technique. ASME J. Vib. & Acoust 114(1), 101–105 (1992)
780. Paoli, L.A.: An existence result for vibrations with unilateral constraints: case of a non–smooth set of constraints. Math. Models & Methods Appl. Sci. 10(6), 815–831 (2000)
781. Paoli, L.: Time discretization of vibro–impact. Phil. Trans. Royal Soc London, Series A (Math., Phys. & Eng. Sci.) 359(1789), 2405–2428 (2001)
782. Paoli, L.: An existence result for non–smooth vibro–impact problems. J. Diff. Eq. 211, 247–281 (2005)

783. Paoli, L., Schatzman, M.: Resonance in impact problems. Math. & Comp. Modeling 28, 385–406 (1998)
784. Paoli, L., Schatzman, M.: Penalty approximation for dynamical systems submitted to multiple non–smooth constraints. Multibody System Dyn. 8(3), 347–366 (2002)
785. Paoli, L., Schatzman, M.: Numerical simulation of the dynamics of an impacting bar. Comp. Methods Appl. Mech. & Eng. 196(29–30), 2839–2851 (2007)
786. Papalou, A., Masri, S.F.: Response of impact dampers with granular materials under random excitation. Earthquake Eng. & Struct. Dyn. 25(3), 253–267 (1996a)
787. Papalou, A., Masri, S.F.: Performance of particle dampers under random excitation. ASME J. Vib. & Acoust 118(4), 614–621 (1996b)
788. Papalou, A., Masri, S.F.: An experimental investigation of particle dampers under harmonic excitation. J. Vib. & Cont. 4, 361–379 (1998)
789. Park, S.W., Lee, J.K., Oh, S.H., Song, M.J., Kwin, S.M.: Whipping Analysis of Ship Hulls Considering Slamming Impact Loads. In: Proc. Int. Offshore and Polar Eng. Conf., Honolulu, HI, pp. 2799–2805 (2003)
790. Park, W.H.: Mass–spring–damper response to repetitive impact. ASME J. Eng. Indust. 89, 587–596 (1967)
791. Parui, S., Banerjee, S.: Border collision bifurcations at the change of state–space Dimension. Chaos 12, 1054–1069 (2002)
792. Pascal, M.: Dynamics and stability of a two–degree–of–freedom oscillator with an elastic stop. J. Comput. Nonlin. Dyn. 1(1), 94–102 (2006)
793. Patrick, S., Jazar, G.N.: Frequency response analysis of piecewise nonlinear vibration isolator. In: Proc. ASME 20th Biennial Conf. Mechanical Vibration and Noise, Long Beach, CA, USA, vol. 1B, pp. 917–924 (2005a)
794. Patrick, S., Jazar, G.N.: Stability analysis of a piecewise nonlinear vibration isolator. In: Proc. ASME Design Engineering Division, Orlando, FL, USA, vol. 118B(2), pp. 1115–1123 (2005b)
795. Paulino, M., Antunes, J., Izquierddo, P.: Remote identification of impact forces on loosely supported tubes: Analysis of multi–supported systems. ASME J. Pressure Vessel Techn. 121(1), 61–70 (1999)
796. Pavlovskaia, E., Wiercigroch, M.: Modeling of vibro–impact system driven by beat frequency. Int. J. Mech. Sci. 45, 623–641 (2003a)
797. Pavlovskaia, E., Wiercigroch, M.: Periodic solution finder for an impact oscillator with a drift. J. Sound & Vib. 267(4), 893–911 (2003b)
798. Pavlovskaia, E., Wiercigroch, M.: Analytical drift reconstruction for visco–elastic impact oscillators operating in periodic and chaotic regimes. Chaos, Solitons & Fractals 19(1), 151–161 (2004)
799. Pavlovskaia, E., Wiercigroch, M., Grebogi, C.: Modeling of an impact system with a drift. Phys. Rev. E (Statistical, Nonlinear, and Soft. Matter Physics) 64(5), 056224/1–9 (2001)
800. Pavlovskaia, E., Wiercigroch, M., Grebogi, C.: Two–dimensional map for impact oscillator with drift. Phys. Rev. E 70(3–2), 036201–1–036201–10 (2004)
801. Pavlovskaia, E., Wiercigroch, M., Woo, K.C., Rodger, A.A.: Modeling of ground moling dynamics by an impact oscillator with a frictional slider. Meccanica 38, 85–97 (2003)
802. Payr, M., Glockner, C.: Oblique frictional impact of a bar: Analysis and comparison of different impact laws. Nonlin. Dyn. 41, 361–383 (2005)

803. Peng, Z., He, Y., Lu, Q., Chu, F.: Feature extraction of the rub–impact rotor system by means of wavelet analysis. J. Sound & Vib. 259(4), 1000–1010 (2003)
804. Pereira, M.S., Nikravesh, P.: Impact dynamics of multibody mechanical systems with frictional contact using joint coordinates and canonical equation of motion. In: Proc. NATO–ASI Conf. Computer Aided Analysis of Rigid and Flexible Mechanical Systems, Troia, Portugal, pp. 505–526 (1993)
805. Perret–Liaudet, J.: Subharmonic resonance of order two on a sphere–plane contact. Comptes Rendus de l'Academie des Sciences, Serie II (Mecanique–Physique–Chimie–Astronomie) 325, 443–448 (1997) (in French)
806. Perret–Liaudet, J.: Superharmonic resonance of order two on a sphere–plane contact. Comptes Rendus de l'Academie des Sciences, Serie II (Mecanique–Physique–Chimie–Astronomie) 326, 787–792 (1998) (in French)
807. Perret–Liaudet, J., Rigaud, E.: Experiments and numerical results on nonlinear vibrations of an impacting Hertzian contact. Part 2. Random excitation. J. Sound & Vib. 265(2), 309–327 (2003)
808. Perret–Liaudet, J., Rigaud, E.: Response of an impacting Hertzian contact to an order–2 subharmonic excitation: Theory and experiments. J. Sound & Vib. 296(1–2), 319–333 (2006)
809. Peterka, F.: Simulation of vibro–impact systems motion by use of analogue computer. Automatizace 9, 235–237 (1965)
810. Peterka, F.: Analysis of the periodic motion of impact dampers. Rev. Roum. Sci. Tech., Ser. Mec. Appl. 16(4), 875–886 (1971)
811. Peterka, F.: Laws of impact motion of mechanical system with one degree of freedom, Part I: Theoretical analysis of n–multiple (1/n)–impact motions. Acta Technica. ČSAV 4, 462–473 (1974a)
812. Peterka, F.: Contribution to the investigation of impact dampers with free additional mass. Zagadnienia Drgan Nieliniowych/Nonlinear Vibration Problems, 315–325 (1974b)
813. Peterka, F.: Existence and stability of self–excited vibrations with impacts. Mechanism & Mach. Theory 13(1), 75–83 (1978)
814. Peterka, F.: Flow–induced impact oscillation of a spherical pendulum. J. Fluids & Struct. 5, 627–650 (1991)
815. Peterka, F.: Dynamics of impact motions of beam between stops and of a spherical pendulum in a channel with a flow of liquid. Adv. Mech. 15(3–4), 93–121 (1992a)
816. Peterka, F.: Mathematical model of bi–planar vibration of beams with oblique impacts. In: ASME WAM Proc. 3rd Int. Symp. on Fluid–Induced Vibration and Noise (FIVN), Anaheim, California, vol. 5, pp. 85–97 (1992b)
817. Peterka, F.: Impact vibration of a tube in the channel with axial flow. In: Proc. 2nd Int. Conf. EAHE, Pilsen, Czech Republic, pp. 334–339 (1994)
818. Peterka, F.: Impact interaction of two heat exchanger tubes. In: Bearman, P.W. (ed.) Proc. 6th Int. Conf. Flow–Induced Vibration, London, Belkema, Rotterdam, pp. 393–400 (1995)
819. Peterka, F.: Bifurcations and transition phenomena in an impact oscillator. Chaos, Solitons & Fractals 7(10), 1635–1647 (1996a)
820. Peterka, F.: Dynamics of mechanical systems with impacts and dry friction. In: Půst, L., Peterka, F. (eds.) Proc. EUROMECH–2nd European Nonlinear Oscillations Conf., vol. 1, pp. 33–44 (1996b)

821. Peterka, F.: Dynamics of the impact oscillator. In: Proc. IUTAM Symp. CHAOS 1997 Application of Nonlinear and Chaotic Dynamics in Mechanics, Ithaca, New York (1997)
822. Peterka, F.: Dynamics of oscillator with soft impacts. In: Proc. ASME 18th Biennial Conf. Mechanical Vibration and Noise, Pittsburgh, PA, September 9–12, vol. 6C, pp. 2639–2645 (2001a)
823. Peterka, F.: Application of double impact oscillator in forming. In: Proc. ASME 15th Reliability, Stress Analysis, and Failure Prevention Conf. and Int Issues in Engineering Design, Pittsburg, PA, pp. 175–182 (2001b)
824. Peterka, F.: Behavior of impact oscillator with soft and preloaded stop. Chaos, Solitons & Fractals 18(1), 79–88 (2003)
825. Peterka, F., Blazejczyk–Okolewska, B.: Some aspects of the dynamical behavior of the impact damper. J. Vib. & Cont. 11(4), 459–479 (2005)
826. Peterka, F., Cipera, S.: Transition to chaos and basins of attraction in the impact oscillator. In: Proc. 1997 1st Int. Conf. Control of Oscillations and Chaos, St. Petersburg, Russia, vol. 3, pt. 3, pp. 512–515 (1997)
827. Peterka, F., Formanek, P.: Simulation of motion with strong nonlinearities. In: Proc. 1st Joint Conf. of Int. Simulation, San Diego, CA, August 22–25, pp. 137–141 (1994)
828. Peterka, F., Kotera, T.: Laws of impact motion of mechanical systems with one degree of freedom – 5. Regions of existence and stability and domains of attraction of different kinds of impact motion. Acta. Technica. CSAV (Ceskoslovensk Akademie Ved.) 27(1), 92–117 (1982)
829. Peterka, F., Kotera, T.: Different ways from periodic to chaotic motion in mechanical vibro–impact systems. In: Proc. 9th World Congress of the Theory of Machines and Mechanisms, Politecnico of Milano, Italy, August 30-September 2 (1995)
830. Peterka, F., Kotera, T., Cipera, S.: Explanation of appearance and characteristics of intermittency chaos of the impact oscillator. Chaos, Solitons & Fractals 19(5), 1251–1259 (2004)
831. Peterka, F., Szollos, O.: Simulation of two–mass–system motion with plastic impacts and dry friction. In: Proc. European Simulation MultiConf: Modelling and Simulation (ESM 1995), Prague, Czech Republic, June 5–7, 1995, pp. 257–258 (1995)
832. Peterka, F., Szollos, O.: The stability analysis of a symmetric two–degree–of–freedom system with impacts. In: EUROMECH 2nd European Nonlinear Oscillation Conf., Prague, vol. 1 (1996)
833. Peterka, F., Vacik, J.: Laws of impact motion of mechanical systems with one degree of freedom em dash 3: Statistical characteristics of beat motions. Acta Technica ČSAV (Ceskoslovensk Akademie Ved) 2(2), 161–184 (1981)
834. Peterka, F., Vacik, J.: Hybrid simulation of transversal vibrations of a beam between stops. Automatizace 25(4), 109–113 (1982) (in Czech)
835. Peterka, F., Vacik, J.: Transition to chaotic motion in mechanical systems with impacts. J. Sound & Vib. 154(1), 95–115 (1992)
836. Pfeiffer, F.: Mechanische systems mit unstetigen Ubergängen. Ing.–Archiv. 54(3), 232–240 (1984)
837. Pfeiffer, F., Glocker, C.: Multibody Dynamics with Unilateral Contacts. Wiley, New York (1996)
838. Pfeiffer, F., Glocker, C.: Contact in multibody systems. J. Appl. Math. Mech. 64(5), 773–782 (2000)

839. Pfeiffer, F., Kunert, A.: Rattling models from deterministic to stochastic processes. Nonlin. Dyn. 1, 63–74 (1990)
840. Piccoli, H.C., Weber, H.I.: Experimental observation of chaotic motion in a rotor with rubbing. Nonlin. Dyn. 16, 55–70 (1998)
841. Piiroinen, P.T., Budd, C.J.: Corner bifurcations in non–smoothly forced impact oscillators. Physica D 220(2), 127–145 (2006)
842. Piiroinen, P.T., Virgin, L.N., Champneys, A.R.: Chaos and period–adding; experimental and numerical verification of the grazing bifurcation. J. Nonlin. Sci. 14(4), 383–404 (2004)
843. Pilipchuk, V.N.: Analytical study of vibrating systems with strong nonlinearities by employing saw–tooth time transformations. J. Sound & Vib. 192, 43–64 (1996)
844. Pilipchuk, V.N.: Application of special non–smooth temporal transformations to linear and nonlinear systems under discontinuous and impulsive excitation. Nonlin. Dyn. 18(3), 203–234 (1999a)
845. Pilipchuk, V.N.: An explicit form general solution for oscillators with a non–smooth restoring force, $x + sign(x)f(x) = 0$. J. Sound & Vib. 226(4), 795–798 (1999b)
846. Pilipchuk, V.N.: Non–smooth spatio–temporal transformation for impulsively forced oscillators with rigid barriers. J. Sound & Vib. 237(5), 915–919 (2000)
847. Pilipchuk, V.N.: Impact modes in discrete vibrating systems with rigid barriers. Int. J. Nonlin. Mech. 36(6), 999–1012 (2001)
848. Pilipchuk, V.N.: Some remarks on non–smooth transformations of space and time for vibrating systems with rigid barriers. J. Appl. Math. & Mech. (PMM) 66(1), 33–40 (2002)
849. Pilipchuk, V.N.: Oscillators with a generalized power–form elastic term. J. Sound & Vib. 270(1–2), 470–472 (2004)
850. Pilipchuk, V.N.: Strongly nonlinear vibrations of damped oscillators with two non–smooth limits. J. Sound & Vib. 302(1-2), 398–402 (2007)
851. Pilipchuk, V.N.: Transient mode localization in coupled strongly nonlinear exactly solvable oscillators. Nonlin. Dyn. 51(1–2), 245–258 (2008)
852. Pilipchuk, V.N., Ibrahim, R.A.: The dynamics of a nonlinear system simulating liquid sloshing impact in moving structures. J. Sound & Vib. 205(5), 593–615 (1997)
853. Pilipchuk, V.N., Ibrahim, R.A.: Application of the Lie group transformations to nonlinear dynamical systems. ASME J. Appl. Mech. 66(2), 439–447 (1999)
854. Pilipchuk, V.N., Ibrahim, R.A.: Dynamics of a two–pendulum model with impact interaction and an elastic support. Nonlin. Dyn. 21, 221–247 (2000)
855. Pilipchuk, V.N., Volkova, S.A., Starushenko, G.A.: Study of a nonlinear oscillator under parametric impulsive excitation using a non–smooth temporal transformation. J. Sound & Vib. 222(2), 307–328 (1999)
856. Pinnington, R.J.: Collision dynamics of two adjacent oscillators. J. Sound & Vib. 268(2), 343–360 (2003a)
857. Pinnington, R.J.: Energy dissipation prediction in a line of colliding oscillators. J. Sound & Vib. 268(2), 361–384 (2003b)
858. Plaut, R.H., Farmer, A.L.: Large motions of a moored floating breakwater modeled as an impact oscillator. Nonlin. Dyn. 23(4), 319–334 (2000)
859. Popp, K.: Non–smooth mechanical systems: An overview. Forschung im Ingenieurwesen/Engineering Research 64(9), 223–229 (1998)

860. Popp, K.: Non–smooth mechanical systems. J. Appl. Math. 64, 765–772 (2000)
861. Popplewell, N., Bapat, C.N., McLachlan, K.: Stable periodic vibro–impacts of an oscillator. J. Sound & Vib. 87(1), 41–59 (1983)
862. Popplewell, N., Liao, M.: A simple design procedure for optimum impact dampers. J. Sound & Vib. 146(3), 519–526 (1991)
863. Popplewell, N., Semercigil, S.E.: Performance of the bean bag impact damper for a sinusoidal external force. J. Sound & Vib. 133(2), 193–223 (1989)
864. Popprath, S., Ecker, H.: Nonlinear dynamics of a rotor contacting an elastically suspended stator. J. Sound & Vib. 308, 767–784 (2007)
865. Potthast, C., Twiefel, J., Wallaschek, J.: Modeling approaches for an ultrasonic percussion drill. J. Sound & Vib. 308, 405–417 (2007)
866. Powell, G., Schricker, V., Row, D., Hollings, J., Sause, R.: Ice–structure interaction of an offshore platform. In: Civil Engineering in the Arctic Offshore, Proc. Conf. Arctic 1985., San Francisco, CA, pp. 230–238 (1985)
867. Privalov, E.A.: One form of the non–smooth–transformation method for use in analysis of vibro–impact systems. Mech. Solids 27(4), 34–37 (1992)
868. Privalov, E.A.: Modification of the method of non–smooth transformations for vibration–impact systems with bilateral motion constraint. Mech. Solids 28(3), 99–101 (1993)
869. Proppe, C.: Exact stationary probability density functions for nonlinear systems under Poisson white noise excitation. Int. J. Nonlin. Mech. 38, 557–564 (2003)
870. Pun, D., Lau, S.L., Cao, D.Q.: Forced vibration analysis of a multi–degree impact vibrator. J. Sound & Vib. 213(3), 447–466 (1998)
871. Půst, L.: Equivalent coefficient of restitution. Eng. Mech. 5(5), 303–318 (1998)
872. Půst, L., Peterka, F.: Impact oscillator with Hertz's model of contact. Meccanica 38(1), 99–114 (2003)
873. Půst, L., Peterka, F., Stépán, G., Tomlinson, G.R., Tondl, A.: Nonlinear oscillations in machines and mechanisms theory. Mechanism & Mach. Theory 34, 1237–1253 (1999)
874. Qian, D., Sun, X.: Invariant tori for asymptotically linear impact oscillators. Science in China, Series A (Mathematics, Physics, Astronomy) 49(5), 669–687 (2006)
875. Qian, D., Torres, P.J.: Periodic motions of linear impact oscillators via the successor map. SIAM J. Math. Anal. 36(6), 1707–1725 (2005)
876. Qiao, N., Lin, W., Qin, Q.: Bifurcations and chaotic motions of a curved pipe conveying fluid with nonlinear constraints. Comp. & Struct. 84, 708–717 (2006)
877. Qin, W., Chen, G., Meng, G.: Nonlinear responses of a rub–impact overhung. Chaos, Solitons & Fractals 19, 1161–1172 (2004)
878. Qiu, J., Feng, Z.C.: Parameter dependence of the impact dynamics of thin plates. Comp. & Struct. 75, 491–506 (2000)
879. Rajaraman, R., Dobson, I., Jalali, S.: Nonlinear dynamics and switching time bifurcations of a thyristor controlled reactor circuit. IEEE Trans. Circuits Syst. I 43(12), 1001–1006 (1996)
880. Ramachandran, S., Lesieutre, G.A.: Numerical investigation of particle impact dampers. Proc. SPIE – The Int. Society for Optical Engineering 5386(1), 466–477 (2004)

881. Ramachandran, S., Lesieutre, G.A.: Dynamics and performance of a vertical impact damper. In: Proc. 46th AIAA/ASME/ASCE/AHS/ASC Structures, Structural Dyn. & Materials Conf., Austin, TX, pp. 6425–6432 (2005)
882. Ramachandran, S., Lesieutre, G.A.: Dynamics and performance of a harmonically excited vertical impact damper. ASME J. Vib. & Acoust. 130(2), 021008–1–11 (2008)
883. Rand, R.H., Moon, F.C.: Bifurcations and chaos in forced zero–stiffness impact oscillator. Int. J. Nonlin. Mech. 25(4), 417–432 (1990)
884. Rao, A., Gupta, G., Eisinger, F., Hibbit, H., Steininger, D.: Computer modeling of vibration and wear multispan tubes with clearances at tube supports. In: Proc. BHRA Conf. Flow–Induced Vibrations, Bowness–on–Windmere, U.K., K3, pp. 449–465 (1987)
885. Rega, G., Lenci, S.: Non–smooth dynamics, bifurcation and control in an impact system. Syst. Anal. Model Sim. 43(3), 343–360 (2003)
886. Remington, P.J.: Wheel/rail squeal and impact noise: What do we know? What don't we know? Where do we go from here? J. Sound & Vib. 116(2), 339–353 (1987)
887. Richards, E.J.: On the prediction of impact noise, Part III: Energy accountancy in industrial machines. J. Sound & Vib. 76, 187–232 (1981)
888. Richards, E.J., Carr, I.: On the prediction of impact noise, X: The design and testing of a quietened drop hammer. J. Sound & Vib. 104(1), 137–164 (1986)
889. Richards, E.J., Lenzi, A.: On the prediction of impact noise, VII: The structural damping of machinery. J. Sound & Vib. 97(4), 549–586 (1984)
890. Richards, E.J., Westcott, M.E., Jeyapalan, R.K.: On the prediction of impact noise, I: Acceleration noise. J. Sound & Vib. 62(4), 547–575 (1979)
891. Rigaud, E., Perret–Liaudet, J.: Experiments and numerical results on nonlinear vibrations of an impacting Hertzian contact, Part 1: Harmonic excitation. J. Sound & Vib. 265(2), 289–307 (2003)
892. Rogers, R.J., Pick, R.J.: On the dynamic spatial response of a heat exchanger tube with intermittent baffle contacts. Nuclear Eng. Design 36(1), 81–90 (1976)
893. Rogers, R.J., Pick, R.J.: Factors associated with support plate forces due to heat exchanger tube vibratory contact. Nuclear Eng. Design 44, 247–252 (1977)
894. Rook, T.E., Singh, R.: Dynamic analysis of a reverse–idler gear pair with concurrent clearances. J. Sound & Vib. 182(2), 303–322 (1995)
895. Rosellini, L., Hersen, F., Clanet, C., Bocquet, L.: Skipping stones. J. Fluid Mech. 543, 137–146 (2005)
896. Roslyakov, V.P., Nakhtigal, N.G.: Choice of parameters of vibration protecting system with nonlinear characteristics. Mech. & Electrification of Agric. 10, 36–37 (1975)
897. Rossikhin, Y.A., Shitikova, M.V.: The impact of elastic bodies upon a Timoshenko beam. In: Modeling and Optimization of Distributed Parameter Systems, pp. 370–374. Chapman Hall, London (1996)
898. Routroy, B., Dutta, P.S., Banerjee, S.: Border collision bifurcations in n–dimensional piecewise linear discontinuous map. In: Proc. National Conference on Nonlinear Systems and Dynamics, Chennai, India, p. 149 (2006)
899. Rumyantsev, V.V.: Collision with obstacle of a body containing a viscous fluid. J. Appl. Math. (PMM) 33(5), 876–881 (1969)

900. Rusakov, I.G., Kharkevich, A.A.: Forced vibration of systems impacting against a stop (in Russian) Zhurnal Tekh Fyz (J Technical Physics) XII, No. 12, 715–721 (1942)
901. Sachse, R., Reum, G.: Beiträge zur theorie des kurbelfeder–hammers (Teil I). Wiss. Z. Th Karl–Marx Stadt, vol. VII(2) (1965)
902. Sadek, M.M.: Impact dampers for controlling vibration in machine tools. Machinery 102, 152–161 (1972)
903. Sadek, M.M., Mills, B.: The application of the impact damper to the control of machine tool chatter. In: Proc. 7th Int. Machine Tool Design and Research Conf., Birmingham, UK, pp. 243–257 (January 1966)
904. Sadek, M.M., Mills, B.: Effect of gravity on the performance of an impact damper: part 1. Steady–state motion. J. Mech. Eng. Sci. 12(4), 268–277 (1970)
905. Sadek, M.M., Williams, C.J.H., Mills, B.: Effect of gravity on the performance of an impact damper: part 2. Stability of vibrational modes. J. Mech. Eng. Sci. 12(4), 278–287 (1970)
906. Saeki, M.: Impact damping with granular materials. In: Proc. ASME Int. Mech. Eng. Cong. & Expos, Aerospace Division AD, Adaptive Structures and Material Systems, New York, NY, vol. 64, pp. 381–386 (2001)
907. Saeki, M.: Impact damping with granular materials in a horizontally vibrating system. J. Sound & Vib. 251(1), 153–161 (2002)
908. Saito, S.: Calculation of nonlinear unbalance response of horizontal Jeffcott rotors supported by ball bearings with radial clearance. In: ASME Design Eng. Tech. Conf., Cincinnati, OH, Paper 85–DET–33, 5 pages (1985)
909. Saito, T., Matsuda, T., Endo, M.: Damping effect of an impact damper for pillar bodies. Kikai Gakkai Ronbunshu, C Hen/Trans JSME, Part C 60(571), 824–830 (1994) (1st Report, Examination by free and random vibration responses)
910. Salapaka, A., Dahleh, M., Mezic, I.: On the dynamics of a harmonic oscillator undergoing impacts with a vibrating platform. Nonlin. Dyn. 24(4), 333–358 (2001)
911. Salvaggio, M.A., Rojansky, M.: The importance of wave–driven icebergs impacting an offshore structure. In: Proc Offshore Technology Conf. OTC 1986, Paper OTC 5086, vol. 1, pp. 29–38 (1986)
912. Sampaio, R., Soize, C.: On measures of nonlinearity effects for uncertain dynamical systems–Application to a vibro–impact system. J. Sound & Vib. 303(3-5), 659–674 (2007)
913. Sanky, G.O.: Some experiments on a particle or 'shot' damper. Memorandum, Westinghouse Research Labs (1955)
914. Sanz, A.S., Miret–Artés, S.: Quantum trajectories in elastic atom–surface scattering: Threshold and selective adsorption resonances. J. Chemical Physics 122(1), 014702(1–12) (2005)
915. Sato, T., Takase, M., Kaiho, N., Makino, T., Tanaka, K.: Vibration reduction of pantograph–support system using an impact damper (influence of curve track). In: Proc. Int. Conf. Noise & Vib Engineering, ISMA, Leuven, Belgium, pp. 1669–1676 (2002)
916. Sauve, R.G., Teper, W.W.: Impact simulation process equipment tubes and support plates – a numerical algorithm. ASME J. Pressure Vessels Tech. 109(1), 70–79 (1987)

917. Schatzman, M.: Uniqueness and continuous dependence on data for one–dimensional impact problems. Math. & Comp. Model 28, 1–18 (1998)
918. Semercigil, S.E., Popplewell, N., Tyc, R.: Impact damping of random vibrations. J. Sound & Vib. 122(1), 178–184 (1988)
919. Semercigil, S.E., Collette, F., Huynh, D.: Experiments with tuned absorber–impact damper combination. J. Sound & Vib. 256(1), 179–188 (2002)
920. Semler, C., Li, G.X., Païdoussis, M.P.: The nonlinear equations of motion of pipes conveying fluid. J. Sound & Vib. 169, 577–599 (1994)
921. Senator, M.: Existence and stability of periodic motions of a harmonically forced impacting system. J. Acoust Soc. Amer. 47(5), pt.2, 1390–1397 (1970)
922. Sharif–Bakhtar, M., Shaw, S.W.: The dynamic response of a centrifugal pendulum vibration absorber with motion–limiting stops. J. Sound & Vib. 126(2), 221–235 (1988)
923. Sharkovsky, A.N., Chua, L.O.: Chaos in some 1–D discontinuous maps that appear in the analysis of electrical circuits. IEEE Trans. Circuits Syst. I 40(10), 722–731 (1993)
924. Shaw, J., Shaw, S.W.: The onset of chaos in a two–degree–of–freedom impacting system. ASME J. Appl. Mech. 46, 168–174 (1989)
925. Shaw, S.W.: A Periodically Forced Piecewise Linear Oscillator, Ph.D. Dissertation, Cornel University (1983)
926. Shaw, S.W.: Forced vibration of a beam with one–sided amplitude constraint: Theory and experiment. J. Sound & Vib. 99(2), 199–212 (1985a)
927. Shaw, S.W.: Dynamics of harmonically excited systems having rigid amplitude constraints, Part 1– Subharmonic motions and local bifurcations. ASME J. Appl. Mech. 52, 453–458 (1985b)
928. Shaw, S.W.: Dynamics of harmonically excited systems having rigid amplitude constraints, Part 2– Chaotic motions and global bifurcations. ASME J. Appl. Mech. 52, 459–464 (1985c)
929. Shaw, S.W., Haddow, A.G., Hsieh, S.R.: Properties of cross–well chaos in an impacting systems. Phil. Trans. Royal Soc. London Series A–Mathematical Physical and Engineering Sciences 347(1683), 391–410 (1994)
930. Shaw, S.W., Holmes, P.J.: A periodically forced piecewise linear oscillator. J. Sound & Vib. 90, 129–155 (1983a)
931. Shaw, S.W., Holmes, P.J.: A periodically forced impact oscillator with large dissipation. ASME J. Appl. Mech. 50, 849–857 (1983b)
932. Shaw, S.W., Holmes, P.J.: A periodically forced linear oscillator with impacts: Chaos and long period motions. Phys. Rev. Lett. 51, 623–626 (1983c)
933. Shaw, S.W., Pierre, C.: The dynamic response of tuned impact absorbers for rotating flexible structures. ASME J. Comput. Nonlin. Dyn. 1(1), 13–24 (2006)
934. Shaw, S.W., Rand, R.H.: The transition to chaos in a simple mechanical system. Int. J. Nonlin. Mech. 24, 41–56 (1989)
935. Shen, X., Jia, J., Zhao, M.: Numerical analysis of a rub–impact rotor–bearing system with mass unbalance. J. Vib. & Cont. 13(12), 1819–1834 (2007)
936. Sherif, H.A.: Effect of Contact Stiffness on the Establishment of Self–Excited Vibrations. Wear 141, 227–234 (1991a)
937. Sherif, H.: On the design of anti–squeal friction pads for disc brakes. Soc. Aut. Eng. Trans. 100(Sect 6), 678–686 (1991b)
938. Sherif, H.: Parameters affecting contact stiffness of nominally flat surfaces. Wear 145(1), 113–121 (1991c)

939. Shin, Y.S., Jendrzejczyk, J.K., Wambsganss, M.W.: Vibration of a heat exchanger tube with tube/support impact. ASME Paper No. 77–JPGC–NE–5, 13 p. (1977)
940. Shin, Y.S., Sass, D.E., Jendrzejczyk, J.A.: Vibro–impact responses of a tube with tube–baffle interaction. In: Proc. ASME Pressure Vessels and Piping Conf. Proceedings, Paper No. 78–PVP–20, ASMSA4, 8 p. (1978)
941. Shu, Z.Z., Shen, X.Z.: Theoretical analysis of complete stability and automatic vibration isolation of impacting and vibrating systems with double masses. Chinese J. Mech. Eng. 26(3), 50–57 (1990) (in Chinese)
942. Sijin, Z., Libiao, Z., Qishao, L.: A map method for grazing bifurcation in linear vibro–impact system. Acta Mechanica Sinica 39(1), 132–136 (2007) (in Chinese)
943. Silberschmidt, V.V., Casas–Rodriguez, J.P., Ashcroft, I.A.: Impact fatigue in adhesive joints. Proc. IMech. E Part C: J. Mechanical Engineering Science 222, 1981–1994 (2008)
944. Sims, N.D., Amarasinghe, A., Ridgway, K.: Particle dampers for work–piece chatter mitigation. ASME Manufacturing Engineering Division MED 16–1, 825–832 (2005)
945. Sin, V.T.W., Wiercigroch, M.: Symmetrically piecewise linear oscillator: Design and measurement. Proc. Inst. Mech. Eng. C 213, 241–249 (1999)
946. Singh, R., Xie, H., Comparin, R.J.: Analysis of an automotive neutral gear rattle. J. Sound & Vib. 131, 177–196 (1989)
947. Sinha, S.K.: Nonlinear dynamic response of a rotating radial Timoshenko beam with periodic pulse loading at the free end. Int. J. Nonlin. Mech. 40, 113–149 (2005)
948. Skipor, E., Bain, L.J.: Application of impact damping to rotary printing equipment. ASME J. Mech. Design 102, 338–343 (1980)
949. Slade, K.N., Virgin, L.N., Bayly, P.V.: Extracting information from interimpact intervals in a mechanical oscillator. Phys. Revs. E 56(3), 3705–3708 (1997)
950. Sokolov, I.J., Babitsky, V.I., Halliwell, N.A.: Autoresonant vibro–impact system with electromagnetic excitation. J. Sound & Vib. 308, 375–391 (2007)
951. Song, L.L., Iourtchenko, D.V.: Analysis of stochastic vibro–impact systems with inelastic impacts. Mech. Solids 41(2), 146–154 (2006)
952. Sorokin, V.N.: The Theory of Speech Production, Radio I Svyaz, Moscow (1985) (in Russian)
953. Sosnovskiy, L.A., Sherbakov, S.S.: Vibro–impact in rolling contact. J. Sound & Vibr. 308(3-5), 489–503 (2007)
954. Spencer, M., Siegmund, T., Mongeau, L.: Determination of superior surface strains and stresses, and vocal fold contact pressure in a synthetic larynx model using digital image correlation. J. Acoust. Soc. Amer. 123(2), 1089–1103 (2008)
955. Steindl, A., Troger, H.: Flow–induced bifurcations to 3–dimensional motions of tubes with an elastic support. In: Besserling, J.F., Eckhaus, W. (eds.) Trends in Applications of Mathematics to Mechanics, pp. 128–138. Springer, Berlin (1988)
956. Steindl, A., Troger, H.: Nonlinear Stability and Bifurcation Theory. Springer, Berlin (1991)
957. Steindl, A., Troger, H.: Nonlinear three–dimensional oscillations of elastically constrained fluid conveying tubes with perfect broken O(2)–symmetry. Nonlin. Dyn. 7, 165–193 (1995a)

958. Steindl, A., Troger, H.: One and two–parameter bifurcations to divergence and flutter in the three–dimensional motions of a fluid conveying viscoelastic tube with D4–symmetry. Nonlin. Dyn. 8(1), 161–178 (1995b)
959. Steindl, A., Troger, H.: Heteroclinic cycles in the three–dimensional post–bifurcation motion of O(2)–Symmetrical fluid conveying tubes. Appl. Math. & Comput. 78, 269–277 (1996)
960. Stensson, A., Nordmark, A.B.: Chaotic vibrations of a spring–mass system with unilateral displacement limitation: Theory and experiments. Phil. Trans. Royal Soc. London A 337, 439–448 (1994)
961. Stensson, A., Nordmark, A.B.: Existence of periodic solutions in grazing bifurcations of impacting mechanical oscillators. J. Tech. Phys. 37(3–4), 531–534 (1996)
962. Stepanov, P.T.: Investigation of vibro–impact mechanical systems. Novosibirsk, NETI, pp. 67–71 (1979)
963. Stinner, A.: Physics and the Dumpsters. Phys. Educ. 24, 260–267 (1989)
964. Stronge, W.J.: Impact Mechanics. Cambridge University Press, Cambridge (2000)
965. Su, T.C., Kang, S.Y.: Analysis of liquid impact of moving containers. Developments in Appl. Mech. 12 (1984a)
966. Su, T.C., Kang, S.Y.: Numerical simulation of liquid sloshing. In: Boresi, A.P., Chong, K.P. (eds.) Mechanical and Civil Engineering. ASCE, vol. 2, pp. 1069–1072 (1984b)
967. Su, T.C., Kang, S.Y.: Analysis and testing of the large amplitude liquid sloshing in rectangular containers. In: Proc. ASME Pressure Vessels and Piping Conf., Seismic Engineering Piping Systems, Tanks and Power Plant Equip, New York, N. Y., PVP. ASME, vol. 108, pp. 149–154 (1986)
968. Su, T.C., Wang, Y.: Numerical simulation of three–dimensional large amplitude liquid sloshing in rectangular containers subjected in vertical excitation. In: ASME Pressure Vessels and Piping Division, Sloshing and Fluid Structure Vibration, Honolulu, HI, PVP, vol. 157, pp. 115–126 (1989)
969. Su, T.C., Wang, Y.: Numerical simulation of three–dimensional large amplitude liquid sloshing in cylindrical tanks subjected to arbitrary excitations. In: ASME Pressure Vessels and Piping Conf., Flow–Structure Vibration and Sloshing, Nashville, TN, PVP, vol. 191, pp. 127–148 (1990)
970. Sugiyama, Y., Tanaka, Y., Kishi, T., Kawagoe, H.: Effect of a spring support on the stability of pipes conveying fluid. J. Sound & Vib. 100, 257–270 (1985)
971. Sun, H., Faltinsen, O.M.: Water impact of horizontal circular cylinders and cylindrical shells. Appl. Ocean Res. 28(5), 299–311 (2006)
972. Sung, C.K., Yu, W.S.: Dynamics of a harmonically excited impact damper: bifurcations and chaotic motion. J. Sound & Vib. 158, 317–329 (1992)
973. Svahn, F., Dankowicz, H.: Energy transfer in vibratory systems with friction exhibiting low–velocity collisions. J. Vib. & Cont. 14(1–2), 255–284 (2008)
974. Swamidas, A.S.J., Arockiasamy, M.: Iceberg impact forces on gravity platforms. In: Proc. 3rd Cold Regions Engineering specialty Conf., Edmonton, vol. 1, pp. 431–458 (1984)
975. Swamidas, A.S.J., Arockiasamy, M.: Structural integrity of semi–submersibles and gravity platforms to berg–bit/iceberg impact. In: Proc. 18th Annual Offshore Technology Conf., Houston, OTC, vol. 5087, 39–49 (1986)

976. Tan, X., Rogers, R.: Dynamic friction modeling in heat exchanger tube simulation. In: ASME Proc. Pressure Vessels and Piping Conf., Flow–Induced Vibration, Montreal, Canada, 328, pp. 347–358 (1996)
977. Tanagasawi, O., Theodossiades, S., Rahnejat, H.: Lightly loaded lubricated impacts: Idle gear rattle. J. Sound & Vib. 308, 418–430 (2007)
978. Tanaka, H., Ushio, T.: Analysis of border–collision bifurcations in a flow model of a switching system. IEICE Trans. Fund Elect, Comm. & Comp. Sci. E85-A(4), 734–739 (2002)
979. Tanaka, N., Kikushima, Y.: Study of the dynamic damper with a preview action (experiment of the dynamic damper with a preview action). JSME Int. Journal 30(268), 1631–1637 (1987)
980. Tanaka, N., Kikushima, Y.: Impact vibration control using a semi–active damper. J. Sound & Vib. 158(2), 277–292 (1992)
981. Tang, D.M., Dowell, E.H.: Chaotic oscillations of a cantilevered pipe conveying fluid. J. Fluids & Struct. 2, 263–283 (1988)
982. Tang, D., Dowell, E.H.: Flutter and limit cycle oscillations of a wing–store model with free-play. J. Aircraft 43(2), 487–503 (2006a)
983. Tang, D., Dowell, E.H.: Experimental and theoretical study of gust response for a wing–store model with free-play. J. Sound & Vib. 295, 659–684 (2006b)
984. Tang, D., Kholodar, D., Dowell, E.H.: Nonlinear aeroelastic response of airfoil section with control surface free-play to gust loads. AIAA Journal 38(9), 1543–1557 (2000)
985. Tangasawi, O., Theodossiades, S., Rahnejat, H., Kelly, P.: Nonlinear vibro–impact phenomenon belying transmission idle rattle. Proc. IMech. E Part C: J. Mechanical Engineering Science 222, 1909–1923 (2008)
986. Tao, C., Jiang, J.J.: Mechanical stress during phonation in a self–oscillating finite–element vocal fold model. J. Biomechanics 40(10), 2191–2198 (2007)
987. Tao, C., Jiang, J.J., Zhang, Y.: Simulation of vocal fold impact pressures with a self–oscillating finite–element model. J. Acoust Soc. Amer. 119(6), 3987–3994 (2006)
988. Teng, X., Wierzbicki, T.: Interactive failure of two impacting beams. J. Eng. Mech. 129(8), 918–926 (2003)
989. Teng, X., Wierzbicki, T.: Multiple impact of beam–to–beam. Int. J. Impact Eng. 31(2), 185–219 (2005)
990. Theodossiades, S., Natsiavas, S.: Periodic and chaotic dynamics of motor–driven gear–pair system with backlash. Chaos, Solitons & Fractals 12, 2427–2440 (2001)
991. Thomas, M.D., Knight, W.A., Sadek, M.M.: The impact damper as a method improving cantilever boring bars. ASME J. Eng. Indust. 97(3), 859–866 (1975)
992. Thomsen, J.J., Fidlin, A.: Discontinuous transformations and averaging for vibro–impact analysis. In: Proc. XXI Int. Cong. Theor. Appl. Mech. (ICTAM 2004) Abstract Book and CDROM ID:SM25L–12694 (2004) (CD–ROM)
993. Thomsen, J.J., Fidlin, A.: Near–elastic vibro-impact analysis by discontinuous transformations and averaging. J. Sond & Vib. 311, 386–407 (2008)
994. Thompson, J.M.T.: Complex dynamics of compliant off–shore structures. Proc. Royal Society of London Series A, Mathematical and Physical Sciences 387, 407–427 (1983)
995. Thompson, J.M.T.: Chaotic behavior triggering the escape from a potential well. Proc. Royal Society of London A 421, 195–225 (1989)

996. Thompson, J.M.T., Ghaffari, R.: Chaos after period doubling bifurcations in the resonance of an impact oscillator. Phys. Lett 91A, 5–8 (1982)
997. Thompson, J.M.T., Ghaffari, R.: Chaotic dynamics of an impact oscillator. Phys. Rev. A 27(3), 1741–1743 (1983)
998. Thompson, J.M.T., McRobie, F.A.: Indeterminate bifurcations and the global dynamics of driven oscillators. In: Kreuzer, E., Schmidt, G. (eds.) Proc. 1st European Nonlinear Oscillations Conf., Hamburg, August 16–20, 1993, pp. 107–128. Academie Verlag, Berlin (1993)
999. Thota, P.: Analytical and Computational Tools for the Study of Grazing Bifurcations of Periodic Orbits and Invariant Tori, PhD Dissertation, Virginia Tech., Blacksburg, Virginia (2007)
1000. Thota, P., Dankowicz, H.: Continuous and discontinuous grazing bifurcations in impacting oscillators. Physica D 214, 187–197 (2006a)
1001. Thota, P., Dankowicz, H.: Analysis of grazing bifurcations of quasi–periodic system attractors. Physica D 220(2), 163–174 (2006b)
1002. Thota, P., Zhao, X., Dankowicz, H.: codimension–two grazing bifurcations in single–degree–of–freedom impact oscillator. ASME J. Comput. & Nonlin. Dyn. 2, 328–335 (2006)
1003. Titze, I.R. (ed.): Vocal Fold Physiology: New Frontiers in Basic Science. Singular Publication Group, San Diego (1993)
1004. Titze, I.R.: Mechanical stress in phonation. J. Voice 8, 99–105 (1994)
1005. Todd, M.D., Virgin, L.N.: An experimental impact oscillator. Chaos, Solitons & Fractals 8(4), 699–714 (1997)
1006. Tokumaru, H., Kotera, T.: On impact–damper for concentrated–mass–continuum system. Bull JSME 13, 59 (1970)
1007. Tolstoi, D.M.: Significance of the Normal–Degree–of–Freedom and Natural Normal Vibrations in Contact Friction. Wear 10, 199–213 (1967)
1008. Tolstoi, D.M., Borisova, G.A., Grigorova, S.R.: Role of Interinsic Contact of Oscillations in Normal Direction during Friction. In Nature of the Friction in Solids, Nauka i Tekhnica, Minsk (1971)
1009. Torvik, P.J., Gibson, W.: Design and effectiveness of impact dampers for space applications. ASME Design Engineering Division (Publication) DE, vol. 5, 65–74 (1987)
1010. Toulemonde, C., Gontier, C.: Multiple degree of freedom impact oscillator. European J. Mech., A/Solids 16(5), 879–904 (1997)
1011. Toulemonde, C., Gontier, C.: Sticking motions of impact oscillators. European J. Mech., A/Solids 17(2), 339–366 (1998)
1012. Trindade, M.A., Wolter, C., Sampaio, R.: Karhunen–Loeve decomposition of coupled axial/bending vibrations of beams subject to impacts. J. Sound & Vib. 279, 1015–1036 (2005)
1013. Tse, C.K.: Complex Behavior of Switching Power Converters. CRS Press, Boca Raton (2003)
1014. Tufillaro, N.B.: Braid analysis of a bouncing ball. Phys. Rev. E 50, 4509–4522 (1994)
1015. Tufillaro, N.B., Abbott, T.A., Reilly, J.P.: An Experimental Approach to Nonlinear Dynamics and Chaos. Addison–Wesley, Reading (1992)
1016. Tufillaro, N.B., Albano, A.M.: Chaotic dynamics of a bouncing ball. Amer. J. Physics 54(10), 939–944 (1986)
1017. Tufillaro, N.B., Mello, T.M., Choi, Y.M., Albano, A.M.: Period doubling boundaries of a bouncing ball. J. de Physique 47, 1477–1482 (1986)

1018. Tung, P.C.: Dynamics of a nonharmonically forced impact oscillator. JSME Int. Journal, Series 3: Vibration, Control Engineering, Engineering for Industry 35(3), 378–386 (1992)
1019. Tung, P.C., Shaw, S.W.: The dynamics of an impact print hammer. ASME J. Vib., Acoust, Stress & Rel. Design 110, 193–200 (1988a)
1020. Tung, P.C., Shaw, S.W.: A method for the improvement of impact printer performance. ASME J. Vib., Acoust, Stress & Rel. Design 110, 528–532 (1988b)
1021. Twizell, E.H., Jin, L., Lu, Q.S.: A method for calculating the spectrum of Lyapunov exponents by local maps in non–smooth impact–vibrating systems. J. Sound & Vib. 298(4–5), 1019–1133 (2006)
1022. Valente, A.X.C.N., McClamroch, N.H., Mezic, I.: Hybrid dynamics of two coupled oscillators that can impact a fixed stop. Int. J. Nonlin. Mech. 38, 677–689 (2003)
1023. Van de Vorst, E.L.B., Heertijes, M.F., Van Campen, D.H., de Kraker, A., Fey, R.H.B.: Experimental and numerical analyses of the steady state behavior of a beam system with impact. J. Sound & Vib. 212, 321–336 (1998)
1024. Van de Wouw, N., Van de Bosch, H.L.A., de Kraker, A., Van Campen, D.H.: Experimental and numerical analysis of nonlinear phenomena in a stochastically excited beam system with impact. Chaos, Solitons & Fractals 9(8), 1409–1428 (1998)
1025. Van de Wouw, N., de Kraker, A., Van Campen, D.H., Nijmeijer, H.: Nonlinear dynamics of a stochastically excited beam system with impact. Int. J. Nonlin. Mech. 38, 767–779 (2003)
1026. Van den Berg, J., Zantema, J.T., Doornebal, P.: On the air resistance and Bernoulli effect of the human larynx. J. Acoust. Soc. Amer. 29, 626–631 (1957)
1027. Van Hirtum, M., Ruty, N., Pelorson, X., Lopez, I.: Experimental validation of some additional issues in physical vocal folds models. Acta Acustica United With Acustica suppl.1,S 53 (2005)
1028. Veluswami, M.A., Crossley, F.R.E.: Multiple impacts of a ball between two plates, part 1: Some experimental observations. ASME J. Eng. Indust. 97, 820–827 (1975)
1029. Veluswami, M.A., Crossley, F.R.E., Horvay, G.: Multiple impacts of a ball between two plates. Part 2: Mathematical modeling. ASME J. Eng. Indust. 97, 835–838 (1975)
1030. Vento, M.A., Antunes, J., Axisa, F.: Tube/support interaction under simulated fluidelastic instability: Two–dimensional experiments and computations of the nonlinear responses of a straight tube. In: Proc. ASME Pressure Vessels and Piping Conf., Cross–Flow Induced Vibration of Cylinder Arrays, Anaheim, CA, PVP, vol. 242, pp. 151–166 (1992)
1031. Veprik, A.M., Babitsky, V.I.: Vibration protection of sensitive electronic equipment from harsh harmonic vibration. J. Sound & Vib. 238(1), 19–30 (2000)
1032. Veprik, A.M., Babitsky, V.I.: nonlinear correction of vibration protection system containing tuned dynamic absorber. J. Sound & Vib. 239(2), 335–356 (2001)
1033. Veprik, A.M., Krupenin, V.L.: On the solution of systems containing distributed impact elements. Machine Sciences 6 (1985) (in Russian)

1034. Veprik, A.M., Krupenin, V.L.: On the resonance oscillation of a system with a distributed impact element. Machine Science 6, 39–47 (1988) (in Russian)
1035. Veprik, A.M., Meromi, A., Leschecz, A.: Novel technique of vibration control for split Sterling crycooler with linear compressor. In: Proc. SPIE 11th Ann. Int. Symp. on Aerospace/Defense Sensing, Simulation and Controls AeroSense, Orlando, FL, USA. Infrared Technology and Applications XXIII, vol. 3061, p. 640 (1997)
1036. Vér, I.L., Ventres, C.S., Myles, M.M.: Wheel/rail noise, Part III: Impact noise generation by wheel and rail discontinuities. J. Sound & Vib. 46, 395–417 (1976)
1037. Verdolini, K., Hess, M.M., Titze, I.R., Bierhals, W., Gross, M.: Investigation of vocal fold impact stress in human subjects. J. Voice 13, 184–202 (1999)
1038. Vinogradov, O.C.: Simulation methodology of vessel–ice floes interaction problems. In: Proc. Int. Symp. on Offshore Mechanics and Arctic Engineering (OMAE), Tokyo, Japan, vol. 4, pp. 601–606 (1986)
1039. Virgin, L.N., Begley, C.J.: Grazing bifurcations and basins of attraction in an impact–friction oscillator. Physica D 130(1–2), 43–57 (1999)
1040. Virgin, L.N., Begley, C.J.: Nonlinear features in the dynamics of an impact–friction oscillator. In: Proc. AIP Conf., Ambleside, UK, vol. 502, pp. 469–475 (2000)
1041. Voronina, S., Babitsky, V., Meadows, A.: Modeling of autoresonant control of ultrasonic transducer for machining applications. Proc. IMech E Part C: J. Mechanical Engineering Science 222, 1957–1974 (2008)
1042. Vuorio, J., Riska, K., Varsta, P.: Long–term measurements of ice pressure and ice–induced stresses on the icebreaker Sisu in Winter. Winter Navigation Research Board, Report 28, Helsinki (1979)
1043. Wadham–Gagnon, M., Païdoussis, M.P., Semler, C.: Dynamics of cantilevered pipes conveying fluid, Part 1: Nonlinear equations of motion of three–dimensional motion. J. Fluids & Struct. 23, 545–567 (2007)
1044. Wagg, D.J.: A note on using the collocation method for modelling the dynamics of a flexible continuous beam subject to impacts. J. Sound & Vib. 276(3–5), 1128–1134 (2004a)
1045. Wagg, D.J.: Rising phenomena and multi–sliding bifurcation in a two–degree–of–freedom impact oscillator. Chaos, Solitons & Fractals 22(3), 541–548 (2004b)
1046. Wagg, D.J.: Periodic sticking motion in a two–degree–of–freedom impact oscillator. Int. J. Nonlin. Mech. 40(8), 1076–1087 (2005)
1047. Wagg, D.J.: Multiple non–smooth events in multi–degree–of–freedom vibro–impact systems. Nonlin. Dyn. 43, 137–148 (2006)
1048. Wagg, D.J.: A note on coefficient of restitution models including the effects of impact induced vibration. J. Sound & Vib. 300(3), 1071–1078 (2007)
1049. Wagg, D.J., Bishop, S.R.: A note on modeling multi–degree–of–freedom vibro–impact systems using coefficient of restitution models. J. Sound & Vib. 236, 176–184 (2000)
1050. Wagg, D.J., Bishop, S.R.: Chatter, sticking and chaotic impacting motion in a two–degree of freedom impact oscillator. Int. J. Bifurc & Chaos Appl. Sci. Eng. 11, 57–71 (2001)
1051. Wagg, D.J., Bishop, S.R.: Application of non–smooth modeling techniques to the dynamics of a flexible impacting beam. J. Sound & Vib. 256(5), 803–820 (2002)

1052. Wagg, D.J., Bishop, S.R.: Dynamics of a two degree of freedom vibro–impact system with multiple motion limiting constraints. Int. J. Bifurc. & Chaos Appl. Sci. Eng. 14, 119–140 (2004)
1053. Wagg, D.J., Karpodinis, G., Bishop, S.R.: An experimental study of the impulse response of a vibro–impacting cantilever beam. J. Sound & Vib. 228, 243–264 (1999)
1054. Walker, J.S., Soule, T.: Chaos in a simple impact oscillator: the Bender bouncer. Amer. J. Phys. 64(4), 397–409 (1996)
1055. Wang, C., Kim, J.: New analysis method for a thin beam impacting against a stop based on the full continuous model. J. Sound & Vib. 191, 809–823 (1996)
1056. Wang, C., Kim, J.: Dynamic analysis of a thin beam impacting against a stop of general three–dimensional geometry. J. Sound & Vib. 203(2), 237–249 (1997)
1057. Wang, L., Ni, Q.: A note on the stability and chaotic motions of a restrained pipe conveying fluid. J. Sound & Vib. 297, 1079–1083 (2006)
1058. Wang, L., Ni, Q., Huang, Y.Y.: Flow–induced vibration of a nonlinearly restrained curved pipe conveying fluid. China Ocean Eng. 18(3), 347–356 (2004)
1059. Wang, L., Xu, W., Li, Y.: Impulsive control of a class of vibro–impact systems. Physics Letters A 372(32), 5309–5313 (2008a)
1060. Wang, L., Xu, W., Li, Y.: Dynamical behavior of a controlled vibro–impact system. Chinese Physics B 17(7), 2446–2450 (2008b)
1061. Warburton, G.B.: Discussion of the theory of the acceleration damper. ASME J. Appl. Mech. 24, 322–324 (1957)
1062. Watanabe, T.: Forced vibration of continuous system with nonlinear boundary conditions. J. Mechanical Design 100, 487–491 (1978)
1063. Watanabe, T.: Steady impact vibrations of continuous elements (case of colliding one in a half cycle). Bull JSME 24 (1981)
1064. Watanabe, T., Shibata, H.: On nonlinear vibration of a beam –response of a beam with a gap at one end. Report of the Institute of Industrial Science, The University of Tokyo 36(1), 1–25 (1991)
1065. Watanabe, T., Shibata, H., Maezawa, S.: Steady impact vibration of a body having hysteresis collision characteristics. Bull JSME 22(167), 655–660 (1979)
1066. Weaver, D.S., Schneider, W.: The effect of flat bar supports on the cross–flow induced response of heat exchanger U–tubes. ASME J. Pressure Vessel Techn. 105, 775–781 (1983)
1067. Wei, Y., Tang, X., Hogat, S.: Nonlinear Response of Rotor to Stator Rubs. Far East Levingston Shipbuilding Ltd. (Singapore), 10 pages (May 1997)
1068. Wen, G.L.: codimension–two Hopf bifurcation of a two–degree–of–freedom vibro–impact system. J. Sound & Vib. 242, 475–485 (2001)
1069. Wen, G.L., Xie, J.H.: Period–doubling bifurcation and non–typical route to chaos of a two–degree–of–freedom vibro–impact system. ASME J. Appl. Mech. 68, 670–674 (2001)
1070. Wen, G.L., Xie, J.H., Xu, D.: Onset of degenerate Hopf bifurcation of a vibro–impact oscillator. ASME J. Appl. Mech. 71, 579–581 (2004)
1071. Wen, J.M., Feng, Q.: Two–stage stochastic model on rattling vibration with amplitude modulation. Shock & Vib. 11(5–6), 693–702 (2004)
1072. Whiston, G.S.: Remote impact analysis by use of propagated acceleration signals. I. Theoretical methods. J. Sound & Vib. 97(1), 35–51 (1984)

1073. Whiston, G.S.: The vibro–impact response of a harmonically excited and preloaded one–dimensional linear oscillator. J. Sound & Vib. 115, 303–319 (1987a)
1074. Whiston: On the statistics of chaotic vibro–impact. J. Sound & Vib. 116, 598–603 (1987b)
1075. Whiston, G.S.: Global dynamics of a vibro–impacting linear oscillator. J. Sound & Vib. 118, 395–429 (1987c)
1076. Whiston, G.S.: Singularities in vibro–impact dynamics. J. Sound & Vib. 152(3), 427–460 (1992)
1077. Wiercigroch, M., Sin, V.T.W.: Experimental study of a symmetrical piecewise based–excited oscillator. ASME J. Appl. Mech. 65, 657–663 (1998)
1078. Wiercigroch, M., Sin, V.T.W., Li, K.: Measurement of chaotic vibration in a symmetrically piecewise linear oscillator. Chaos, Solitons & Fractals 9(1–2), 209–220 (1998)
1079. Wiesenfeld, K., Tufillaro, N.B.: Suppression of period doubling in the dynamics of a bouncing ball. Physica D 26, 321–335 (1987)
1080. Woo, K.C., Rodger, A.A., Neilson, R.D., Wiercigroch, M.: Application of the harmonic balance method to the ground moling machines operating in periodic regimes. Chaos, Solitons & Fractals 11, 2515–2525 (2000)
1081. Woo, K.C., Rodger, A.A., Neilson, R.D., Wiercigroch, M.: Phase shift adjustment for harmonic balance method applied to vibro–impact systems. Meccanica 41, 269–282 (2006)
1082. Wood, L.A., Byrne, K.P.: Analysis of a random repeated impact process. J. Sound & Vib. 78(3), 329–345 (1981)
1083. Wood, L.A., Byrne, K.P.: Experimental investigation of a random repeated impact process. J. Sound & Vib. 85(1), 53–69 (1982)
1084. Woolston, D.S., Runyan, H.W., Andrews, R.E.: An investigation of effects of certain type of structural nonlinearities on wing and control surface flutter. J. Aeronautical Sci. 24, 57–63 (1957)
1085. Wu, S.C., Yang, S.M., Haug, E.J.: Dynamics of mechanical systems with Coulomb friction, stiction, impact, and constraint addition–deletion. Mechanism & Mach. Theory 21(5), 417–425 (1986)
1086. Xie, J.H.: codimension–two bifurcations and Hopf bifurcations of an impacting vibrating system. J. Appl. Math. Mech. (PMM) 17, 65–75 (1996)
1087. Xie, J.H., Ding, W.: Hopf–Hopf bifurcation and invariant torus T2 of a vibro–impact system. Int. J. Nonlin. Mech. 40, 531–543 (2005)
1088. Xie, J.H., Ding, W., Dowell, E.H., Virgin, L.N.: Hopf–flip bifurcation of high dimensional maps and application to vibro–impact systems. Acta Mechanica Sinica (English Series) 21(4), 402–410 (2005)
1089. Xie, J.H., Wen, G.L., Xiao, J.: Determining bifurcation parameters of two–degree–of–freedom vibro–impact system. J. Vib. Eng. 14, 285–290 (2001)
1090. Xu, Z., Wang, M.Y., Chen, T.: Particle damping for passive vibration suppression: numerical modelling and experimental investigation. J. Sound & Vib. 279(3–5), 1097–1120 (2005)
1091. Yagci, B., Iannacci, M., Ozdoganlar, O.B., Wickert, J.A.: On the repetitive impact dynamics of one and two degree of freedom systems. In: Proc. Int. Mechanical Engineering Congress and Exposition, Orlando, Florida, November 5–11 (2005)

1092. Yamagata, Y., Higuchi, T.: A micro-positioning device for precision automatic assembly using impact force of piezoelectric elements. In: Proc. 1995 IEEE Int. Conf. Robotics and Automation, Nagoya, Japan, vol. 1, pp. 666–671 (1995)

1093. Yamamoto, Y., Akamatsu, M., Tsutsumi, M.: Vibration and noise reduction of C–frame turret punch press by applying turret as impact mass damper. Nippon Kikai Gakkai Ronbunshu, C Hen/Trans JSME, Part C 62(602), 4026–4031 (1996) (in Japanese)

1094. Yamauchi, S.: The nonlinear vibration of flexible rotors, 1st report: Development of a new analysis technique. Nippon Kikai Gakkai Ronbunshu JSME 49(446), Series C, 1862–1868 (1983)

1095. Yang, D.C.H., Lin, J.Y.: Hertzian damping, tooth friction and bending elasticity in gear impact dynamics. ASME J. Mechanisms, Transmission & Automotive Design 109, 189–196 (1987)

1096. Yang, M.Y.: Development of Master Design Curves for Particle Impact Dampers, PhD Thesis, The Pennsylvania State University, Department of Mechanical Engineering, University Park, PA (2003)

1097. Yang, M.Y., Lesieutre, G.A., Hambric, S.A., Koopmann, G.H.: Development of a design curve for particle impact dampers. Proc. SPIE – Int. Soc. Optical Eng. 5386(1), 450–465 (2004)

1098. Yang, Z.C., Zhao, L.C.: Analysis of limit cycle flutter of an airfoil in incompressible flow. J. Sound & Vib. 123(1), 1–13 (1988)

1099. Yashin, A.F., Kim, L.I., Nikiforov, I.S.: Nonlinear vibrations of systems with an arbitrary polynomial restoring force. In: Proc. Mechanics of Deformed Bodies and Structural Calculations, Novosibirsk, NIIZhT, pp. 136–143 (1975)

1100. Yasuda, K., Kamiya, K., Kato, R.: Vibration suppression by an elastic impact damper. In: Proc. ASME Design Eng. Tech. Conf. 19th Bien Conf. Mech. Vib. & Noise, vol. 5 B, pp. 1047–1054 (2003)

1101. Yau, C.H., Bajaj, A.K., Nwokah, O.D.I.: Active control of chaotic vibration in a flexible pipe conveying fluid. J. Fluids & Struct. 9, 99–122 (1995)

1102. Ye, Z., Birk, A.M.: Fluid pressure in partially liquid–filled horizontal cylindrical vessels undergoing impact acceleration. ASME J. Pressure Vessel Technology 116, 449–459 (1994)

1103. Yetisir, M., McKerrow, E., Pettigrew, M.J.: Fretting wear damage of heat exchanger tubes: a proposed damage criterion based on tube vibration response. ASME J. Pressure Vessel Technology 120(3), 297–305 (1998)

1104. Yetisir, M., Weaver, D.: The dynamics of heat exchanged U–bend tubes with flat–bar supports. ASME J. Pressure Vessel Techn. 108, 406–412 (1986)

1105. Yigit, A.S., Christoforou, A.P.: Coupled axial and transverse vibrations of oil well drill–strings. J. Sound & Vib. 195(4), 617–627 (1996)

1106. Yigit, A.S., Christoforou, A.P.: Coupled torsional and bending vibrations of drill–strings subject to impact with friction. J. Sound & Vib. 215(1), 167–181 (1998)

1107. Yin, X.C., Qin, Y., Zou, H.: Transient responses of repeated impact of a beam against a stop. Int. J. Solids & Struct. 44(22–23), 7323–7339 (2007)

1108. Yokomichi, I., Aisaka, M., Araki, Y., Inoh, T.: Stationary random oscillations of engine mounting system with motion–limiting stops. Nippon Kikai Gakkai Ronbunshu, C Hen/Trans JSME, Part C 63(608), 1074–1080 (1997) (in Japanese)

1109. Yokomichi, I., Araki, Y., Jinnouchi, Y., Nishida, E., Inoue, J.: Impact damper with granular materials for multibody system (Damping performance of plural dampers). Nippon Kikai Gakkai Ronbunshu, C Hen/Trans. JSME, Part C 59(566), 2991–2995 (1993) (in Japanese)
1110. Yokomichi, I., Araki, Y., Jinnouchi, Y., Inoue, J.: Impact damper with granular materials for multibody system. ASME J. Pressure Vessel Technology 118(1), 95–103 (1996)
1111. Yokomichi, I., Muramatsu, H., Araki, Y., Jinnouchi, Y., Nishida, E.: Impact damper with granular materials applied to self–induced vibration. Nippon Kikai Gakkai Ronbunshu, C Hen/Trans. JSME, Part C 61(587), 2769–2775 (1995) (in Japanese)
1112. Yokomichi, I., Miyauchi, M., Araki, Y.: On shot impact dampers applied to self–excited vibrations. Int. J. Acoustics & Vib. 6(4), 193–199 (2001)
1113. Yongyong, H., Xinyun, Y., Felei, C.: Acoustic emission characteristics of rub–impact for rotor–bearing system based on wavelet scalogram. Chinese J. Mech. Eng. 43(6), 149–153 (2007) (in Chinese)
1114. Yoshitake, Y., Sueoka, A.: Quenching of self–excited vibrations by an impact damper. Nippon Kikai Gakkai Ronbunshu, C Hen/Trans. JSME, Part C 60(569), 50–56 (1994) (in Japanese)
1115. Yuan, G.H.: Shipboard Crane Control, Simulation Data Generation and Border Collision Bifurcations, Ph.D. Thesis, University of Maryland, College Park (1997)
1116. Yuan, G.H., Banerjee, S., Ott, E., Yorke, J.A.: Border–collision bifurcations in the buck converter. IEEE Trans. Circuits & Systems–I 45, 707–716 (1998)
1117. Yuan, Z., Chu, F., Hao, R.: Simulation of rotor's axial rub–impact in full degrees of freedom. Mechanism & Mach. Theory 42, 763–775 (2007)
1118. Yudenko, V.Y., Vorobev, V.A.: Investigation into vibratory impact loading in disc friction clutches and brakes. Soviet Engineering Research 2(1), 34–37 (1982)
1119. Yue–Gang, L., Zhi–hao, J., Chang–li, L., Bang–Chun, W.: Effects of nonlinear friction force on chaotic motion of rotor–bearing system at rub–impact faults. J. Northeastern University (Natural Science) 24(9), 843–846 (2003)
1120. Yue, Y., Xie, J.H.: Symmetry of the Poincaré map and bifurcations of a two–degree–of–freedom vibro–impact system. Zhendong Gongcheng Xuebao/Journal of Vibration Engineering 21(4), 376–380 (2008) (in Chinese)
1121. Zavala–Río, A., Brogliato, B.: On the control of a one degree–of-freedom juggling robot. Dynamics and Control 9, 67–90 (1999)
1122. Zentner, I., Poirion, F.: Non–smooth dynamics and stability of mechanical systems featuring a free–play nonlinearity. Int. J. Nonlin. Mech. 41(6,7), 796–806 (2006)
1123. Zesch, W., Buchi, R., Codourey, A., Siegwart, R.: Inertial drives for micro– and nano–robots: two novel mechanisms. In: Proc. SPIE – The Int. Society for Optical Engineering, Philadelphia, PA, October 25, vol. 2593, pp. 80–88 (1995)
1124. Zhang, H.Z., Cheng, G.M., Zhao, H.W., Zeng, P., Yang, Z.G.: Two–dimensional precise actuator driven by piezoelectric bimorph impact. Jilin Daxue Xuebao (Gongxueban)/J. Jilin. University (Engineering and Technology Edition) 36(1), 67–71 (2006) (in Chinese)

1125. Zhang, W.M., Meng, G.: Stability, bifurcation and chaos of a high–speed rub–impact rotor system in MEMS. Sensors & Actuators A 127, 163–178 (2006)
1126. Zhang, W.M., Meng, G., Feng, J.Z.: Nonlinear dynamic characteristics of a micro–rotor system with rub–impact in MEMS. In: Proc. Asia–Pacific Microwave Conf. (APMC), Suzhou, China, vol. 2, p. 1606528 (2005a)
1127. Zhang, W.M., Meng, G., Chen, D., Zhou, J.B., Chen, J.Y.: Nonlinear dynamics of a rub–impact micro–rotor system with scale–dependent friction model. J. Sound & Vib. 309(3–5), 756–777 (2008)
1128. Zhang, W.M., Zhang, W., Turner, K.L.: Nonlinear dynamics of micro impact oscillators in high frequency MEMS switch application. In: TRANSDUCERS 2005. The 13th Int Conf Solid–State Sensors, Actuators and Microsystems. Digest of Technical Papers (IEEE Cat. No. 05TH8791), pt. 1, vol. 1, pp. 768–771 (2005b)
1129. Zhang, Y.P., Luo, G.W.: An introduction to recent researches into dynamics of vibro–impact system contacting a single stop. Current Topics in Acoustical Research 3, 137–140 (2003)
1130. Zhao, X., Dankowicz, H.: Unfolding degenerate grazing dynamics in impact actuators. Nonlinearity 19(2), 399–418 (2006a)
1131. Zhao, X., Dankowicz, H.: Control of Impact Microactuators for Precise Positioning. ASME J. Comput. & Nonlin. Dyn. 1(1), 65–70 (2006b)
1132. Zhao, X., Dankowicz, H., Reddy, C.K., Nayfeh, A.H.: Modeling and simulation methodology for impact microactuators. J. Microelectromechanical Systems 14, 775–784 (2004)
1133. Zhao, X., Reddy, C.K., Nayfeh, A.H.: Nonlinear dynamics of an electrically driven impact micro–actuator. Nonlin. Dyn. 40, 227–239 (2005)
1134. Zhiqing, C., Pilkey, W.D.: Application of wavelets to optimal shock and impact isolation. Trans. Nanjing University of Aeronautics and Astronautics 18, 1–10 (2001)
1135. Zhong, Z.H.: Finite Element Procedures for Contact–Impact Problems. Oxford University Press, Oxford (1993)
1136. Zhou, T., Rogers, R.J.: Simulation of two–dimensional squeeze film and solid contact forces acting on a heat exchanger tube. J. Sound & Vib. 203(4), 621–639 (1997)
1137. Zhuravlev, V.F.: Investigation of certain vibro–impact systems by the method of non–smooth transformations. Izvestiva AN SSSR Mehanika Tverdogo Tela (Mechanics of Solids) 12, 24–28 (1976)
1138. Zhuravlev, V.F.: The application of monomial lie groups to the problem of asymmetrically integrating equations of mechanics. J. Appl. Math. Mech. (PMM) 50, 346–352 (1986)
1139. Zhuravlev, V.F., Klimov, D.M.: Applied Methods in Vibration Theory, Nauka, Moscow (1988)
1140. Zhusubalyev, Z., Mosekilde, E.: Bifurcations and Chaos in Piecewise–Smooth Dynamical Systems. World Scientific, Singapur (2003)
1141. Zhusubalyev, Z., Mosekilde, E.: Quasi-periodicity and torus birth bifurcations in non–smooth systems. In: Int. Conf. Physics and Control (IEEE Cat. No. 05EX1099), Saint Petersburg, Russia, August 24–26, pp. 429–433. IEEE, Piscataway (2005)
1142. Zhusubalyev, Z., Mosekilde, E., Maity, E., Mohanan, S., Banerjee, S.: Border collision route to quasi-periodicity: numerical investigation and experimental confirmation. Chaos 16(2), 23122–1–11 (2006)

1143. Zhusubalyev, Z., Soukhoterin, E.A., Mosekilde, E.: Border–collision bifurcations and chaotic oscillations in a piecewise–smooth dynamical system. Int. J. Bifurcation and Chaos in Applied Sciences and Engineering 11(12), 2977–3001 (2001)
1144. Zhusubalyev, Z., Soukhoterin, E.A., Mosekilde, E.: Complexity and chaos in piecewise–smooth dynamical systems. In: Proc. 2003 Int. Conf. Physics and Control, Piscataway, NJ, vol. 4, pp. 1159–1164 (2003a)
1145. Zhusubalyev, Z.T., Soukhoterin, E.A., Mosekilde, E.: Border–collision bifurcations on a two–dimensional torus. Chaos, Solitons & Fractals 13, 1889–1915 (2003b)
1146. Zimmerman, R.L., Celaschi, S., Neto, L.G.: The electronic bouncing ball. American J. Physics 60(4), 370–375 (1992)
1147. Zuyev, A.K., Nikitin, A.A.: Vibration protecting mechanism for a riveting hammer. In: Proc. Dynamics of Vibro–Impact Mechanical Systems, Novosibirsk, NETI, pp. 50–52 (1973)

Index

Printed in the United Kingdom by
Lightning Source UK Ltd., Milton Keynes
139186UK00008B/132/P